PHYSICAL CHEMICAL ASPECTS OF CELL SURFACE EVENTS IN CELLULAR REGULATION

DEVELOPMENTS IN CELL BIOLOGY

Volume 1—Development and Differentiation in the Cellular Slime Moulds, edited by P. Cappuccinelli and J. M. Ashworth, 1977

Volume 2—Biomathematics and Cell Kinetics, edited by A. J. Valleron and P. D. M. Macdonald, 1978

Volume 3—Developmental Biology of Acetabularia, edited by S. Bonotto, V. Kefeli and S. Puiseux-Dao, 1979

Volume 4—Physical and Chemical Aspects of Cell Surface Events in Cellular Regulation, edited by Charles DeLisi and Robert Blumenthal, 1979

PHYSICAL CHEMICAL ASPECTS OF CELL SURFACE EVENTS IN CELLULAR REGULATION

Proceedings of An International Conference on Physical Chemical Aspects of Cell Surface Events in Cellular Regulation, NIH, Bethesda, Maryland, USA on October 19 –21, 1978

Editors:

CHARLES DeLISI and ROBERT BLUMENTHAL
Cancer Biology and *Diagnosis, National Cancer Institute*

ELSEVIER/NORTH-HOLLAND
NEW YORK • AMSTERDAM • OXFORD

Published by:

Elsevier North Holland, Inc.
52 Vanderbilt Avenue, New York, New York 10017

Sole distributors outside USA and Canada:

Elsevier/North-Holland Biomedical Press
335 Jan van Galenstraat, PO Box 211
Amsterdam, The Netherlands

Library of Congress Cataloging in Publication Data

International Conference on Physical Chemical Aspects of
Cell Surface Events in Cellular Regulation, National Institutes of Health, 1978.
Physical chemical aspects of cell surface events in cellular regulation.
(Developments in cell biology ; 4)

Bibliography: p.
Includes index.
1. Cellular control mechanisms—Congresses. 2. Cell membranes—Congresses. I. DeLisi, Charles, 1941- II. Blumenthal, Robert. III. Title. IV. Series. [DNLM: 1. Cell membrane—Physiology—Congresses. W1 DE997VN v. 4 / QH601 I633p 1978
QH604.I57 1978 574.8'76 78-26955
ISBN: 0-444-41607-2 (series)
ISBN: 0-444-00311-8 (volume)

Manufactured in the United States of America

Contents

Preface

Physical chemical studies of hormones, neurotransmitters and antigens interacting with plasma membrane receptors, and the dynamical events which follow, are becoming increasingly more abundant and more quantitative. Their analysis is leading to new insights and a deeper understanding of the role which physical processes play in regulating the initial phases of the biochemical sequences leading to various cellular responses, such as growth, secretion, endocytosis and desensitization.

The rapid expansion in this area can be atrributed to a number of technical advances including the refinement of techniques for isolating and purifying membrane receptors, and the development of new methodologies, both biophysical and biochemical, for following molecular motions upon and within the cell. In many instances the data now available are sufficiently extensive and accurate to allow application of the mathematical tools of physical chemistry which have been so successfully applied in the physical sciences.

This volume collects the proceedings of an NIH sponsored international workshop on the relationship between the physical chemistry of cell surface events and biological activity. Both the scientific content of the meeting and the composition of the audience were unusual. The stress was on concepts and data analysis, and consequently on "clean" cellular systems for which extensive quantitative information is available. The participants were drawn from several disciplines including endocrinology, immunology and neurobiology, with representation from experimental as well as mathematically oriented researchers.

The topics have been organized in a hierarchal fashion, beginning with the thermodynamics and kinetics of receptor ligand interactions and moving through topics of increasing complexity toward cell-cell interactions. The presentations display a level of sophistication in analytical techniques which is unprecedented in cell biology, and illustrate how new insights may be achieved, and current hypotheses rigorously tested, by an interplay between appropriate experimental approaches and mathematical modeling. We hope that the book will be valuable to researchers entering the field, as a representative sample of some of the more experimentally developed cell surface systems; and to quantitatively oriented biologists, as an indication of the role which mathematical methods are beginning to play in this area.

Charles DeLisi

Robert Blumenthal

Bethesda

October, 1978

PHYSICAL CHEMICAL ASPECTS
OF CELL SURFACE EVENTS IN CELLULAR REGULATION

I
MOLECULAR INTERACTION IN CELL MEMBRANES

MOLECULAR INTERACTIONS IN CELL MEMBRANES

ROBERT BLUMENTHAL
Laboratory of Theoretical Biology, NCI, NIH, Bethesda, MD 20014

Ladies and Gentlemen,

I would like to thank those who have traveled long distances to join us in this big event. Participants have come from Israel, France, Belgium and England. I would also like to thank those participants who have come from various corners of this country, and those who have made it all the way from Buildings 10 or 37. I should not forget to acknowledge our gratitude for the financial support provided by NCI, NIAMDD, NIDR, and FIC.

This workshop came out of the Laboratory of Theoretical Biology, and I am surprised to see that there is so much interest in theoretical biology. It is an interdisciplinary meeting bringing together experimentalists from the fields of neurobiology, endocrinology and immunology with theoreticians who represent a physical chemical and/or mathematical approach to the analysis of biological activity and physical events involving cell surfaces. It is certainly hazardous to organize a meeting bringing together theoreticians and experimentalists. I have been to other meetings, at which leading scientists from the two cultures were present. The results were disastrous because they did not know how to communicate with each other. In this workshop, on the other hand, we have selected only theoreticians who had gotten their hands dirty... at least by shaking those of the experimentalist ... and experimentalists who do not have a fit when they see a differential equation.

The workshop was motivated last year by discussions between Charles DeLisi and myself, about physical chemical aspects of receptor-ligand interactions. We felt that the time was ripe for bringing experimentalists and theoreticians together to formulate rigorous models of the physical chemical aspects of cell surfaces in regulation of the immune response. We held a planning session with, among others, Bill Paul and Henry Metzger, more or less to assess whether this was feasible. The upshot of the planning session was that although a great deal is known about the structure, sequence and physical chemical properties of immunoglobulin molecules which function as B cell receptors, very little is known about how they transmit information through the membrane. On the other hand, physical chemical studies of neurotransmitters and hormones interacting with plasma membrane receptors, and the dynamical events which follow, are becoming increasingly more abundant and more quantitative. Their analysis is leading to new insights into the role which physical processes play in regulating the initial phases of biological responses, such as signal transduction, secretion, endocytosis and desensitization. The rapid expansion in this area can be attributed to a number of technical advances including the refinement of techniques for isolating and purifying membrane receptors, and the development of new methodologies, both biophysical and biochemical, for following molecular motions upon and within the cell. In many instances the data now available are sufficiently extensive and accurate to allow application of the mathematical tools which have been so successfully applied in the physical sciences. With the advice and guidance of our collegues at NIH, in various areas of receptorology, we therefore decided to consider first the better defined systems in neurobiology and endocrinology hoping that whatever is learned from the interaction of chemical stimuli impinging upon a receptor in these systems, can be applied to immune triggering at the cell surface. One criterion used in selecting the topics was that the systems to be analyzed are "clean" cellular systems for which extensive quantitative information is available.

As you can see from the program, we designed the workshop such that the systems to be discussed are in a hierarchical order from least to most complex and, in general, from best to least understood. This morning we will discuss molecular interactions on the cell surface, without yet being directly concerned with the response to the stimulus. In the afternoon, we will consider the mobility of cell surface receptors, a matter which may be of considerable interest, in relation to function. It is most appropriate to have Michael Edidin, who was the first to observe mobility of all surface receptors, lead this discussion. It is still uncertain, for example, what role (if any) is played by patching and capping in the regulation of cellular events. Originally patching and capping were only shown in the context of cell surface immunology, induced by bivalent antibodies, surface immunoglobulins, or lectins. Recently with the development of fluorescently labeled hormones (insulin and epidermal growth factor) and of refined image intensification techniques it turns out that polypeptide hormones also patch and cap. Since it has been shown that polypeptide hormones also enter cells, the question naturally arise "what are they doing there".

Tomorrow we will consider the coupling between the binding of a ligand to a cell surface receptor and a cellular event, namely production of CAMP. The glucagon receptor, the B-adrenergic receptor and the vasopressin receptor seem to be the best worked out systems to study in this context. This discussion will be lead by a young, promising scientist, named Jesse Roth, who is relatively new in the field and therefore not burdened by the current dogmas. Tomorrow afternoon we will go into more complex, but experimentally well understood receptor-mediated endocytosis of low density lipoproteins, and receptor-mediated exocytosis of mast cell granules. Henry Metzger will lead that discussion. Finally, Saturday morning we will discuss the even more complex issues which initally motivated this workshop. The relationship between immunological triggering on the cell surface and the consequent cellular events. We hope that by the end of the morning, under Martin Raff's direction, we will better understand how to approach these systems.

The ground rules for this workshop are the following: the announced speakers will give 20-30 minute presentations. We hope that the lectures will be conceptual, and since this is an interdisciplanary group, we will not go too much into nitty gritty details. With luck we will also avoid long debate over parochial issues of interest only to those actively investigating a particular system. We had hoped that the audience would not be so large that it is difficult to interrupt for quick clarifications, but we will have to see how it goes. After each talk there will be some time for discussion and then hopefully two hours for a general discussion.

Now, let me take up my role of Chairman of this morning's session: As stated before we will first consider molecular interactions. It is most appropriate to have Jean-Pierre Changeux lead off this conference, because he is the one who originally formulated ideas about the regulatory functions of biological membrane proteins. The ideas originated from considering regulatory functions of allosteric proteins. Jean-Pierre combines the skills of the biochemist with those of the theoretical biologist. He has a subtle intuition about complexities of biological systems. When I started a post-doctoral fellowship with Jean-Pierre about ten years ago, the situation appeared quite simple. The neurotransmitter acetylcholine interacts with the acetylcholine-receptor, provokes an allosteric conformational change, and thereby activates an ionophoric channel. All one had to do was to isolate the receptor, put it back into lipid model membrane, and study the response in terms of a permeability change. Now, ten years later, Jean-Pierre will show how complex this relatively simple system is. He has dissected the system into a regulatory (Ach binding) protein and an ionophoric protein, and will present the evidence that those proteins are associated with their putative roles.

The fact, that the transduction step is mediated by a membrane permeability change induced by a transport molecule associated with the receptor is obvious in the case of receptors for neurotransmitters. The interaction directly causes a depolarization or hyperpolarization of the post synaptic membrane due to changes in ionic permeability. Evidence is, however, mounting that the action of receptors for hormones or antigens is also associated with an ionic permeability change, leading to the notion of metallic ions as "second messengers".

On the other hand, data on the action of active transport systems such as the Na^{+}+K^{+} ATPases and Ca^{2+} ATPases suggest a more complex picture: ions are bound to receptor sites (or "gates") in the transport molecule, which then undergoes an energy-mediated transduction, facilitating the passage of that ion through a membrane channel, and thereby promoting active accumulation in or extrusion from the cell.

In view of the fact that transport molecules are so intimately involved in receptor action it is important to understand their mechanisms of action, either as separate entities or preferably in conjunction with the receptor or recognition part of the system.

Ion transport in conjunction with receptor action in reconstituted systems could appropriately be the topic of a separate workshop, but here we will have the opportunity to consider two aspects. First Bill Catteral has been studying the excitable Na^{+} ionophore, and has shown that the receptor paradign applies: allosteric interactions associated with permeability changes, with ionophoric and regulatory sites for drugs. Second Adil Shamoo will discuss the receptor concept of active transport and will show how he takes the system apart and reassembles it. Catteral and Shamoo will each spend ten minutes discussing those topics.

Allosteric considerations will be extended to negative cooperativity in the insulin receptor by Pierre de Meyts. We seem to know a lot about the insulin receptor -- but not what it is really doing. The notions of cooperativity and heterogeneity of binding sites will be further elaborated upon by Candace Pert, with particular reference to opiate-receptor binding. I would at this point like to say a word about terminology: when cooperativity is mentioned, most people have a specific model in mind thanks to the work of Monod, Wyman and Changeux, and of Koshland and his coworkers. For the purposes of this meeting we would like to give it a more phenomenological definition, namely binding of one ligand to a multivalent receptor enhances or decreases the probability for the next one binding. As we will see at this workshop various mechanisms can account for that phenomenon.

Before giving Jean-Pierre the floor, I would like to conclude this introduction with the following story, which may be appropriate to this meeting of theorists and experimentalists. Mort Weisiger, the creator of Superman, once got a letter from a group of MIT students who had consulted with Einstein about the speed with which Superman was flying. Einstein replied that Weisiger had Superman flying twice the speed of light -- which is impossible. Weisiger showed the letter to his friend Isaac Asimov who commented: "Superman has got the facts, Einstein the theory."

I do not wish to imply that we have gathered a workshop of Einsteins and Supermen, but I do hope that towards the end of the meeting we will at least know who has the facts and who has the theory. The workshop will be successful if the two are not found to be incompatible.

STRUCTURAL AND FUNCTIONAL PROPERTIES OF THE ACETYLCHOLINE-REGULATOR

A. SOBEL, J. HOFLER, T. HEIDMANN & J.P. CHANGEUX
Neurobiologie Moléculaire et Laboratoire Associé Centre National de la Recherche Scientifique, Interactions Moléculaires et Cellulaires, Institut Pasteur, 75015 Paris, France

Acetylcholine (ACh) triggers two distinct reactions in the post-synaptic membrane of the neuromuscular junction: the rapid opening of cation selective gates or "activation" and a slow (second, minutes) decrease of the response amplitude upon prolonged exposure to the agonist or "desensitization". To account for these two processes, it was potulated (1) that the elementary membrane unit carrying the receptor site for acetylcholine and the site for ion translocation is made up of, at least, two distinct elements of structure: the acetylcholine receptor protein (AChR) sensu stricto and the acetylcholine ionophore which might equally be part of the same polypeptide chain or carried by distinct subunits.

By analogy with the known behavior of regulatory enzymes, it was further postulated (see 2) that the ACh-Regulator exists under at least three discrete and interconvertible conformational states. These states referred to as resting R, active A and desensitized D would differ by a) the affinity of the acetylcholine receptor site for cholinergic ligands: the affinity for agonists increase from R (low affinity) to A (medium affinity) to D (high affinity); on the other hand, the affinity for antagonists is lower in the A than in the R or D states; "non-exclusive" binding of a given ligand to more than one state is expected to occur; b) the affinity of the site for local anesthetics: in the D and A states the affinity is higher than in the R state; c) the ion gate is open only in the A state.

In the membrane at rest, the R state is favored; the fraction of A state is negligible, that of the D state small but not negligible, at least in *vitro*. In the presence of a given concentration of ligand, the equilibrium is shifted in favor of the state to which it binds preferentially.

Agonists will shift the equilibrium towards the D state, antagonist to either the R or D state and local anesthetics (and multivalent cations) to either the D or A state. Upon rapid change of ligand concentration and depending on the nature of the ligand, some states can be transiently populated due to preferential pathways for the interconversion. The "activation" reaction is viewed as a transient population of the A state by acetylcholine and the "desensitization" as the final equilibration in the D state.

This model accounts qualitatively for the *in vivo* responses of "activation" and "desensitization"; it also fits quantitatively the *in vitro* data presently available with subsynaptic membrane fragments purified from the electric organ of *Torpedo*.

I - STRUCTURE OF THE ACh-REGULATOR IN ACh-RECEPTOR RICH MEMBRANE FRAGMENTS

The early studies on the ACh-regulator have been focussed mainly, though not exclusively, on the purification and characterization of the ACh-receptor protein sensu stricto after detergent extraction from its membrane environment (ref. in 2, 3, 4, 5). A possible draw back of this approach is that some critical elements of the ACh-regulator are lost during purification. This is why a method of purification of membrane fragments containing the ACh-receptor protein with a particularly high specific activity has been developed (6, 7). In these ACh-receptor rich membranes, which are available in hundreds of mg quantities, the ACh-receptor protein represents up to 40% of the proteins (7). Examined by electron miscroscopy (see 8) they disclose 9nm centered particles, which have the same size and shape as the rosettes observed in preparations of ACh-receptor protein purified in the presence of detergent. These particles are present on the etched surfaces and therefore are exposed to the surface of the membrane fragments. Similar images have been recently reported on the cleft surface of the subsynaptic membrane in situ after rapid freezing (Salpeter and Heuser unpublished). In the fracture plane but after fixation "internal" particles are observed with a different shape but a similar distribution and size as the ACh-receptor rosettes. They are attached mostly to the cytoplasmic face of the membrane. Internal particles of a smaller size are noticed also on the cleft surface of the fracture plane, in the absence of fixation (8). An important but still pending question is the significance of these "internal" particles. If they belong to the ACh regulator, do they represent the ACh receptor sensu stricto, the ACh-ionophore or their complex?

The external rosettes and the internal particles often make regular arrays (see 8) both *in vitro* and *in situ*. Adequate *in vitro* treatment of the AChR-rich membranes (9) lead to an increase of the surface of these arrays which can be analysed by optical diffraction and give a pseudo-hexagonal lattice organisation (at the present level of 3 nm resolution). Accordingly, the 9 nm repeated unit would possess a 6-fold symmetry, however an asymmetrical organisation of this unit has been suggested by others (10). A direct observation of the subunit structure of the rosette is needed to solve the dilemma.

Polyacrylamide gel electrophoresis of ACh-receptor rich membranes in the presence of sodium dodecyl sulfate yields several bands after Coomassie blue staining (see 11, 12). With the membrane preparations from *T. marmorata* their apparent mol. w. are respectively 40, 43, 50 and 66 thousand daltons (7). Only the 40,000 daltons band is labelled by Karlin's covalent affinity labeling reagent MPTA and therefore belongs to the ACh receptor protein *sensu stricto*. The 50 and 66 thousand daltons bands almost completely disappear in the purest membrane fractions and, therefore, may represent contaminants. This interpretation is supported by the observation that on a Sepharose column, in the presence of non-denaturing detergent, the different bands do not follow the same elution pattern and therefore should belong to distinct macromolecules. A particular susceptibility of the high mol. w. bands to proteolysis which, according to some of the authors, would explain the variability of the gel patterns (for a discussion see 2), as yet, cannot be excluded.

Dissolution of the ACh-rich membranes in non-denaturing detergent followed by centrifugation or gel filtration results in the separation of two main protein components: the α-toxin labelled ACh-receptor protein and another macromolecule almost exclusively made up of the 43,000 daltons chain: the "43 K protein" (13, 14).

The most purified preparations of ACh-receptor protein give in SDS gels almost exclusively the 40,000 daltons chain and only serine as N-terminal aminoacid. The mol. w. of the receptor protein is close to 250,000 daltons (see 15, 16). If the molecule is composed of only the chain with an apparent mol. w. of 40,000 daltons, then, it should be a regular hexamer, but this particular quaternary structure deserves confirmation. Cyanogen bromide split of the purified receptor protein labelled by MPTA gives a MPTA-labelled hydrophobic peptide which can be purified by Sepharose filtration in the presence of 50% acetic acid. It migrates slightly more slowly than the α-chain of insulin and has an apparent mol. w. close to 2,500 daltons. Its sequence starts by NH_2-Met-Lys-Gly...

The 43 K protein most often forms high molecular weight aggregates under conditions where the receptor protein remains in solution. It can however be obtained in a soluble form in the presence of non-denaturing detergents with an apparent molecular weight (on Sepharose columns) of about 100,000. Its aminoacid composition is different from that of the ACh-receptor protein and from that of actin which makes a closely located but distinct protein band in

SDS polyacrylamide gels. The 43 K protein also possesses highly reactive sulphydryl groups. Antibodies directed against the AChR protein do not cross react with the 43 K protein. Without ambiguity the 43 K protein is a macromolecule entirely different from the ACh-receptor protein sensu stricto. Densitometric scans of SDS polyacrylamide gels of highly purified ACh-receptor rich membranes suggest that in these membranes the two proteins are present in equivalent amounts.

II - ALLOSTERIC INTERACTION BETWEEN AChR AND IONOPHORE IN AChR-RICH MEMBRANE FRAGMENTS

The particularly high density of AChR-rosettes (10-20,000/μm^2) in the AChR-rich membrane fragments is encountered in situ only in the area of the cytoplasmic membrane which underly the nerve terminals. Thus, they most likely derive from the subsynaptic membrane and therefore are exposed to ACh in vivo. Using the $^{22}Na^+$ efflux assay of Kasai and Changeux (17), one can show that these membrane fragments, which make close vesicles or miscrosacs, are still functional in vitro. In the presence of carbamylcholine the rate of passive transport of $^{22}Na^+$ increases and this effect is blocked by snake venom α-toxins and other cholinergic antagonists. Moreover, prolonged exposure of the membrane fragments to carbamylcholine causes a time dependent decrease of the response amplitude. The cholinergic agonists still trigger on the AChR-rich membranes, the two characteristic reactions of "activation" and "desensitization" (18, 19). The AChR-rich membranes therefore contain the

ACh-ionophore in a functional state. In addition, its "coupling" with the AChR protein closely resembles that found in vivo.

The permeability response of the AChR-rich microsacs is blocked by a series of compounds, which strictly differ, by their mode of action, from the classical "competitive" antagonists (for a review see 2). They are, for instance, amine local anesthetics like tetracaine, prilocaine, di- and tri- methisoquin or quinacrine, detergents like triton X-100 or cholate and the frog toxin histrionicotoxin (20, 21). All of them block the response to cholinergic agonists in a "non competitive" manner and in a concentration range often several orders of magnitude lower than that at which they bind to the AChR site (22, 23). Recent electrophysiological studies carried out mostly on skeletal muscle show that they interfere with the mechanism of ion translocation which is under the command of the AChR-site (24, 25). Most likely these non-competitive blocking agents bind directly to the ion gate and therefore are potential labels of the ACh-ionophore. Another interesting property of these compounds is that they markedly enhance, both in vivo and in vitro, the rate of desensitization of the permeability response to cholinergic agonists (26).

In vitro, their effect is not limited to the blocking of ion transport. In the domain of concentration where they alter the permeability response to cholinergic agonists they also modify the binding of these agonists to the AChR-site. Paradoxically, all of them (27), including histrionicotoxin (23), increase the affinity of the AChR-site for cholinergic agonists (but also for some antagonists) and change the shape of the binding curve of radioactive ACh to the membrane bound AChR-site from a sigmoid into an hyperbola (27). The data have been interpreted in terms of allosteric interactions between the AChR-site and the binding site for local anesthetics (27).

New insights on this interaction have been recently brought about by the introduction of a fluorescent cholinergic ligand, C_5DAChol (28), which behave as a full agonist both on E. electricus electroplaque (28) and on T. marmorata microsacs, with an apparent dissociation constant ($\sim 1\ \mu M$) and a Hill coefficient (~ 2.0) close to those found for ACh. The interactions of C_5DAChol with the AChR-rich microsacs can be followed by fluorescence spectroscopy under conditions of energy transfer from proteins (28) (λ_{ex} = 290 nm, λ_{em} = 557 nm). At equilibrium, the binding of C_5DAChol inferred from the fluorescence experiments closely parallels that found by following the binding of radioactive ACh: high affinity ($K_D \sim 50$ nM) and enhanced binding by local anesthetics (28).

In vivo, the genesis of the endplate potential elicited by ACh release from the nerve endings takes place within a millisecond. In vitro, this time range becomes accessible to direct measurement by rapidly mixing the AChR-rich membrane fragments with C_5DAChol in a stopped-flow apparatus (29, 30, 31). Under these conditions a fluorescence intensity signal is recorded in the second to millisecond time range which disappears after preincubation of the AChR-rich membrane fragments with the α-toxin from N. nigricollis and therefore results exclusively from the interaction of C_5DAChol with the AChR-site. The quantitative analysis of the traces at various concentrations of C_5DAChol reveals a minimum of three relaxation processes:

1) a "rapid" process takes place in the millisecond time range; its apparent rate constant varies linearly with C_5DAChol concentration and its amplitude corresponds to approx. 20% of the total signal; it is interpreted as the result of C_5DAChol binding to a population of AChR-sites with a <u>high</u> affinity for C_5DAChol ($K_D \sim 3$ nM) which represents approx. 20% of the total population of AChR-site and preexists to agonist binding in the membrane at rest.

2) an "intermediate" relaxation process is recorded in the second to millisecond time range and becomes important at high C_5DAChol concentration; the signal reaches its full amplitude around 10^{-4}M C_5DAChol; the apparent rate constant again increases linearly with the concentration of C_5DAChol and the data are accounted for, in first approximation, by a bimolecular binding reaction to a state (or states) of the AChR-site of lower affinity ($\sim 1\ \mu$M),

3) a "slow" relaxation process takes place in the second range; it strikingly differs from the other two in several instances: its apparent rate constant increases with C_5DAChol concentration but reaches a plateau at high C_5DAChol concentration indicating that an isomerisation process takes place; its amplitude also reaches a plateau but decreases at high C_5DAChol concentration; it is accounted for by an isomerisation between the above mentioned low and high affinity states of the AChR-protein.

The experimental data are fitted, in first approximation, by a simplified version of the three-state model presented in the

Introduction. It postulates that the ACh-regulator exists under two interconvertible states of affinity for cholinergic ligands (low R, high D) which preexist to C_5DAChol binding. Binding to the R state would take place in the concentration range where C_5DAChol triggers the permeability change and therefore would correspond to the "activation" process. The isomerisation to the high affinity D state occurs in the second range and therefore would coincide with the "desensitization" reaction.

The interpretation of the "slow" process as being part of the desensitization reaction is supported by the observation that local anesthetics, compounds known to accelerate desensitization *in vivo*, modify this slow process. Preequilibration of AChR-rich membranes with local anesthetics or histrionicotoxin increases the amplitude of the rapid relaxation process which, as already mentioned, is proportional to the fraction of AChR-sites in a high affinity state for agonists. Expectedly, and in agreement with other observations done with different methods, the local anesthetics (but not histrionicotoxin) increase the apparent rate constant of the slow transition leading to this high affinity state.

Local anesthetics also accelerate the "intermediate" relaxation process suggesting that this intermediate process includes an isomerisation step towards a third state to which the local anesthetics would also bind. This state would then correspond to an "active" state of the ACh-regulator. It might be identical to that revealed in the presence of the fluorescent local anesthetic: quinacrine (32, 33). The fast kinetic data, available at present, therefore can be accounted for, at least in first approximation, by the three-state model presented in the Introduction.

Additional information on the "allosteric" site to which local anesthetics bind has been obtained with radioactive local anesthetics (34) or histrionicotoxin (21, 35). Quaternary trimethisoquin, which acts as a potent non competitive blocking agent on E. electricus electroplaque, binds at equilibrium to a limited population of sites on T. marmorata AChR-rich membrane fragments (A. Sobel unpublished). Similar conclusion was reached previously with the fluorescent local anesthetic quinacrine (32). These sites are not blocked by the snake venom α-toxins and carbamylcholine enhances the binding of radioactive trimethisoquin or histrionicotoxin to these sites. This confirms that the local anesthetic binding site is distinct from the AChR-site and further indicates that the allosteric interactions between these two classes of sites are, indeed, reciprocal. In the presence of carbamylcholine the number of trimethisoquin binding sites ranges between the half and the same number as that of α-toxin binding sites. High and low affinity states of the ACh-regulator for local anesthetics also exist.

The structural specificity of the local anesthetic binding sites followed by displacement of radioactive trimethisoquin is that expected for the in vivo pharmacology of the non-competitive blocking agents: it binds the amine local anesthetics, di- and trimethisoquin, histrionicotoxin and the detergent Triton X-100 in the adequate range of concentrations.

III - RECONSTITUTION OF A FUNCTIONAL ACh-REGULATOR

Knowing the specificity of the local anesthetic binding site it becomes possible to test if the purified AChR-protein or the 43 K protein carries this site. Since detergents block the site, the experiment has to be done in the absence of detergent. Histrionicotoxin is also an excellent tool to check the specificity of the site. With native AChR-rich membrane fragments equilibration with quinacrine, histrionicotoxin causes a decrease of fluorescence intensity recorded under conditions of energy transfer from proteins (14). The K_{app} for this effect is close to 1 μM a value which is also found by following the displacement of radioactive trimethisoquin. Interestingly, when the same experiment is carried out with the purified AChR and 43 K-protein after elimination of the detergent, only this last protein exhibits a quinacrine signal sensitive to histrionicotoxin but at variance with what is observed with the native membranes this signal is no longer enhanced by cholinergic agonists. The 43 K-protein therefore possesses a site for local anesthetics and histrionicotoxin. However, the conclusions drawn from these experiments are limited for the following reasons:

1) inferences from fluorescence experiments about binding are always indirect;

2) since the regulation by carbamylcholine is lost with the purified protein the specificity of the site binding quinacrine or even histrionicotoxin cannot be ascertained with the same security as with the native membrane;

3) both receptor and 43 K protein have been solubilized by detergent and, as a consequence, their binding properties can be altered.

Recently, Neubig et al. (39) reported that the 43 K protein can be separated from the ACh-receptor rich membrane by alkaline extraction in the absence of detergent and without significant loss of α-toxin binding. Interestingly, the alkaline extracted membranes still bind the radioactive local anesthetic {^{14}C}-meproadifen in a manner similar to that found with the native membranes. We have confirmed that the pH 11 treated membranes still bind {^{3}H}- trimethisoquin with the same stoichiometry as the native membranes and that carbamylcholine still regulates this binding (Sobel unpublished). Since in the alkali treated membranes only the receptor peptides persist, the site for the pharmacological action of local anesthetics must be carried out by one of these peptides.

Several attempts have been made in the past to reconstitute a functional receptor-channel complex (or ACh-"regulator") from purified components of the excitable microsacs (for a review see 40). Although the results of these experiments are certainly correct, their lack of reproductibility indicates that some basic elements of the system are not under control. Our approach has been to try to "reconstitute", first, the several features of the kinetics of interaction of C_5DAChol with the ACh-receptor site and the effect of local anesthetics on these kinetics. In a first attempt (37), only the slow change of affinity for agonists was recovered. Recently, we have developed a method which leads to the recovery of the normal kinetics of interaction of C_5DAChol with the ACh-receptor site as well as the effect of local anesthetics on these kinetics (36) after complete solubilization of the receptor protein.

The method is quite simple (see 41). The membrane fragments are solubilized by cholate. Under these conditions, the sensitivity to local anesthetics is lost and the receptor protein is stabilized under an "intermediate" state of affinity, different from any of the affinity states present in the native membrane. Moreover the receptor protein is in a 9 S soluble form. Elimination of the detergent by dilution leads to the reaggregation of the receptor protein and to the "reconstitution" altogether of the three kinetic processes of C_5DAChol binding and of the effect of local anesthetics on these processes.

To test if the 43 K protein is necessary for the recovery of these interactions, the same experiment was carried out with alkali treated ACh-receptor-rich microsacs which do not contain the 43 K protein. Qualitatively, the results are the same (Heidmann & Sobel unpublished). This confirms that the 43 K protein is not involved in the allosteric interaction between ACh-receptor site and local anesthetic binding site. The product of this reconstitution constitutes the best starting material for the recovery of the permeability response.

IV - CONCLUSIONS

If the AChR-protein becomes a rather well characterized molecular entity (sequencing has started), the identity of the ACh-ionophore remains to be stablished without ambiguity. Good evidence exists for conformational transitions of the membrane bound ACh-regulator; but the detailed correlation in the milli-second to minute time range of agonist binding to the AChR-site and ion translocation through the ACh-ionophore deserves further investigations.

ACKNOWLEDGEMENT

This work was supported by grants from the Muscular Dystrophy Association of America, the College de France, the Delegation Generale a la Recherche Scientifique et Technique, the Centre National de la Recherche Scientifique, the Institut National de la Sante et de la Recherche Medicale and the Commissariat a l'Energie Atomique.

REFERENCES

1. Changeux, J.P., Podleski, T.R. & Meunier J.C. (1969), J. Gen. Physiol., 54, 225-244 S
2. Heidmann, T. & Changeux, J.P. (1978), Ann. Rev. Biochem., 47, 371-441.
3. Changeux, J.P. (1975), In "Handbook of Psychopharmacology" (S. Snyder & L. Iversen, ed.) 6, 235-301.
4. Karlin, A. (1977), In "Pathogenesis of human muscular dystrophies" (L.P. Rowland, ed.) Excerpta medica, Amsterdam, Oxford, 73-84.
5. Fulpius, B.W. (1976), In "Motor Innervation of muscle" (S. Thesleff, ed.), Academic Press, London.
6. Cohen, J.B., Weber, M., Huchet, M. & Changeux, J.P. (1972), FEBS Lett. 26, 43-47.
7. Sobel, A., Weber, M. & Changeux, J.P., (1977), Eur. J. Biochem., 80, 215-224.
8. Cartaud, J., Benedetti, L., Sobel, A. & Changeux, J.P. (1978), 29, 313-337.
9. Brisson, A. (1978), J. Cell. Biology (in press).
10. Ross, M.J., Klymkowsky, M.W., Agard, D.A. & Stroud, R.M. (1977), J. Mol. Biol., 116, 635-659.
11. Raftery, M.A., Vandlen, R.L., Reed, K.L. & Lee, T. (1976), In "The Synapse". Cold Spring Harbor Symposia on quantitative Biology, 40, 193-202.
12. Karlin, A., Weill, C., McNamee, M. & Valderrama, R. (1976), In "The Synapse" Cold Spring Harbor Symposia on quantitative Biology, 40, 203-210.
13. Sobel, A., Heidmann, T. & Changeux, J.P. (1977), C.R. Acad. Sci. Paris, Seance du 17 Oct. 1977.
14. Sobel, A., Heidmann, T., Hoffler, J. & Changeux, J.P. (1978), Proc. Natl. Acad. Sci. USA, 510-514.
15. Meunier, J.C., Olsen, R.W. & Changeux, J.P. (1972), FEBS Lett. 24, 63-68.
16. Reynolds, J.A. & Karlin, A. (1978), Biochemistry, 17, 2035-2038.
17. Kasai, M. & Changeux, J.P. (1971), J. Memb. Biology, 6, 1-80.
18. Sugiyama, H., Popot, J.L. & Changeux, J.P. (1976), J. Mol. Biol. 106, 485-496.
19. Popot, J.L., Sugiyama, H. & Changeux, J.P. (1976), J. Mol. Biol. 106, 469-483.
20. Daly, J.W., Karle, J., Myers, C.Q., Tokuyama, T., Waters, J.A. & Witkop, B., (1971), Proc. Natl. Acad. Sci. USA, 68, 1870-1875.
21. Eldefrawi, A.T., Eldefrawi, M.E., Albuquerque, E.X., Oliveira, A.C., Mansour, N., Adler, M., Daly, J.W., Brown, G.B., Burgermeister, W.B., & Witkop, B. (1977), Proc. Natl. Acad. Sci. USA, 74, 2172-2176.
22. Weber, M. & Changeux, J.P. (1974), Mol. Pharmacol., 10, 1-40.
23. Kato, G. & Changeux, J.P. (1976), Mol. Pharmacol., 12, 92-100.
24. Adams, P.R., (1977), J. Physiol. London, 268, 291-318.
25. Neher, E. & Steinbach, J.H. (1978), J. Physiol. London (in press).

26. Magazanik, L. & Vyskogil, F. (1973), In "Drug Receptors", (H. P. Rang, ed.), Mac Millan London, 105-120.
27. Cohen, J., Weber, M. & Changeux, J.P., (1974), Mol. Pharmacacol. 10, 904-932.
28. Waksman, G., Fournie-Zaluski, M.C., Roques, B., Heidmann, T., Grunhagen, H.H. & Changeux, J.P. (1976), FEBS Lett. 67, 335-342.
29. Heidmann, T., Iwatsubo, M. & Changeux, J.P. (1977), C.R. Acad. Sci. Paris, 284 D, 771-774.
30. Heidmann, T. & Changeux, J.P. (1978), Eur. J. Biochem. (submitted).
31. Heidmann, T. & Changeux, J.P. (1978), Eur. J. Biochem. (submitted).
32. Grunhagen, H.H. & Changeux, J.P. (1976), J. Mol. Biol., 106, 497-535.
33. Grunhagen, H.H. Iwatsubo, M. & Changeux, J.P. (1977), Eur. J. Biochem., 80, 225-242.
34. Cohen, J.B. (1977), In "Molecular Specialization and Symmetry in Membranes" (edit. by Solomon, A.K.) Harvard University Park press, Cambridge (in press).
35. Elliott, J. & Raftery, M.A. (1977), Biochem. Biophys. Res. Commun., 77, 1347-1353.
36. Heidmann, T., Sobel, A. & Changeux, J.P. (1978), FEBS Lett. 94 (2), 397-411.
37. Briley, M. & Changeux, J.P. (1978), Eur. J. Biochem., 84, 429-439.
38. Sugiyama, H. & Changeux, J.P. (1975), Eur. J.Biochem., 55, 505-515.
39. Neubig, R.R., Krodel, E.K., Boyd, N. & Cohen, J.B. (1979). Proc. Nat. Acad. Sci. USA (in press).
40. Briley, M. & Changeux, J.P. (1977), Int. Rev. Neurobiol. 20, 31-63.
41. Sugiyama, H. & Changeux, J.P. (1975), Eur. J. Biochem. 55, 505-515.

DISCUSSION

DR. SINGER: The molecular weight of actin is 43K. Could the 43K protein be actin?

DR. CHANGEUX: We have done three series of experiments: 1) we compared on SDS gels pure actin from beef with the 43K fragment from *Torpedo*. They do not comigrate. 2) we purified actin from *Torpedo*. It does not comigrate either with the 43K protein. 3) the aminoacid composition of the 43K from *Torpedo* is clearly different from that of beef actin.

It is possible nevertheless that the 43K protein is related to some protein from the cytoskeleton and is involved either in the clustering of the receptor protein in the subsynaptic membrane or in its protection against degradation ("stabilization") (or in both).

At the present stage of knowledge, we do not know the exact function of the 43K protein. It is clear that this protein is strongly associated with the acetylcholine receptor (everybody who has done this type of experiment agrees on this point). During the purification of the receptor-rich membranes by sucrose gradient centrifugation, one always find that the 43K protein follows closely the 40K chain of the receptor protein.

Whether the 43K protein is related to actin or to a derivative of actin, we do not know but we cannot eliminate this possibility. One should remember that there is a collagen moiety on acetylcholine esterase. Therefore, I do not react negatively to the idea that there is an actin moiety in the 43K protein which could be involved in the anchoring of the receptor protein to some subcellular structure. But, of course, we have no proof of this.

DR. HECHTER: What is the difference in K_D between the high affinity and the low affinity states.

DR. CHANGEUX: The binding constant to the low affinity state is difficult to determine but we can infer it from the kinetic experiments. The apparent K_D for the low affinity state is in the 1 μM - 10 μM range, that for the high affinity state in the 10 nM range.

DR. HECHTER: The second question has to do with the paper that came out by Raftery* where the conclusions seem to be that the 65K fragment is very important.

DR. CHANGEUX: Yes. And that is the reason why I emphasized there was a difference among groups about the significance of the 50K and 65K bands.

DR. HECHTER: I would really like to know, what are the obvious deficiencies in this paper to that conclusion?

DR. CHANGEUX: I don't think there is any difference in the real data; most of the people in the field find the same gel patterns. The 50 and 65K bands are present in small amounts in most of the preparations tested.

The question still pending is whether or not they are integral parts of the receptor protein or of the receptor-ionophore complex. In other words, what is the physiological significance of these bands.

Raftery claims that when he adds carbamylcholine, the labelling of the 65K by his covalent reagent increases. As far as we are concerned, we have preparations which contain only five to ten percent of the protein as the 65K pro-

---

*Witzemann, V. and Raftery M. Biochemistry 17: 3598 (1978).

tein. Others, like Karlin and his co-workers* have found that the 40K and the 65K bands can be cross-linked. So, I wish to point out that several groups attribute a role to the 65K protein in Acetyl choline receptor function. Also Jon Lindstrom** at the Salk Institute claims (but this needs to be confirmed) that he can induce autoimmune paralysis in animals with some of the chains different from the 40K chain.

On the other hand, a cross-reaction exists between species, (*Torpedo*, *Electrophorus*, and mammals) only at the level of the 40K and never with the other chains.

DR. BERNHARD: In the stopped-flow experiments, you had three processes, a very fast one, a very slow one, which I think had a plausible molecular interpretation, and an intermediate one in the 100 millisecond range. How do you interpret that kind of number?

DR. CHANGEUX: That is the most interesting part as far as we are concerned. The "slow" process, which is accelerated by local anaesthetics can be correlated with the desensitization of the ion fluxes, the high affinity state being a "desensitized" state. The "intermediate" process is clearly complex. The quinacrine experiments tell us that in this time range, a reaction takes place which is triggered by the agonist. What Thierry Heidmann has done in my laboratory is the analysis of the effect of local anaesthetics on the "intermediate" reaction monitored with the fluorescent agonist C_5DAChol. As I said, local anaesthetics affect this reaction which means that they bind to a low affinity state of the ACh-regulator in addition to the high affinity state. The idea is that a third state, distinct from the low and high affinity states, exists which would be, then, the active state. This is something we are trying to dissect now. Your question is, of course, very pertinent.

---

*Hamilton, S., McLaughlin, M., Karlin, A. 1978. Fed. Proc. 37, 529.

**Lindstrom, J. 1976 in Receptors and Recognition P. Cuatrecasas S, M. Greaves eds. London, Chapman and Hall p. 1-45.

ALLOSTERIC PROPERTIES OF VOLTAGE-SENSITIVE SODIUM CHANNELS

WILLIAM A. CATTERALL
Department of Pharmacology, University of Washington, Seattle, WA 98195

Work in my laboratory over the last few years has focussed on the mechanism of action of neurotoxins which alter the properties of voltage-sensitive sodium channels. This short paper focuses on the allosteric interactions among different neurotoxins and on the voltage dependence of binding of these neurotoxins in electrically excitable neuroblastoma cells.

Neurotoxins which alter the properties of voltage-sensitive sodium channels interact with three separate receptor sites in neuroblastoma cells (Table 1).

TABLE 1
NEUROTOXIN RECEPTOR SITES ASSOCIATED WITH VOLTAGE-SENSITIVE SODIUM CHANNELS

Toxin Receptor Site	Ligands
1	Saxitoxin Tetrodotoxin
2	Veratridine Batrachotoxin Aconitine Grayanotoxin
3	Scorpion Toxin Sea Anemone Toxin

The four lipid soluble neurotoxins veratridine, batrachotoxin, aconitine and grayanotoxin cause persistent activation of voltage-sensitive sodium channels in neuroblastoma cells at the resting membrane potential.[1-3] They interact with a common receptor site in causing persistent activation.[2-4] We have measured persistent activation of sodium channels in neuroblastoma cells by determining the increase in sodium permeability caused by toxins in $^{22}Na^{+}$ flux experiments carried out at constant membrane potential.[1-4] In this experimental system, batrachotoxin is a full agonist activating greater than 95% of the sodium channels whereas aconitine, veratridine, and grayanotoxin are partial agonists. The action of all four of these toxins is reversible. The fraction of sodium channels activated at saturation (P_∞) and the concentrations of toxins required for 50% activation ($K_{0.5}$) are presented in Table 2.

TABLE 2

ACTIVATION OF SODIUM CHANNELS BY NEUROTOXINS

Toxin	Concentration for 50% Activation	Fractional Activation at Saturation
Batrachotoxin	7.0×10^{-7}	.95
Veratridine	2.9×10^{-5}	.08
Grayanotoxin	1.2×10^{-3}	.51
Aconitine	3.6×10^{-6}	.02

Data are derived from computed fits of titration curves to a simple hyperbolic saturation curve of the form $P(A) = P_\infty A/(K_{0.5} + A)$ where P_∞ is the fraction of ion channels activated at saturation, $K_{0.5}$ is the concentration of toxin required for 50% maximal activation, and A is the concentration of toxin activator. The complete data are published in reference 3.

Saxitoxin and tetrodotoxin are noncompetitive inhibitors of the activation of sodium channels by the lipid soluble toxins.[2] These two toxins bind to a common receptor site in neuroblastoma cells. Neither the lipid-soluble toxins nor the polypeptides scorpion toxin or sea anemone toxin affect ^{3}H-saxitoxin binding.[5]

The polypeptides scorpion toxin and sea anemone toxin have no effect on sodium permeability when tested alone, but cause a marked enhancement of activation of sodium channels by the lipid-soluble toxins.[3,4,6,7] The effect of different fixed concentrations of scorpion toxin on activation of sodium channels by veratridine and batrachotoxin is illustrated in Figure 1. With the full agonist batrachotoxin, scorpion toxin causes a shift of the concentration-effect curves to the left with no change in the fraction of ion channels activated at saturation. With the partial agonist veratridine (and also with aconitine and grayanotoxin), scorpion toxin causes both an increase in P_∞ and a decrease $K_{0.5}$. In all cases, the concentration-effect curves are simple hyperbolae. Similar results are obtained with sea anemone toxin.[7] Therefore, there are heterotropic cooperative interactions between the receptor site for the lipid-soluble toxins and the receptor site for the polypeptide toxins but no homotropic cooperative interactions are observed.

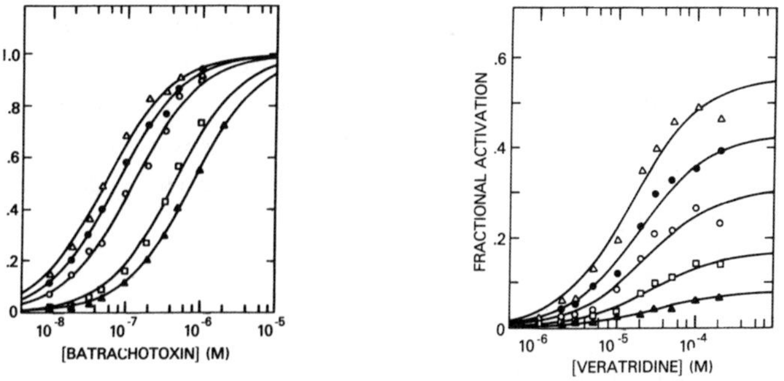

Fig. 1. Cooperative interaction between veratridine, batrachotoxin and scorpion toxin. N18 cells were incubated with the indicated concentrations of veratridine or batrachotoxin and 0 (Δ), 3 (◘), 10 (0), 30 (●), or 100 (▲) nM scorpion toxin for 30 min at 36° in sodium-free medium. $^{22}Na^+$ uptake was then measured for 30 sec at 36° in medium containing 5 mM Na^+ and the same toxin concentrations. Ordinates are the fraction of the total sodium channels activated. Taken from reference 3 with permission of the American Society of Biological Chemists.

An allosteric model has been developed which provides a unified mechanism for activation of sodium channels by lipid soluble toxins and for the heterotropic cooperative interaction between the polypeptide toxins and the lipid-soluble toxins.[3] The smooth curves in Figure 1 are theoretical curves derived from a computer fit to the allosteric model described below. The good agreement between the data points and the theoretical curves support the validity of the allosteric model.

The allosteric model makes three assumptions. (a) Sodium channels exist in two states with respect to ion transport activity, active (conducting) and inactive (non conducting). (b) Alkaloid toxins alter sodium channel properties by binding with greater affinity to active state(s) of sodium channels. These toxins activate sodium channels by shifting a pre-existing equilibrium between active and inactive states according to the law of mass action. (c) Scorpion toxin and sea anemone toxin, by binding to their receptor sites, reduce the energy required to activate sodium channels and alter the equilibrium constant between the active and inactive states. These assumptions are identical to those made by Monod et al.[8] in describing heterotropic cooperativity in allosteric enzymes.

In terms of this simple model, the ability of grayanotoxin, veratridine, aconitine, and batrachotoxin to activate sodium channels is due to their ability to selectively bind to the active state of sodium channels. This selective binding is most conveniently represented by the ratio of dissociation constants for binding to the inactive and active states (K_T/K_R). These values are presented in Table 3. The full agonist batrachotoxin binds 100,000 fold better to the active state of sodium channels whereas aconitine binds only 140 fold better.

TABLE 3

SELECTIVE BINDING OF NEUROTOXINS TO THE ACTIVE STATE OF SODIUM CHANNELS

Toxin	Binding Selectivity (K_T/K_R)
Batrachotoxin	1.4×10^5
Veratridine	6.2×10^2
Grayanotoxin	1.3×10^4
Aconitine	1.4×10^2

The data are derived from computed fits of titration curves to the allosteric model described in the text. The complete data are published in reference 3.

The success of this simple model in fitting our toxin activation data suggests that there are fundamental similarities between allosteric enzymes and sodium channels. Our current data are most consistent with the view that receptor sites for grayanotoxin and the alkaloid toxins and the receptor site for scorpion toxin and sea anemone toxin reside on separate subunits of a multi-subunit complex. Each toxin causes a conformational change in the subunit to which it binds and influences the state of the other subunit through protein-protein interactions.

The heterotropic cooperative effect of scorpion toxin and sea anemone toxin on toxin activation of sodium channels is highly dependent on membrane potential.[4,7,9] As neuroblastoma cells are progressively depolarized from -41 mV to 0 mV, concentration-effect curves for scorpion toxin action are shifted progressively to the right (Figure 2). These results suggest that depolarization reduces the affinity of the scorpion toxin receptor site for its ligands.

^{125}I-labelled scorpion toxin derivatives can be prepared without loss of biological activity.[9,10] These derivatives bind to a single class of receptor sites in electrically excitable neuroblastoma cells and extensive studies have demonstrated that these sites are associated with voltage-sensitive sodium channels.[9,10] Binding of scorpion toxin to these receptor sites is dependent on membrane potential. Binding data at -41 mV

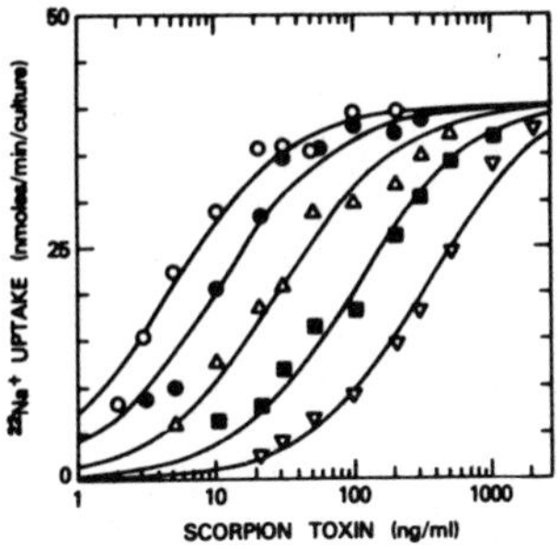

Fig. 2. Effect of depolarization with K^+ on activation of Na^+ channels by scorpion toxin. N18 cells were incubated for 20 min at 36° with the indicated concentrations of scorpion toxin in Na^+-free medium containing 5 mM (O), 10 mM (●), 25 mM (Δ), 60 mM (■), or 135 mM (∇) K^+ and choline Cl such that $[K^+] + [choline^+] = 135$ mM. Cells were rinsed to remove unbound scorpion toxin and excess K^+ and the initial rate of $^{22}Na^+$ uptake was measured in medium containing 5 mM K^+. Curves are least-squares fits to the equation $v = V_\infty A/(K_D + A)$ for $V_\infty = 40.2$ nmol/min per culture and K_D = 4.8 ng/ml at 5 mM K^+, 10.0 ng/ml at 10 mM K^+, 28.6 ng/ml at 25 mM K^+, 98.3 ng/ml at 60 mM K^+, and 331 ng/ml at 135 mM K^+. Taken from reference 9 with permission of the National Academy of Science.

and -28 mV are presented as a Scatchard plot in Figure 3. Depolarization causes an increase in the K_D with no effect on the number of receptor sites. The dependence of K_D for monoiodo and diiodo scorpion toxin on membrane potential is presented in Figure 3. The K_D is increased 10 fold for each 31 mV depolarization.

These results suggest that the scorpion toxin receptor site undergoes a conformational change to a low affinity state on depolarization. Scorpion toxin therefore provides a molecular probe of a voltage-dependent change of state of the sodium channel. Recently, the first clamp analyses of membrane currents in neuroblastoma cells have been made.[11,12] The voltage dependence of activation and inactivation of sodium currents in neuroblastoma cells is compared to the voltage dependence of inhibition of scorpion toxin binding in Figure 4. Inactivation of sodium channels is half-maximal at -65 mV and nearly complete at -40 mV (●, Figure 4). In contrast, activation of the sodium current begins at -40 mV and is half-maximal at -28 mV or -21 mV for the two sets of voltage clamp data available. The voltage dependence of inhibition of scorpion toxin binding

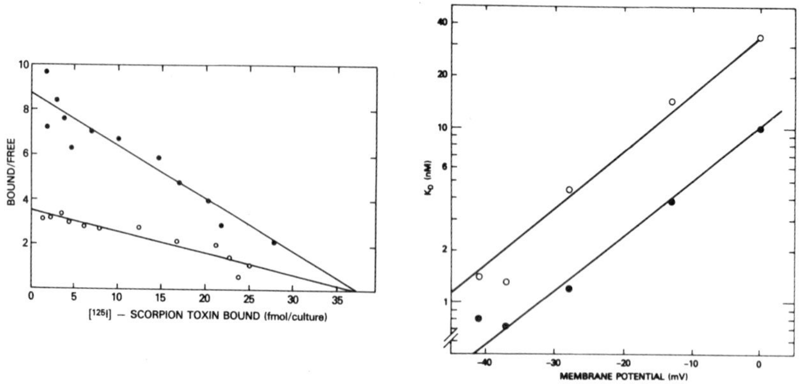

Fig. 3. Effect of membrane potential on scorpion toxin binding. Left. Specific binding of [^{125}I] diiodo-scorpion toxin was measured in the presence of 5 mM K^+ (●) or 25 mM K^+ (0). The data are presented as a Scatchard plot. Right. Specific binding of [^{125}I] monoiodoscorpion toxin (●) and [^{125}I] diiodoscorpion toxin (0) was measured at different membrane potentials: 5 mM K^+, -41 mV; 10 mM K^+, -37 mV; 25 mM K^+, -28 mV; 60 mM K^+, -13 mV; and 135 mM K^+, 0 mV. Dissociation constants were calculated and plotted as a function of average membrane potential. Taken from reference 10 with permission of the American Society of Biological Chemists.

parallels the voltage dependence of activation of sodium channels (0, Figure 4). The close correspondence between these two sets of data indicate that the scorpion toxin receptor site is located on a region of the sodium channel complex which undergoes a voltage-dependent conformational change leading ultimately to activation of the ion channel. Scorpion toxin therefore provides a probe of the voltage sensor which initiates activation of the sodium channel.

Our studies of interaction of neurotoxins with voltage-sensitive sodium channels leads to a view of the sodium channel as a macromolecular membrane complex containing multiple components which interact allosterically and respond to changes in membrane potential with specific conformational changes. Solubilization and isolation of the specific neurotoxin receptor sites should lead to further understanding of the structure and function of the voltage-sensitive sodium channel.

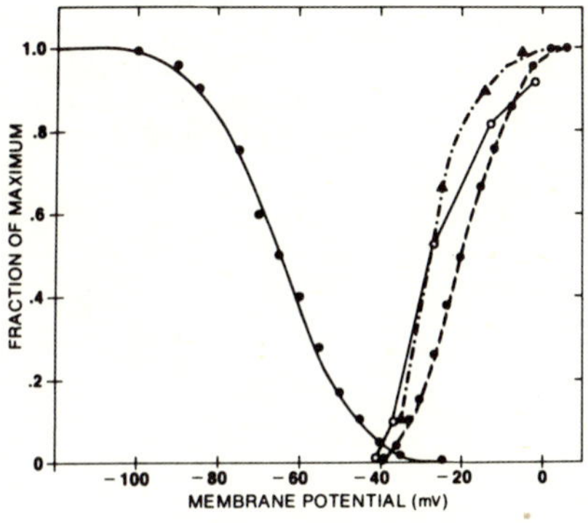

Fig. 4. Comparison of the voltage dependence of scorpion toxin binding and sodium conductance activation and inactivation. Scorpion toxin binding was measured at five different membrane potentials as described in Methods. Fractional inhibition of specific scorpion toxin binding is plotted vs. membrane potential (0). Values of sodium conductance calculated from data in the literature are presented as the fraction of the maximum conductance observed in each experiment. (●——●), inactivation of the peak sodium conductance during voltage steps from the potentials indicated on the abscissa to -10 mV from the data of Moolenaar and Spector [11] on clone N1E-115. (●---●), activation of the peak sodium conductance during voltage steps from -85 mV to the potentials indicated on the abscissa from the data of Moolenaar and Spector[11] on clone N1E-115. (▲---▲), activation of the peak sodium conductance during voltage steps from -100 mV to the potentials indicated on the abscissa from the data of Veselovsky et al.[12] on clone N18.

REFERENCES

1. Catterall, W.A., and Nirenberg, M. (1973) Proc. Nat. Acad. Sci. U.S.A. 70, 3759-3763
2. Catterall, W.A. (1975) J. Biol. Chem. 250, 4053-4059
3. Catterall, W.A. (1977) J. Biol. Chem. 252, 8669-8676
4. Catterall, W.A. (1975) Proc. Nat. Acad. Sci. U.S.A. 72, 1782-1786
5. Catterall, W.A., and Morrow, C.S. (1978) Proc. Nat. Acad. Sci. U.S.A. 75, 218-222
6. Catterall, W.A. (1976) J. Biol. Chem. 251, 5528-5536
7. Catterall, W.A., and Beress, L. (1978) J. Biol. Chem. 253, 7393-7396
8. Monod, J., Wyman, J., and Changeux, J.P. (1965) J. Mol. Biol. 12, 88-118
9. Catterall, W.A., Ray, R., and Morrow, C.S. (1976) Proc. Nat. Acad. Sci. U.S.A. 73, 2682-2686

10. Catterall, W.A. (1977) J. Biol. Chem. 252, 8660-8668
11. Moolenaar, W.H., and Spector, I. (1978) J. Physiol. 278, 265-286
12. Veselovsky, N.S., Kostyuk, P.G., Krishtal, D.A., Naumor, A.P., and Pidoplichko, V.I. (1977) Neurofysiology 9, 641-643

INTERACTION BETWEEN THE BENZODIAZEPINE RECEPTOR, THE GABA RECEPTOR AND A COMMON ANION CHANNEL

Tommaso Costa[1]

David Rodbard[1]

Candace B. Pert[2]

[1]Endocrinology and Reproduction Research Branch
National Institute of Child Health and Human Development
and
[2]Section on Biochemistry and Pharmacology, Biological Psychiatry Branch, National Institute of Mental Health

National Institutes of Health, Bethesda, Md. 20014

ABSTRACT

Benzodiazepines (BZ) interact with brain membrane receptors with high affinity, stereospecificity, and saturability. The existence of these sites suggests the presence of a still undiscovered neurotransmitter which serves as a natural substrate for these receptors. The search for this endogenous ligand, which might be expected to mediate anxiety states, has been futile to date, due to the influence of many nonspecific factors on benzodiazepine binding.

We have systematically investigated the role of ions in modulating BZ binding. Iodide, bromide, chlorife (and other ions of similar size and comparable ability to penetrate post-synaptic membranes) at millimolar concentrations show a significant dose-dependent, temperature-sensitive increase of BZ binding.

Further, there is a significant effect of GABA (and of its agonists and antagonists) on benzodiazepine binding:

1) GABA (as well as iodide) increases the affinity of the BZ receptor for BZ; 2) The effects of GABA and iodide are additive; 3) The effect of GABA on BZ binding is mediated by the GABA receptor, since it can be blocked by the reversible; stereospecific antagonist bicuculline; 4) The effect of bicuculline is enhanced by iodide; 5) Preincubation of membranes with Triton-X100 (or washing or dialysis) attenuates the iodide effect, enhances the GABA effect, and reduces the bicuculline effect; 6) Enhancement of BZ binding by iodide in the presence of GABA appears to vary systematically with brain region, being most pronounced in frontal and occipital cortex and barely detectable in cerebellar cortex and spinal cord. These finding suggest a model, wherein the BZ receptor is intimately coupled to an anion (presumeably chloride) channel, which is also directly coupled with and modulated by the GABA receptor. The possibility of differential coupling of several different neurotransmitter receptors to the same chloride ionophore in different brain regions is discussed.

INTRODUCTION

Despite the enormous clinical popularity of the benzodiazepines (BZ) as minor tranquillizers, the mechanism(s) of action of these drugs has remained obscure. Investigations in numerous laboratories have indicated a role of benzodiaepines in virtually every class of neuronal system which has been studied: cholinergic, adrenergic, serotoninergic, dopaminergic, glycinergic, etc.[1,2] However, in no case could a definitive biochemical mechanism of action be established for BZ.

Early studies suggested that the effects of BZ might be mediated through postsynaptic inhibition.[3,4] This led to a series of recent investigations of the ability of BZ and its analogs to lower the cyclic GMP content of cerebellum.[5,6] These studies indicated that BZ exerted a facilitatory effect on the GABA-ergic system. The role of BZ on the GABA-ergic system was further supported by biochemical, pharmacological and behavioral approaches[7-12,50] However, the mechanism(s) of the effect remained highly controversial, with both synergistic and antagonistic effects being reported.

MEMBRANE RECEPTORS FOR BENZODIAZEPINES

The simultaneous discovery of the BZ receptors by Squires et al.[14] and Mohler et al.[15] provided a major breakthrough. Studies with [^{3}H]-diazepam and with analogs of higher affinity and potency (e.g. [^{3}H]-flunitrazepam) have demonstrated stereospecificity, high affinity, saturability, and selective distribution in brain.[14-18] A very high correlation has been established between the affinity of binding, the potency in experimental systems in animals, and the therapeutic dose in man, for more than 20 analogs of BZ.[17,18] The apparent localization of BZ receptors to synaptosomes,[17] and the inability of any of the known or putative neurotransmitters to competitively displace BZ from the receptor, has raised the exciting possibility that all of the effects of BZ may be due to its mimicry or blockade of an as yet unidentified neurotransmitter.[14,20] This problem is now directly analogous to the opiate receptor prior to the discovery of the enkephalins and endorphins, with an "orphan" receptor in search of its natural ligand. Repeated attempts in our laboratory (unpublished data) and several others have failed to isolate this new neurotransmitter, but have led to improved understanding of the BZ receptor, which appears to be influenced by a variety of both specific and non-specific factors (ions, phospholipids, choline, etc.). One group reported a low molecular weight, protease-resistant factor in bovine brain extracts with an ability to displace [^{3}H]-diazepam from rat brain membrane receptors,[21] but this material was later identified as hypoxanthine.[22] A novel class of purine-like compounds which posses anti-anxiety properties is currently under active investigation (A. Lippa, personal communication). Another laboratory has claimed extraction of a "factor" from muscle which also displaces BZ from its receptor,[23] but the specificity of the effect and the nature of the "substance" remains in doubt. The search for the ligand continues, but it is hampered by the lack of an _in vitro_ bioassay and the lack of antagonists to permit the evaluation of specificity;

Studies in our laboratory with a glioma-neuroblastoma hybrid cell line showed a homogeneous class of binding sites for [^{3}H]-flunitrazepam with a K_d of 26 nM which is twenty-fold higher than for the brain membrane receptor (Fig. 1). These low affinity sites

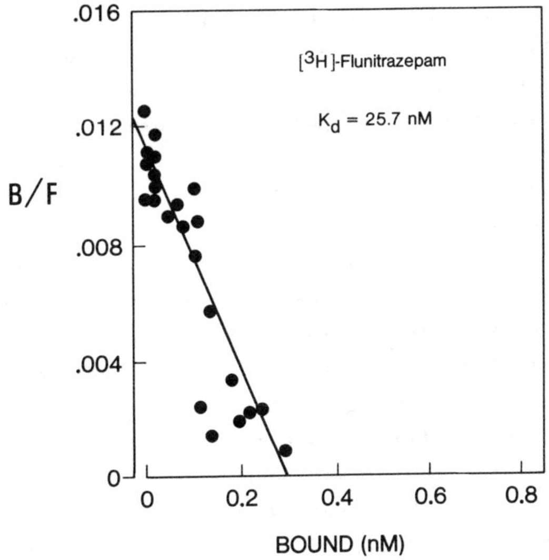

Fig. 1: Scatchard analysis of [^{3}H]-Flunitrazepam binding to NG 108-15 cells. Cells were washed 3 times in Tris-Cl buffer (50mM, pH 7.0) and resuspended at a concentration of 1-3 millions/ml. A fixed amount of tracer (0.3nM) with increasing concentrations of unlabeled drug were used. Binding has been corrected for the presence of 1 μM. Flunitrazepam. The equilibrium constant was computed using "SCATFIT", a nonlinear curve fitting computer program for Scatchard analysis.[62]

resemble those reported by Squires et al.[17] for kidney membranes, and we have observed similar low affinity binding on common fibroblast cell lines (3T12 and primary cultures of fibroblasts). A recent report that BZ receptors are indeed localized mainly to glia is based primarely on negative data, i.e. failure to detect alterations in density of receptors after numerous lesions.[24-26] However, "nervous" mice which have a genetic deficiency of Purkinje cells in the cerebellum, show a decrased number of receptors, again consistent with a neuronal localization.[27,28]

EFFECTS OF IONS ON BRAIN MEMBRANE RECEPTORS

Ionic strength, temperature, and pH are critical variables which must be optimized for each receptor system. Increasing ionic strength usually leads to a reduction of the equilibrium constant of association, although this effect (weakening electrostatic interactions) is usually present only at supra-physiological concentrations and is largely independent of the nature of the ion employed.[29,30] Investigation of the opiate receptor revealed a new kind of highly specific ion effect.[33] The sodium ion profoundly decreases the binding of opiate agonists (alkaloids or peptides), while increasing the binding of opiate antagonists. This effect is dose dependent (with significant activity at 1 mM), markedly temperature dependent, and mimicked only by lithium (K^+, Ru^+, Ce^+ and organic anions are completely inactive). It was suggested[32] that Na^+ may modulate the transition between an "agonist" and "antagonist" form of the receptor, and later, that this may be mediated by direct coupling of the receptor to an ion channel.[33] Zieglgänsberger has demonstrated that the inhibitory effects of opiate agonists on acetylcholine and glutamate-excited neurons is due to a change in membrane Na^+ permeability.[34] Other specific effects of ions suggesting direct linkage of receptor and ionophore have subsequently been reported in several other systems: e.g. Na^+ decreases the binding of alpha-adrenergic agonists, whereas the binding of antagonists are unaffected.[35]

One group of anions with a radius close to that of chloride are known to inhibit the post-synaptic potential in response to electrical stimulation.[43] In the first experiments to demonstrate specific (neurophysiologically relevant) ionic effects on neurotransmitter receptor binding, the same anions were shown to regulate the the affinity of the glycine receptor for [^{3}H]-strychnine.[36] Similarly, anions increase the binding of the GABA-antagonist bicuculline, while having no effect or decreasing the binding of GABA or its "superagonist", muscimol.[37,38] These studies of "specific" ion mediated effects on the receptor have been extremely useful. For instance, they accurately predicted that the endogenous ligand for the opiate receptor would be an agonist, even prior to its isolation.[39] However, such studies provide only "presumptive" evidence: definitive proof of the proposed receptor-ionophore coupling will ultimately require the isolation of all of the individual components, and demonstration of their physico-chemical interaction. Even for the highly purified and relatively abundant nicotinic receptor, such studies are only now becoming available.[40]

EFFECTS OF ANIONS ON THE BENZODIAZEPINE RECEPTOR

Previous studies by other workers[41-13] failed to demonstrate an effect of monovalent ions on BZ binding. The customary use of a Tris-Cl buffer masked such an effect, but use of a Tris-Phosphate or Tris-maleate buffer combined with the use of smaller concentrations of labeled ligand (0.15 - 0.30 nM) permitted us to detect a significant dose-dependent anion effect (Fig. 2) for iodide, bromide, chloride,

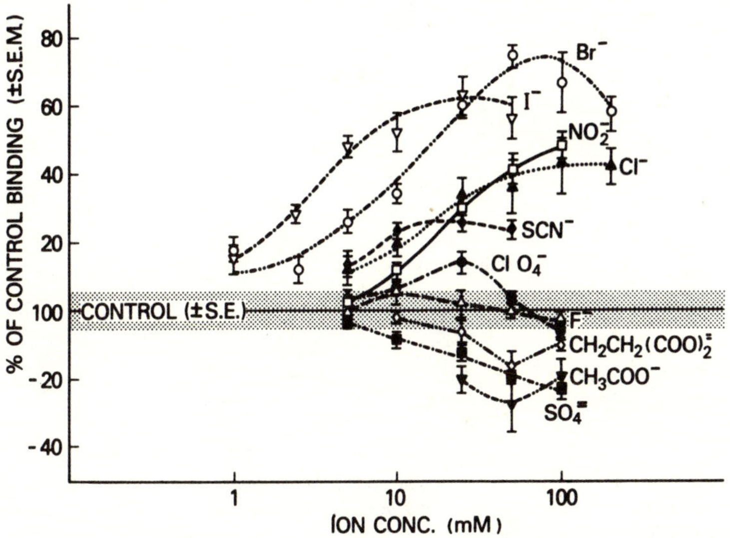

Fig. 2: Effects of anions on [^{3}H]-Diazepam binding. [^{3}H]-Diazepam binding to crude rat brain membranes was carried out as described (12), but 0.3 nM of tracer and Tris Maleate buffer (50 mM pH 7.4 at 4°C) were used. Anions shown were used as ammonium or potassium salts, and did not affect final pH of the incubation mixture even at the highest concentration employed.

nitrite and thiocyanate.[42] There was no direct correlation between the magnitude of the maximal effect and the ED_{50}: Br^- showed the highest effect on the former, while I^- showed the smallest ED_{50} The effect was biphasic in all cases, with reduced binding at higher ionic strenght.

Thirteen other ions (including fluoride) were inactive or decreased the binding of BZ, presumeably through the non-specific effects of high ionic strength.

Scatchard analysis of BZ binding indicated that the effect of I^- or Br^- was due to an increase in the affinity constant, with no significant change in the number of binding sites. Studies of the initial association rate were difficult due to the rapidity of binding ($t_{1/2}$ less than 2 min), but dissociation studies indicated that the change in affinity could be completely accounted for by an increase in k_{off} (Figure 3).

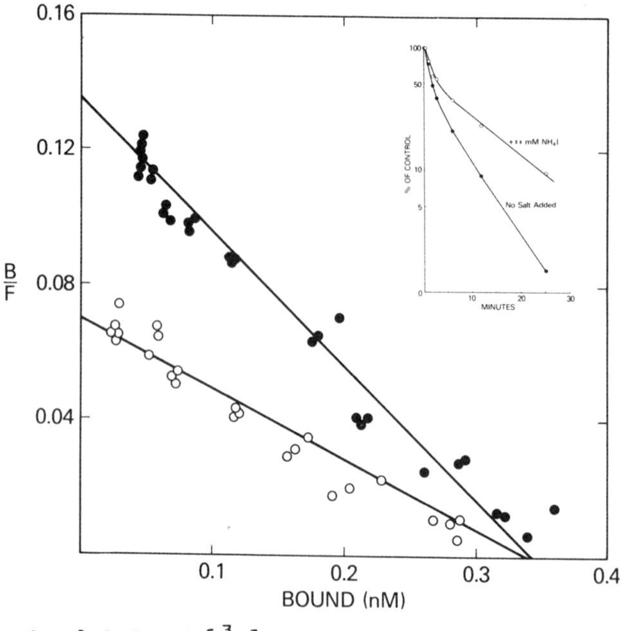

Fig. 3: Scatchard anlyisis of [^{3}H]-Diazepam binding to rat brain membranes in the presence (●) or in the absence (o) of iodide. Experimental conditions as in Fig. 2. A major change in affinity was detected in three separate experiments. The equilibrium constant of association (K) was increased by a factor of 1.4 $\pm$ 0.04 (sem), while the binding capacities were not significantly changed.

The insert shows the dissociation curve for [^{3}H]-Diazepam obtained by a 1000 fold isotopic dilution of the tracer after equilibration for 1 hr. at 0°C. The ratio of $t_{1/2}$ values (1.7) can fully account for the difference in the equlibrium constant without changes in the association rate constant.

Fig. 4: Temperature dependence of the effect of anions on BZ binding. Conditions as in Figs. 1-3, with Tris maleate buffer adjusted to provide the desired pH at different temperatures, to compensate for the shift in pKs. Bromide(●), chloride(▲) and fluoride (■) were present at 100mM.The line with the shaded area represent control binding ± one standard error of the mean.

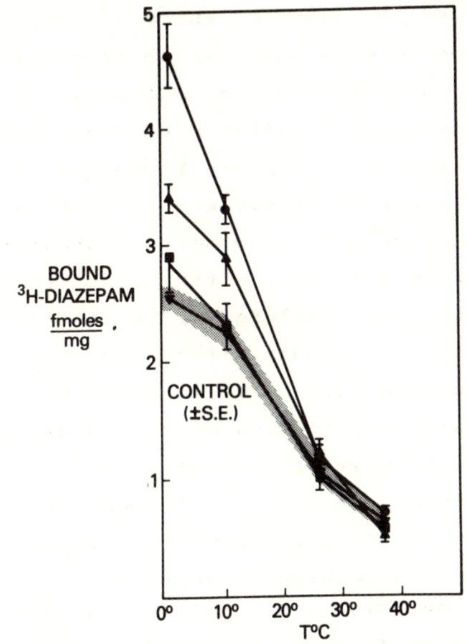

TABLE 1

EFFECTS OF ANIONS ON ^{3}H -DIAZEPAM BINDING AND CORRELATION WITH THEIR PERMEABILITY THROUGH THE CAT MOTONEURON SYNAPSE

ANION	ENHANCEMENT OF BINDING	PERMEABILITY (Araki[43])	ED_{50} (mM)	E_{max}[a] (% over control)
Iodide	+	+	2.8	160
Bromide	+	+	9.0	180
Chloride	+	+	13.0	142
Nitrite	+	+	20.0	148
Thiocyanate	+	+	~10	126
Perchlorate	±	+	----	---
Chlorate	-	+		
Azide	-	+		
Formate	-	+		
Fluoride	-	-		
Oxalate	-	-		
Succinate	-	-		
Acetate	-	-		
Sulfate	-	-		
Lactate	-	-		
Propionate	-	-		
Maleate	-	-		
Phosphate	-	-		

[a]Note: ED_{max} = Maximal Binding relative to control.

The anion effect is strongly temperature dependent (Fig. 4), being maximal at 0°C and undetectable at 37°C. Table I compares the activity of the anions with their activity in the cat motoneuron system used by Araki:[43-45] a significant correlation is observed. Fluoride is the only halide which is inactive in both systems. Three anions (perchlorate, azide, nitrate) which were capable of modulating the postsynaptic response were inactive in terms of BZ binding. This discrepancy is as yet unexplained; conceivably, it may be related to the relatively high chemical reactivity of these species.

The correlation between binding and neurophysiological activity of these ions, though imperfect, encourages us to further explore the possibility of a direct relationship between the BZ receptor and the chloride channel. However, several alternative hypotheses must be considered. First, one must exclude the possibility that the effects observed are due to the chaotropic activities of the ions, or due to effects on water structure (and hence membrane structure). We regard this possibility as unlikely in view of the inactivity of the chlorate anion, which is known to have strong chaotropic activity,[46] and the ability to observe effects in the 1 - 10 mM concentration range. The absence of ion effects at 37°C is puzzling, but we must bear in mind that this may be merely a problem of detection. Conformational interconversion with concurrent alteractions in ionic flux occurs so rapidly at physiological temperatures that artifactually low temperatures (which "freeze" some components of the membrane) are probably required for detection.[47]

THE INTERACTIONS OF BENZODIAZEPINES, GABA, AND IODIDE

Several studies provided indirect evidence for interaction of BZ and GABA[48-50] However, GABA and its analogs do not displace [^{3}H]-diazepam from its receptors _in vitro_ [19] Recently, Tallman _et al._[51] have demonstrated that micromolar concentrations of GABA resulted in a small but statistically significant increase in BZ binding, due to an increase in the affinity constant. This effect was presumeably mediated at the receptor level, because 1) a "superagonist", muscimol, is more potent than GABA itself, and 2) bicuculline (a known antagonist of GABA) could fully reverse the effect of GABA on BZ binding. Amino-oxyacetic acid (AOAA) which increases GABA levels in the synapse by competing for the uptake of GABA by synaptosomes[52] and for the transaminases which inactivate GABA[53] also increases [^{3}H]-Diazepam binding after _in vivo_ administration.[54]

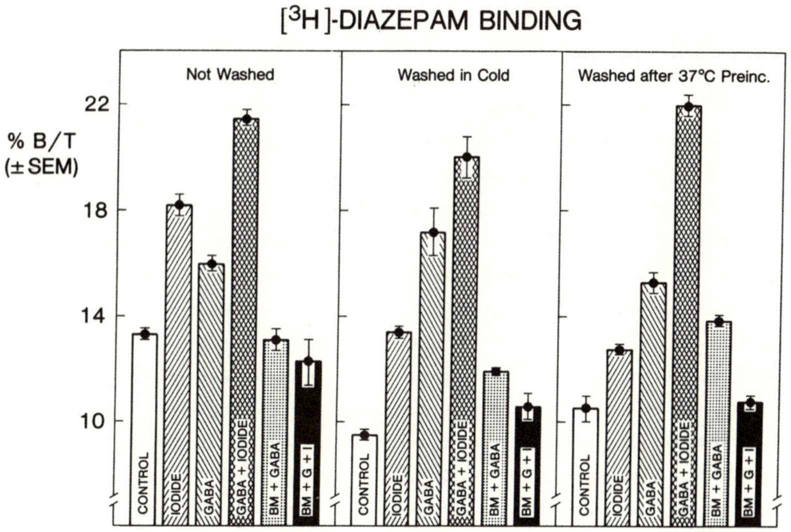

Fig. 5: Effect of GABA, bicuculline and iodide on [^{3}H]-Diazepam binding to rat brain membranes. GABA (50 μM.), iodide (25 mM) and bicuculline methiodide (50 μM) were present in this and in the following experiments. All experiments performed using aliquots of a single membrane preparation.

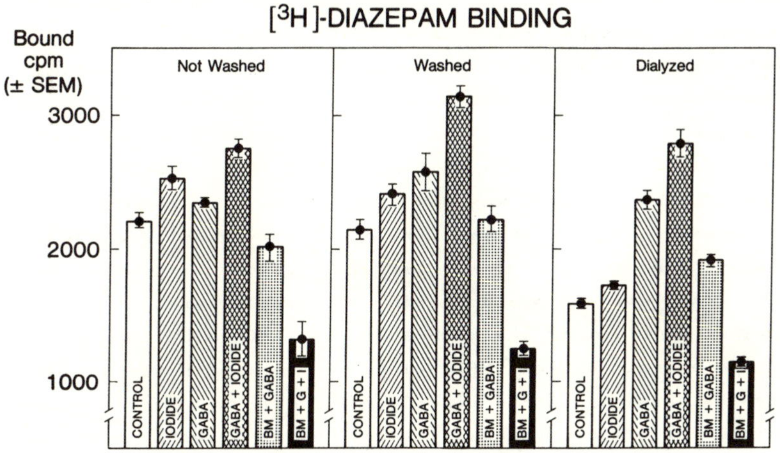

Fig. 6: Effect of various ligands on [^{3}H]-Diazepam binding on different membranes preparations of rat brain. Conditions per Fig. 5. Dialysis was carried out for 1 day at 4°C. against Tris-Maleate buffer (200 fold volume excess). Control membranes were stored for 1 day at 0-4°C.

High-affinity GABA binding is absent from freshly prepared brain membranes, but becomes detectable after freezing and exposure to detergent.[55] This has been attributed to the removal of an endogenous inhibitor of GABA binding. This "inhibitor" is sensitive to trypsin, has an apparent molecular weight of 15000 on gel filtration, and is said to reduce the affinity of GABA binding, in a dose-dependent manner[56] It may also reduce the afffinity of BZ for its receptors, although a dose-dependent effect has not been demonstrated.[57] Guidotti et al. have further proposed that the benzodiazepines are responsible for displacing the inhibitor from the membranes, thereby exposing GABA receptor sites.[57]

With this background, we have investigated the complex interactions of BZ, ions, GABA and GABA antagonists. Figures 5 and 6 show the influence of these factors on [^{3}H]-diazepam binding. Using fresh unwashed membrane preparations, we confirmed the above-noted effects of iodide and of GABA. Further, the effects of iodide and GABA appeared to be additive. Bicuculline inhibited the effect of GABA, and iodide appeared to accentuate the bicuculline effect. All of these effects were statistically significant and reproducible, but the magnitude of the I^- effect was variable between experiments. Accordingly, we examined several details of the procedure for preparing membranes which might affect the concentrations of endogenous ligands, ions, GABA, "GABA inhibitor", etc. Washing of the membrane (three times for one hour at 0-4°C)(Fig. 5, middle panel) always resulted in an inversion of the magnitude of BZ binding in the presence of I^- or GABA alone (N = 4). Binding in the presence of I^- and GABA was essentially unchanged, but the blockade of the GABA effect by bicuculline was now incomplete. The possibility that these effects were due to instability of the membrane (perhaps proteolysis, degradation of ligands) was largely excluded since the above findings were unaffected by preincubation of the membranes at 37°C for 45 min. Further, storage of the membranes at 0-4°C for 24 hours had no effect on the overall pattern (Figure 6, left panel) or on the effect of washing (Fig. 6, middle panel), while dialysis for 24 hours produced a pattern of effects virtually identical to those of washing.

The effects of detergent treatment of membranes are shown in Figure 7. Here the membranes were exposed to Triton X-100 for 30 min, followed by extensive washing (×3) so that the detergent should not interfere with the binding assay per se. Under these conditions, Triton-X100 results in a dose-dependent enhancement of the GABA effect, and a systematic decrease in inhibitory effects of

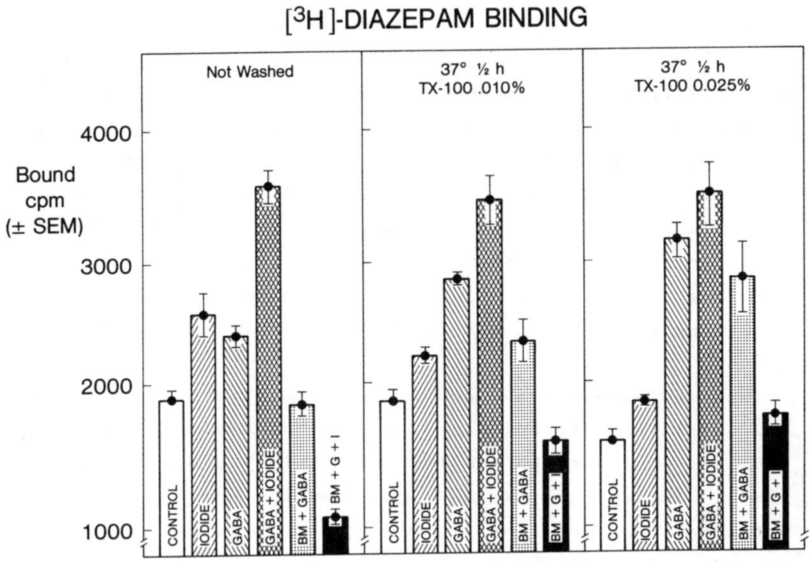

Fig. 7: Effect of various ligands on [^{3}H]-Diazepam binding after Triton treatment of rat brain membranes. Conditions as above, with Triton X-100 in Tris-Maleate buffer incubated with membranes for 45 minutes at 37°C. Membranes were washed 3 times before the incubation to remove most of the Triton.

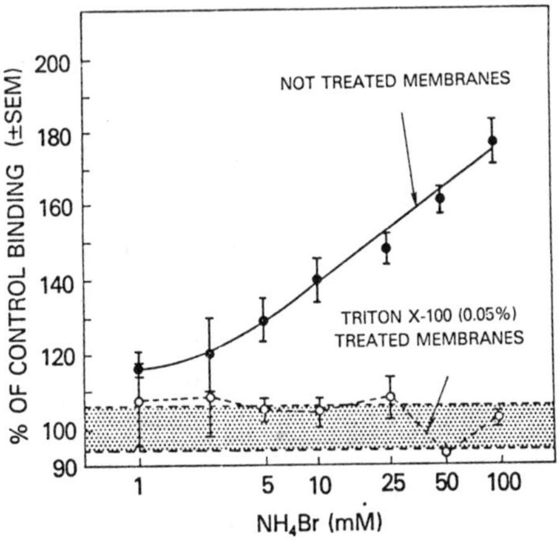

Fig. 8: Effect of increasing concentrations of ammonium bromide on normal or detergent treated membranes. Detergent treatment as before, at a concentration of 0.05% (v/v) Control membranes were not washed.

bicuculline, either in the absence or presence of iodide. Triton also results in a progressive loss of the anion effect when iodide or bromide is present alone, and completely obliterates this effect at concentrations of 0.05% (Figure 8).

When membranes are incubated with Triton and assayed for BZ binding activity without attempting to remove the detergent, one obtains a significant enhancement of binding under all conditions (Figure 9, middle and right panels), although the pattern of effects resembles that for the control membranes with an anion effect which is larger than the GABA effect. Figure 9 (left panel) confirms that Triton treatment followed by washing results in a relative loss of the anion effect.

We have examined the possibility that the benzodiazepine receptor may exist in different forms, or have differential coupling to the GABA receptor through the ion channel in different regions of the brain. Accordingly, we have examined the effects of GABA in the presence and absence of I^- on membranes prepared from four separate regions: frontal and occipital cortex, cerebellar cortex, and spinal cord (Fig. 10). Binding in spinal cord and cerebellum were much lower

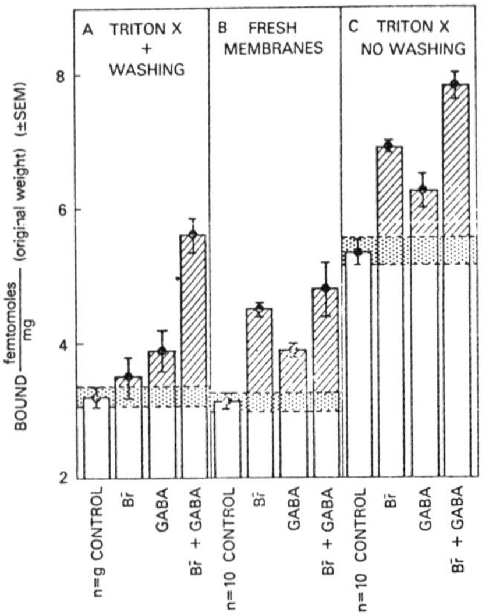

Fig. 9: Effect of washing after Triton X-100 treatment of membranes. Detergent treatment as in fig.7, with and without washing of membranes prior to incubation with [^{3}H]-Diazepam.

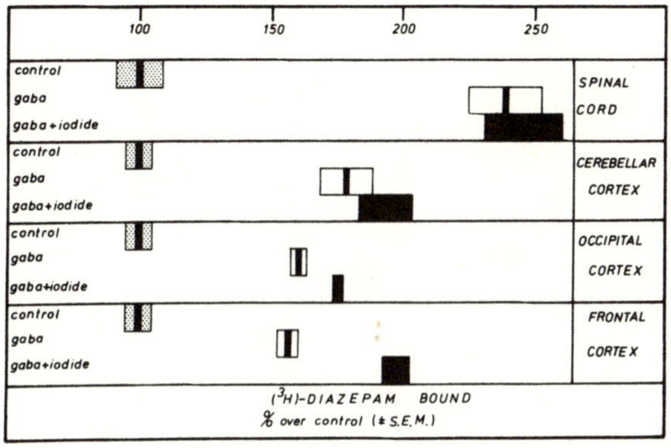

Fig.10. Binding of [^{3}H]diazepam to membranes prepared from different regions of cat brain, in the presence of GABA and in the presence of GABA with 25mM Iodide. All results are expressed relative to the controls. The initial percent binding of ligand varies for the four regions: 0.3%, 5% , 10% and 12% for spinal cord, cerebellar cortex, occipital and frontal cortex, respectively.

than in cerebral cortex; the <u>relative</u> enhancement of binding by GABA appeared higher for spinal cord and cerebral cortex than for cerebral cortex. The relative enhancement of BZ binding is present in all regions, and is inversely related to the initial binding of BZ. The ability of iodide to further increase binding in the presence of GABA is directly proportional to the initial level of binding. The present data suggest that the iodide effect is undetectable in spinal cord, thus implying a qualitative change in behavior. These results suggest, but do not conclusively establish, a qualitative difference in the nature of the BZ receptor in various regions of the brain. Even so it is interesting to speculate that different combinations of neurotransmitters receptors (recognition units) may be coupled to the same ionophore, depending upon the region of the brain being studied (Fig. 11). For example, the observation that [^{3}H]-strychnine binding to the glycine receptor in membrane preparations from spinal cord is inhibited by various BZ analogs with <u>in vitro</u> affinities which parallel <u>in vivo</u> potency,[58] suggests that the glycine receptor may interact with the GABA-BZ-Chloride ionophore in some CNS regions. Neuorophysiological[59,44] and binding studies[36] have established an association between strychnine receptor and a chloride channel. On the basis of neurophysiological studies, Carpenter[60,61] has previously proposed that several recognition units may share the same ionophore with a specific pattern for each neurotransmitter system.

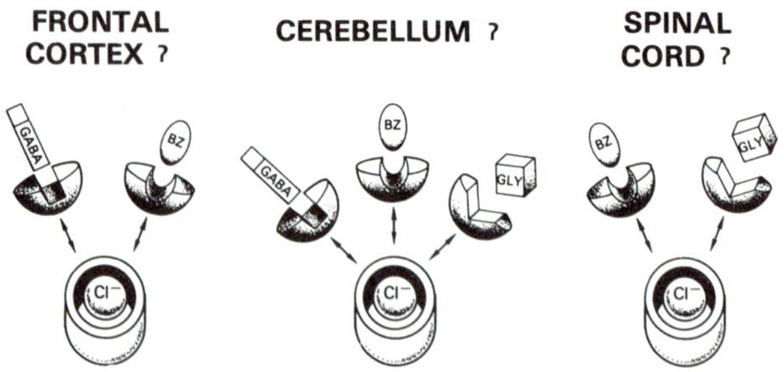

Fig.11. Schematic representation of the interactions between the receptors for benzodiazepines (and the corresponding natural neurotransmitter), GABA and glycine, with a common anion channel. The relative number of receptors, and their ability to interact with the ionophore, differs significantly in various regions of the brain.

CONCLUSION

The benzodiazepine receptor displays many parallels with the opiate receptor, with significant modulation by ions, temperature, and with complex interactions of these factors with the presence of agonists and antagonists of the GABA-ergic system. The present data are consistent with the working hypothesis that the BZ receptors normally interact with an endogenous ligand important in anxiety states, and that both this receptor and the GABA receptor are coupled to a common anion channel. Such a mechanism would be quite novel, and its further investigation is likely to lead to new insights into neural - and membrane - control mechanisms.

ACKNOWLEDGEMENTS

Dr. F. Cortese provided access to a Neuroblastoma-Glyoma Hybrid cell line (NG 108-15). Dr. S. Adamo kindly supplied us with mouse fibroblasts cell lines (3T12 and primary explants).

REFERENCES

1. Garattini, S., Mussini, E. and Randall, L.: (1973). "The Benzodiazepines", Raven Press, N. Y.
2. Costa, E. and Greengard, P.(1975) "Mechanism of action of Benzodiazepines". Raven Press, N.Y.
2. Schmidt, R.E.,Vogel, M.E. and Zimmermann. (1967) Naunyn Schmiedebergs Arch. Pharmacol. Exptl. Pathol. 258, 68-82.
3. Stratten, W.P., and Barnes, C.D. (1971) Neuropharmacol. 10, 685-696.
5. Mao, C.C., Guidotti, A. and Costa, E. (1974) Brain Res. 79, 510-514.
6. Costa, E., Guidotti, A., Mao, C.C. and Suria, A. (1975) Life Sci. 17, 167.
7. Haefely, H., Kulscar, A., Mohler, H., Pieri, L., Polc, P. and Schaffner, R. (1975) in "Ädv. Biochem. Psycopharm. " (E. Costa and P. Greengard eds.) vol. 12 p. 131
8. Curtiss, D.R., Game, C.J.A. and Lodge, D. (1976) Exp. Brain Res. 25, 413.
9. Kozhechkin, S.N. and Ostrovskaya, R.U. (1977) Nature 269, 72.
10. Dray, A. and Stranghan D.W. (1976) J. Pharm. Pharmacol. 28, 314.
11. Gähwiler, B.H. (1976) Brain Res 107, 176.
12. Steiner, F.A. and Felix, D. (1976). Nature 260, 346.
13. Squires, R.F. and Braestrup, C. (1977). Nature 266, 732.
14. Mohler, H. and Okada, T. (1977) Life Sci. 20, 2101-2110.
15. Mohler, H. and Okada, T. (1977) Science 198, 849-851.
16. Speth, R.C., Wastek, G.J., Johnston, P.C. and Yamamura, H. (1978) Life Sci. 22,859-870.
17. Braestrup, C. and Squires, R. F. (1977) Proc. Natl. Acad. Sci. U. S. A. 74, 3805-3809.
18. Mohler, H. and Okada, T. (1978) Life Sci. 22, 985-996.
19. Braestrup, C. and Squires, R. F. (1978) Eur. J. Pharmacol. 48, 263-270.
20. Iversen, L. L. (1977) Nature 266, 678.
21. Marangos, P.J., Paul, S.M., Greenlaw, P., Goodwin, F.K. and Skolnik, P. (1978) Life Sci. 22, 1893-1900.
22. Skolnik, P., Marangos, P.J., Goodwin, F.K., Edwards, M. and Paul ,S. (1978). Life Sci. 23, 1473-1480.
23. Karobath, M., Sperk, G. and Schonbeck, G. (1978) Eur. J. Pharmacol. 49, 323-326.
24. Chang, R.S.L. and Snyder, S.H. (1978) Eur. J. Pharmacol. 48, 213-218.
25. Williamson, M.J., Paul, S.M. and Skolnik, P. (1978) Nature 275, 551-553.
26. Chang, R.S.L. and Snyder, S.H. (1978) (To be published).
27. Lippa, A.S., Sano, M.C., Carpet, J., Klepner, C.A. and Beer, B. (1978) Life Sci. 23, 2213-2218.
28. Paul, S. (personal communication)
29. Cuatrecasas P. (1971) Proc. Natl. Acad. Sci. U. S. A. 68, 1264-1268.
30. Freychet, P., Roth, J. and Neville, D.M. Jr. (1971) Proc. Natl. Acad. Sci. U. S. A. 68, 1833-1837.
31. Pert, C.B. and Snyder, S.H. (1974) Mol. Pharmacol. 10, 868-879.

32. Pert, C.B. and Snyder, S.H. (1975) Neurosci.Res.Bull. 13, 73-79.
33. Pert, C.B. and Garland, B.L. (1978) in "Receptors and Hormone action" (Birnbaumer L. and O'Malley B.W. eds.) pp.535-549. Academic Press. N.Y.
34. Zieglgänsberger, W. and Bayrel, H. (1974) Brain Res. 115, 11-128.
35. Tsai, B.S. and Lefkowitz, R.J. (1978) Mol. Pharmacol. 14, 540-548.
36. Young, A.B. and Snyder, S.H. (1974) Proc. Natl. Acad. Sci. U. S. A. 71, 4002-4005.
37. Enna, S.J. and Snyder, S.H. (1977) Mol. Pharmacol. 13, 452-453.
38. Mohler, H. and Okada, T. (1978) Mol. Pharmacol. 14, 256-265.
39. Pasternak, G.W., Simantov, R. and Snyder, S.H. (1976) Mol. Pharmacol. 12, 504-513.
40. Eldefrawi, M.E. and Eldefrawi, A.T. (1977) in "Receptor and Recognition. Series A, Vol. 4 (Cuatrecasas, P., Greaves, M. F., eds.) pp. 197-258, Chapman and Hall, London.
41. Mackerer, C.R. and Kochman, R.L. (1978) Proc.Soc.Exptl.Biol.Med. 158, 393-397.
42. Costa, T., Rodbard, D. and Pert, C.B. (1979) Nature (In press).
43. Araki T., Ito M.and Oscarsson O. (1962) J. Physiol. 164, 150-156.
44. Eccles, J.C. (1964) in "The Physiology of Synapses" (Academic Press, N. Y.) Proc.Natl.Acad.Sci.U.S.A. 62,1120.
45. Eccles, J.C. (1969) Ann. N. Y. Acad. Sci. 137, 473-495.
46. Hatefy, Y. and Hanstein, W.G. (1969) Proc. Natl. Acad. Sci. U.S.A. 62,1120.
47. Creese, I., Pasternak, G., Pert, C.B., and Snyder, S.H. (1975) Life Sci. 16, 1837.
48. MacDonald, D.W. and Barker, J.L. (1977) Nature 371, 563-564.
49. Choi, D.W., Farb, D.H. and Fischbach, G.D. (1977) Nature 269, 342-344.
50. Gallagher, D.W. (1978) Eur. J. Pharmacol. 49, 133-143.
51. Tallman, J.F., Thomas, J.W. and Gallagher, D.W. (1978) Nature 274, 384-385.
52. Snodgrass, S.R. and Iversen,L.L. (1973) J. Neurochem. 20, 431-439.
53. Collins, G.G.S. (1973) Biochemical Pharmacol. 22, 101-111.
54. Gallagher, D.W., Thomas, J.W., and Tallman, J.F., Biochemical Pharm. 27, 1979.
55. Enna, S.J. and Snyder, S.H. (1975) Brain Res. 93, 168-174.
56. Toffano, A., Guidotti, G. and Costa, E. (1978) Proc. Natl. Acad. Sci. U. S. A. 75, 4024-4028.
57. Guidotti, G., Toffano, G. and Costa,E. (1978) Nature 275, 553-555.
58. Snyder, S.H. and Enna, S.J. (1975) in"Mechanism of action of Benzodiazepines" (Costa E., Greengard P. eds.) Raven Press. pp.81-91.
59. Curtiss, D.R., Duggan, A.W. and Johnston, G.A.R. (1971) Exptl.Brain Res. 12,547-565.
60. Swann, J.W. and Carpenter, D.O. (1975) Nature 258,751-754.
61. Yarowsky, P.J. and Carpenter, D.O. (1978) J.Neurophysiol. 41, 531-541.
62. Rodbard, D. (1973) in "Receptors for Reproductive Hormones" (O'Malley B. W. and Means A. R., eds.) pp. 289-326. Plenum, N.Y.

DISCUSSION

DR. BLUMENTHAL: Thank you, token neuroscientist. I wonder if Jean Pierre could perhaps comment on whether his laboratory has observed the kind of specific ion effects on binding to the acetylcholine receptor that you have observed on the opiate and valium receptors.

DR. CHANGEUX: We have neither a specific sodium nor a specific potassium effect. However, we have observed an effect of calcium on the transition rate.

May I ask a question about the effect of sodium on opiate receptor binding? According to your scheme the antagonists are stabilizing the state which binds sodium. Now, according to Zieglgänsberger, glutamate opens the sodium channel and opiate agonists close it. Therefore, why don't opiate antagonists act as depolarizing agents?

DR. PERT: You are quite right--Dr. Zieglgänsberger has only been able to show excitatory changes with naloxone in morphine-dependent animals. There are two possible explanations for this. It is possible that opiate receptors which are occupied by endogenous opiate peptides are unable to release endorphins easily. There is, in fact, some experimental evidence which suggests that endogenous opiate peptides dissociate extraordinarily slowly from opiate receptors prepared from brain membranes which appear to be about 50% occupied. Release of endogenous ligand from opiate receptors as measured by the increased ability to bind tritiated D-Ala-enkephalin as preincubation time increases proceeds extraordinarily slowly even at physiological temperatures, particularly in the presence of manganese (see Pert and Bowie in Endorphins in Mental Health Research, Usdin, E., Bunney, W.E. and Kline, N.S. (eds.), London, Macmillan Press, 1979). Unoccupied opiate receptors would, according to the model, already be in the antagonist conformation and, thus, would be unaffected by naloxone. It is certainly conceivable that there are states of opiate receptor occupancy by endogenous ligand that are not reversible by naloxone.

Recently work carried out in our laboratory in collaboration with Drs. Duncan Taylor and Agu Pert suggests there are two types of opiate receptors, each with its own characteristic peripheral tissue and brain distribution, and each with its own chemical specificity. We began our investigations as a result of Dr. John Kababian's convincing proposal that dopamine receptors come in two types; Type I is coupled to adenylate cyclase, and the other, Type II, is not. The parathyroid dopamine receptor is the prototype for Type I as characterized by Ed Brown, and the pituitary dopamine receptor is the prototype for Type II. Each receptor has a different drug specificity with certain antipsychotic agents able to antagonize only the Type II receptor, for example. We have been able to use the effects of GTP on binding to assess whether opiate receptors are coupled to adenylate cyclase. Specific effects of guanyl nucleotides on opiate receptor binding were first reported by Art Blume last year in PNAS. These effects are similar to what was originally reported by Dr. Rodbell for the glucagon receptor and later by Dr. Lefkowitz for the beta-adrenergic receptor, and recently by Drs. U'Pritchard on alpha-receptor binding. Similarly, Childers and Snyder have shown that GTP reduces opiate agonist binding but has no effect on opiate antagonist binding in the presence of sodium chloride. We have studied GTP's ability to reverse the inhibition of tritiated diprenorphine binding by met-enkephalin. There is a dramatic difference in GTP's ability to produce this effect from one brain region and tissue to another. Opiate receptors punched from brain regions shown previously to mediate analgesia in the rat such as the periaqueductal gray and pontine reticular formation, are extremely susceptible to reversal of met-enkephalin inhibition by GTP. Other brain regions, including the hypothalamus, amygdala and frontal cortex appear resistant to GTP effects even at concentrations 50-fold higher than the 2-μM concentration required to produce dramatic effects in the analgesic receptor areas and, in fact, most other brain areas. Binding to the guinea pig ileum displays the same type of GTP sensitivity as that displayed by the analgesic receptors. This is consistent with Hans Kosterlitz's reports that the structure-activity relationships of numerous opiate alkaloids and peptides in producing analgesia and inhibition of muscle contraction in the ileum are virtually identical, with morphine being more potent than leu-enkephalin. By contrast, vas deferens opiate receptors seem to display a completely different order of drug potency with the enkephalins, particularly leu-enkephalin, being much more potent than morphine. Vas deferens opiate receptors and neuroblastoma glioma opiate receptors characterized by Klee and Nirenburg show a similar resistence to GTP effects on opiate receptor binding in the higher brain areas. We are suggesting that opiate receptors in the guinea pig ileum, analgesic areas, and most of the brain are Type I opiate receptors, coupled to adenylate cyclase and having a much greater preference for morphine than leucine-enkephalin. Opiate receptors in the vas deferens, neuroblastoma glioma, and frontal cortex, hypothalamus, and amygdala are Type II opiate receptors, uncoupled to adenylate cyclase in the classical way but certainly capable (as Klee and Nirenberg have shown) of regulating cyclase activity by reversing protoglandin-induced activity. We have shown that frontal cortical receptors and neuroblastoma glioma receptors bind tritiated leu-enkephalin and other peptide opiates much more readily than tritiated morphine. At this time, we have some indication that these Type II "weird" opiate receptors are extremely resistant to naloxone which fails to displace tritiated leu-enkephalin binding, even at concentrations in the 10 μM range.

DR. HECHTER: Do you have any evidence that your sodium binding site is related to transport?

DR. PERT: No, in fact we have some evidence to the contrary. Oabain and other manipulations known to effect the sodium pump has no effect on binding.

DR. HECHTER: With respect to peptide agonists and antagonists for the opiate receptor--in other words, for the enkephalins and for endorphins and for analogs thereof which are antagonists, do you get the sodium effect?

DR. PERT: Currently no potent opiate peptide antagonists have been synthesized, or at least published. We tried to create a peptide-opiate antagonist in collaboration with Erhard Gross by placing an N-allyl group on the N-terminal tyrosine of D-Ala-enkephalin in analogy with what has been done to produce opiate alkaloid antagonists. We created a peptide that had a very low affinity, but a small sodium shift indicative of an antagonist. By contrast, all the other opiate peptides which are considered to be agonists have sodium shifts of eight or nine and higher. The opiate peptide antagonist which we made showed a very weak ability to precipitate some opiate withdrawal signs, but otherwise lacked demonstrable antagonist activity.

DR. BLUMENTHAL: Jesse Roth?

DR. ROTH: Candace, as usual, the ideas that you present are very provocative and exciting. I would like to make a small distinction. Clearly, since any ligand-ligand interaction is very sensitive to the ionic environment, are some _in vitro_ ionic manipulations producing major configurational alterations which, while theoretically interesting, could never really occur in a physiological or an _in vivo_ situation? If so, _in vitro_ ionic manipulations would be interesting as probes, but might not necessarily reflect the pathophysiological range of _in vivo_ness. Yet in some cases, such small shifts in ionic environment might, in fact, be regulators.

Let me give an example. In the insulin receptor, if you assume that pH 6.8 to 7.8 are the limits that are acceptable to the living state, since this pH range is associated with major shifts in the affinity and dissociation rate of the insulin receptor for insulin, pH, in fact, is likely to be an important regulator of how insulin reacts with its receptor.

On the other hand, calcium and magnesium, produce very large changes in insulin binding _in vitro_ but these are at such high concentrations that they are interesting in being probes for how the receptors are structured and things that are going on, but they wouldn't be regulators.

Alternatively, if you take the ACTH receptor that is exquisitely sensitive to calcium concentrations that are within the five millimolar range plus or minus a few millimoles, that is the range that occurs in vivo, and, in fact, over that range, major changes in the affinity of ACTH for the receptor are occurring, where with pH you have to get extravagant in order to change the ACTH binding.

As an analogous example, I would cite the angiotensin binding studies that Kevin Catt's group has done, where sodium and potassium within the imaginable range do cause major changes in the binding. I think that it would be useful to make this distinction. I think the sodium effects on the opiate receptor, and certainly the chloride effects on the valium receptor, are interesting probes, but do they occur in vivo?

DR. PERT: Yes, I think the evidence that these highly specific ion effects are biochemical correlates of neurophysiological events is overwhelming. In terms of the anion effects on the valium receptor, they are in the physiological concentration range. True, they are only detectable at the non-physiological temperature of zero degrees, but I would like to make the distinction between being detectable and really happening.

I can envision that valium receptor occupancy alters chloride conductance in an important way in vivo, yet the biochemical correlate of this effect, the affinity enhancement of [^{3}H]valium caused by specific anions is undectable at 37°C. After all, what are we doing? We are stopping a rapidly dissociating receptor complex on a piece of filter paper. Things may oscillate so rapidly at 37°C that they are undectable by biochemical methods yet compatible with ionic fluxes occurring on a microsecond basis.

DR. ROTH: I wasn't trying to make the decision for you about which were probes and which were in vivo regulators, but just trying to emphasize the usefulness of that distinction. And zero degrees, that is an in vivo temperature for a large part of the living world.

(Laughter)

You emphasized how neuroscientists have realized how numerous peptides are acting as neuromodulators in brain. I wanted to raise some studies that Jana Havrankova has been doing in our group with insulin in the brain. She has found that the brain is chock full of insulin receptors in all regions that have been examined. This is work done in collaboration with Michael Brownstein in which we are finding absolutely bona fide insulin receptors all over the brain.

She has also found that the brain is chock full of insulin, and, in fact, the concentrations of insulin in brain tissue are about 10 to 20 times higher than that circulating in the plasma.

Now, even though other GI peptides and other peptides that were thought to exist only outside the nervous system are now being found in the nervous system, the reason I think the insulin receptor is interesting is obviously because I find insulin interesting, but also because of all the peptides that you speak of, the only one whose peripheral function is accepted and agreed upon is insulin.

In other words, we know that insulin exists to regulate peripheral blood glucose, and therefore, the finding of the peptide there in such high concentrations is really extraordinary.

The other thing is that it looks like the receptor and insulin concentrations in brain do not regulate in the same way that the peripheral ones do since jacking up peripheral insulin receptors or peripheral insulin concentrations or the other way around doesn't seem to affect the brain hormones.

I would add that to your list of central nervous system peptides. It is a little bit more perplexing only because I think we feel that we understand insulin's peripheral function more clearly than we do VIP or bombesin or such things.

DR. DE HAEN: I am really very puzzled by a steadily increasing number of peptides in the brain, and in the context Jesse Roth brought up, I would like to suggest that there is some parallelism between peptides in the brain and the fact that you find certain hormones in higher concentration in seminal fluid than in plasma. It has been suggested that in the case of seminal fluid, it arises from a very rapid liquid turnover, and a relatively poor degradation ratio.

I wonder whether the brain has a similar situation: that an enormous amount of fluid is exchanged all the time, but the peptides cannot be degraded as efficiently, and therefore, they accumulate there and get higher than in the plasma. Therefore, I wonder whether they can be physiologically regulated there.

DR. PERT: I don't think the peptides in the brain are like the peptides in seminal fluid at all. I think that there is really an enormous amount of evidence that they are made _in situ_ to serve physiological roles. Their immunohistofluorescent distribution, where they have been studied, are particularly striking. David Jacobowitz, Terry Moody and I have seen bombesin in the boutons and the axons of nerve terminals. This peptide, like others, has its own characteristic distribution.

It has been shown that potassium, electrical stimulation, and procedures which release other neurotransmitters from brain slices release enkephalin and the other peptides that have thus far been studied. Why should we be surprised at how many peptides there are in brain? The brain is going to turn out to be even more complicated than we imagined it at first. I think that almost every peptide receptor that we can characterize in other places is in the brain.

Actually, there is one definite exception so far and that is the peptide receptor found on lymphocytes which Elliot Schiffmann has shown to be responsible for chemotaxis. While we have shown that neutrophils specifically bind the methionyl peptides which elicit a chemotactic response, the brain sadly seems to lack detectable levels of this peptide receptor, at least!

RECONSTITUTION OF THE 20,000 DALTON FRAGMENT OF Ca^{2+}+Mg^{2+}-ATPASE IN PHOSPHATIDYLCHOLINE VESICLES

ADIL E. SHAMOO
Department of Radiation Biology and Biophysics, School of Medicine and Dentistry, University of Rochester, Rochester, New York 14642

ABSTRACT

We present evidence that the 20,000 dalton fragment of (Ca^{2+}+Mg^{2+})-ATPase posseses Ca^{2+}-ionophorous properties in phosphatidylcholine vesicles as has been shown in black lipid membranes. Furthermore, we show that a 12,883 dalton CNBr fragment of the 20,000 dalton fragment posseses the Ca^{2+}-ionophorous characteristic.

INTRODUCTION

Sarcoplasmic reticulum in skeletal muscle plays an essential role in intracellular calcium regulation. Through several reconstitution procedures, Ca^{2+}+Mg^{2+}-ATPase is believed to be the sole protein responsible for the active transport of Ca^{2+} into sarcoplasmic reticulum[1,2]. Thus the enzyme must contain the entire "pumping" machinery. The machinery includes the site of ATP hydrolysis, the ion selective gate, and the channel pathway across the bilayer. For review, see Goldstein and Shamoo[3].

Our recent data[4] show that the intact enzyme Ca^{2+}+Mg^{2+}-ATPase molecule contains the Ca^{2+} conducting pathway (Ca^{2+}-ionophorous property) as an integral part of the enzyme molecule[5]. Furthermore we have shown that controlled tryptic digestion of the enzyme results first in 55,000 (A) and 45,000 (B) dalton fragments, followed by further cleavage of the 55K dalton fragment into 30,000 (A_1) and 20,000 (A_2) dalton fragments[6]. It was also shown that the 55K and 30K dalton fragments are the site of phosphorylation indicative of the ATP hydrolytic site. The intact enzyme, 55K (A), and 20K (A_2) fragments showed similar Ca^{2+}-ionophorous activity when tested on black lipid membranes (BLM). The 45K dalton fragment, however, has been shown to be a relatively non-selective, divalent cation dependent, ionophore[7,8]. We have further shown that the ionophorous property was solely due to the protein and not to the detergents used in the course of isolation and purification procedures of the various fragments[4].

The 20K (A_2) dalton fragment showed the same ionophorous activity in oxidized cholesterol as well as phosphatidylcholine:cholesterol

BLM's[4]. However, the 30K (A_1) fragment known to contain the site of ATP hydrolysis showed no ionophorous activity[9].

The purpose of this paper is to test the effect of the 20K and 30K dalton fragments on the Ca^{2+}-permeability of artificial vesicles (liposomes) formed from phosphatidylcholine. Furthermore, we will examine the ionophorous activity of cyanogen bromide (CNBr) fragments of the 20K dalton fragment on BLM's. The data obtained indicate that the 20K dalton fragment increases the vesicles' Ca^{2+} permeability. Also it was shown that a 12,883 dalton CNBr fragment of the 20K showed an ionophorous activity similar to the 20K in BLM assay.

MATERIALS AND METHODS

Preparation. Sarcoplasmic reticulum (SR) was prepared from rabbit white skeletal muscle by the method of MacLennan[10]. The tryptic 30,000 (A_1) and 20,000 (A_2) dalton fragments were isolated according to our published method[4,6]. The removal of sodium dodecylsulfate (SDS) from the fragments was according to our published method[4]. Care was taken to insure complete removal of SDS from the protein to avoid potassium dodecylsulfate precipitation. This can be done by dialysis of the protein against 2 liters of vesicle formation buffer (see next section). Protein was determined according to the method of Lowry et al.,[11].

Vesicle Formation. A mixture of 20 mg of phosphatidylcholine (Supelco,Bellefonte,Pa.) and 1 mg of protein (either A_1 or A_2) is gently sonicated in 1 ml of a solution containing 2% cholate,0.3M KCl, 20 mM imidazole, pH 7.6 and 0.02% NaN_3 for three minutes. The preparation is centrifuged 15 minutes at 30,000g and dialyzed 18 hours versus 1 liter of vesicle formation buffer without cholate (i.e. 0.3M KCl, 20 mM imidazole pH 7.6 and 0.02% NaN_3). The vesicle preparation is then applied to a Sepharose 4B column at 4°C for elution in the same solution. The Sepharose 4B column separation is used to separate the vesicles from the remainder solution and unreconstituted lipids. Control vesicles were made with the same procedure except protein was not included.

Ca^{2+}-flux assay. One ml of the vesicle preparation is made in 1 mM $CaCl_2$ and 10-20 μl $^{45}Ca^{2+}$. Vesicles are allowed to incubate with tracer 3 to 4 days at 0°C. To initiate efflux 50 μl of the vesicle suspension is added to 2.5 ml of vesicle buffer. At various time points the entire 2.55 ml is pipetted onto a 1 X 3 cm Na-chelex 100 column (Biorad) equilibrated with 0.6M sucrose and further eluted with an additional 2.5 ml of 0.6M sucrose. 200 to 300 μl is then counted in 10 ml of a 50/50 (v/v) toluene/ethanol scintillation cocktail.

Conductance measurement. The lipid bilayer was formed from egg phosphatidylcholine:cholesterol (5:1 mg/mg in n-decane). Conductance and ionic selectivities were all measured and calculated according to our published method[3].

RESULTS AND DISCUSSION

Ca^{2+}-flux. Phosphatidylcholine vesicles have been prepared with and without incorporation of the A_1 and A_2 fragments. Vesicles have been loaded with radioactive tracer both by incubation of the vesicles with tracer and forming the vesicles in the presence of tracer. The rate of efflux was then studied by separating the tracer trapped into vesicles from the remainder by passing the vesicular mixture through the chelex column. We have found the chelex column method to be far superior, easier, and much more consistant than the traditional Millipore filtration methods for measuring fluxes from vesicles. For example, we found that as high as 40% of the total phospholipid in the preparation may travel through 0.45 μM Millipore filters.

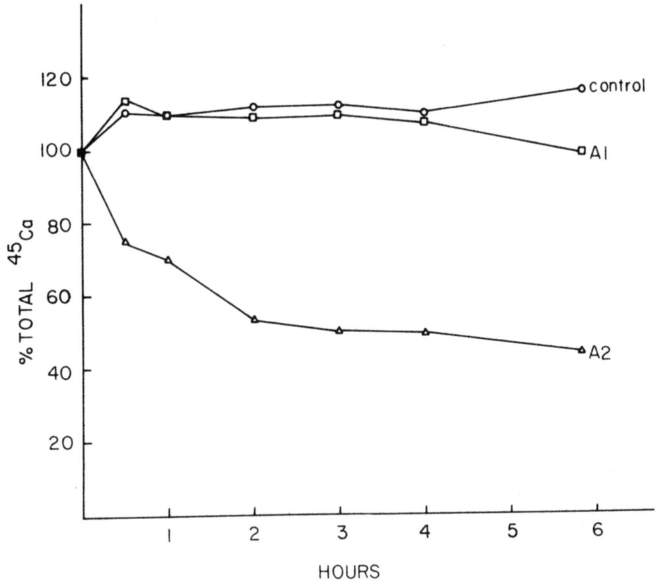

Figure 1. Ca^{2+}-efflux versus time.
The efflux is normalized to the total amount of cpm in the vesicles at t=0. For reconstitution and assay procedures, see the methods section. Control represents vesicles alone. A_1 and A_2 represent vesicles reconstituted with A_1 and A_2, respectively.

Figure 1 shows a representative experiment of ^{45}Ca efflux from vesicles. Vesicles formed with the A_2 fragment lose 50% of their internal Ca^{2+} over a period of two hours. Vesicles formed with the A_1 fragment and the control vesicles are stable over the time period studied.

The levelling off effect of ^{45}Ca efflux from the A_2 vesicles cannot be explained at the present time. It may be due to having two populations of the vesicles, one with and one without the A_2 fragments.

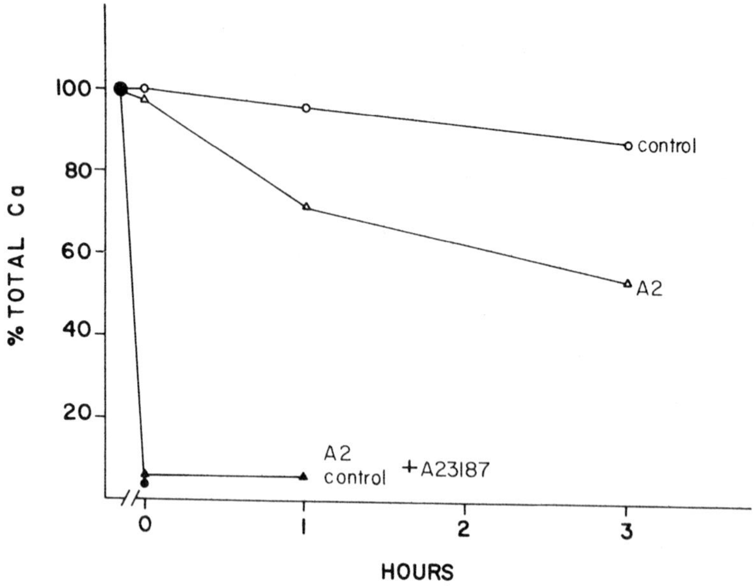

Figure 2. Ca-efflux versus time.

The efflux is normalized to the total amount of cpm in the vesicles at t=0. For reconstitution procedure and assay, see the methods section. Control represents vesicles alone and A_2 represents vesicles reconstituted with A_2. In order to initiate efflux, vesicles are diluted in a buffer. However, the bottom curve represents data with the control or A_2 doped vesicles diluted into a buffer containing A23187 (4.8 μg/ml).

Experiments where vesicles (control or with A_1 or A_2) were diluted into a medium containing the Ca^{2+} ionophore A-23187 show an immediate and total loss of ^{45}Ca from the vesicles (Figure 2). This in itself is taken as presumptive evidence for the existence of vesicles and that Ca^{2+} is not bound but available for the efflux measurement.

Not shown are influx experiments performed on the same type reconstituted vesicles with A_1 and A2 and the corresponding control. ^{45}Ca was present on the outside of the vesicle and the amount trapped inside the vesicle was measured as a function of time using the chelex column procedure. The data indicated the same pattern of permeability where A_2 vesicles showed the fastest uptake as compared to A_1 and control. However, the influx experiments were plagued by the fact that in the presence of 1 mM $CaCl_2$, vesicles may fuse together. This fusion hypothesis may explain why we never reached equilibration even after 3 days.

Preliminary experiments with radioactive tracers of Sr^{2+}, Mn^{2+}, and Zn^{2+} indicate that Sr^{2+} is permeable nearly equally to Ca^{2+} and that Mn^{2+} and Zn^{2+} are impermeable.

Conductance data. Extensive studies of the A_2 fragment in black lipid membranes (BLM's) formed from either oxidized cholesterol or phosphatidylcholine:cholesterol (PC:Chol) (5/1, mg/mg) indicate that A_2 posseses Ca^{2+}-selective ionophorous activity[4,9]. Cyanogen bromide digestion of the A_2 fragment when tested on PC:Chol membranes showed ionophorous activity similar to the A_2 fragment. CNBr fragments isolated by Klip and MacLennan[12] were tested on BLM's. It was found that the C_2 (12,883) fragment has ionophorous activity similar to the CNBr digest of A_2 and similar to the A_2 fragment. However, C_2 was less selective than the A_2 fragment, as A_2 was slightly less selective than the intact enzyme (100K).

The above mentioned data are in good agreement with our previous hypothesis establishing the A_2 fragment as the ion selective gate at the mouth of the B fragment (45K), while the A_1 fragment is the site of ATP hydrolysis.

Since 1974, this author has proposed that active transport processes could occur via a gated channel coupled to an energy transducing system [3,13]. A detailed description of the model for Ca^{2+} transport can be found in our review article[3]. The model consisted of a "channel like" moiety (45K fragment); a selective gate (20K fragment):

and an ATP hydrolyzing site (30K fragment). In this article we have provided further characteristics of the CNBr fragments of the 20K fragment. The C_2 fragment represents the selective gate or the portion which posseses the ionophorous characteristics. The molecular weights of the CNBr fragments are as follows: C_2 (12,883); C_3 (8,115); C_4 (4,800); and C_5 (2,200) daltons[12]. From these and previous evidence we have constructed a more detailed model of active transport of Ca^{2+} by Ca^{2+}+Mg^{2+}-ATPase. The model is still speculative and hypothetical in nature, but it does serve to summarize our data. Figure 3 is the model proposed. The relative location of the C2 fragments is derived from the data of Klip and MacLennan[12].

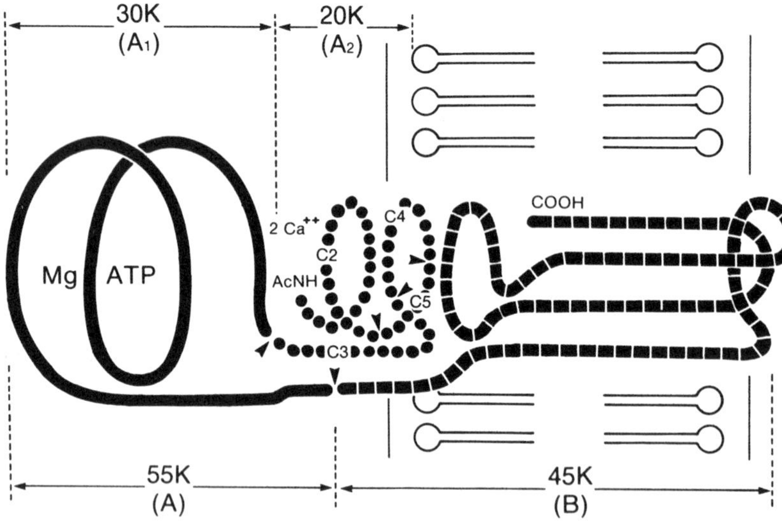

Figure 3. Model for active transport of calcium

In the model, Ca^{2+} is transported from left to right where the 55K fragment represents the portion of the enzyme exposed to the cytoplasm. The 45K fragment represents a large, non-selective channel. The site of phosphorylation is on the 30K fragment near the cleavage point between the 20K and the 30K. The C_2 fragment contains the

AcNH terminal and thus is the furthest from the site of phosphorylation. The peptidic section between C_2 and the 30K portion can allow for 3-dimensional folding to situate C_2 (selective gate) adjacent to the site of phosphorylation. We can speculate that Mg-ATP binds on the 30K site near the C_2 portion. The 2 Ca^{++} are located somewhere in the 3-dimensional space surrounded by the two portions (i.e. C_2 and 30K). ATP hydrolysis then causes the gate (C_2) to translocate Ca^{2+} into the 45K fragment which eventually results in the overall Ca^{2+} movement.

ACKNOWLEDGEMENTS

This paper is based on work performed under contract with the U.S. Department of Energy (DOE) at the University of Rochester Biomedical and Environmental Research Project and has been assigned Report Number UR-3490-1474 . This paper was also supported in part by NIH Grant 1 R01 AM 18892; Program Project Grant ES 10248; the Muscular Dystrophy Association (USA); and the Upjohn Company. A.E.S. is an Established Investigator of the American Heart Association.

REFERENCES

1. Racker, E. 1972. Reconstitution of a calcium pump with phospholipids and a purified Ca^{2+}-ATPase from sarcoplasmic reticulum. J. Biol. Chem. 247: 8198.

2. Racker, E. 1973. A new procedure for the reconstitution of biologically active phospholipid vesicles. Biochem. Biophys. Res. Comm. 55: 224.

3. Shamoo, A.E. and Goldstein, D.A. 1977. Isolation of ionophores from ion transport systems and their role in energy transduction. Biochim. Biophys. Acta. 472: 13.

4. Shamoo, A.E. 1978. Ionophorous properties of the 20,000 dalton fragment of (Ca^{2+}+Mg^{2+})-ATPase in phosphatidylcholine:cholesterol membranes. J. Membrane Biol. 43: 227.

5. Shamoo, A.E. and MacLennan, D.H. 1974. A Ca^{++}-dependent and -selective ionophore as part of the Ca^{++}+Mg^{++}-dependent adenosine triphosphatase of sarcoplasmic reticulum. Proc. Nat. Acad. Sci. USA 71: 3522.

6. Stewart,P.S., MacLennan, D.H. and Shamoo, A.E. 1976. Isolation and characterization of tryptic fragments of the adenosine triphosphatase of sarcoplasmic reticulum. J. Biol. Chem. 251: 712.

7. Shamoo, A.E. and Abramson, J.J. 1977. Ionophore from Ca^{2+}+Mg^{2+}-ATPase. International symposium on "Calcium Binding Proteins and Calcium Function in Health and Disease". (R.H. Wasserman, R.A. Corridino, E. Carafoli, R.H.Kretsinger, D.H. MacLennan, and F.L.Siegel, eds.) pp. 173-80. North Holland, New York.

8. Abramson, J.J. and Shamoo, A.E. 1978. Purification and Characterization of the 45,000 dalton fragment from tryptic digestion of Ca^{2+}+Mg^{2+})-adenosine triphosphatase of sarcoplasmic reticulum. J. Membrane Biol. In Press.

9. Shamoo, A.E., Ryan, T.E., Stewart, P.S. and MacLennan, D.H. 1976. Localization of ionophore activity in a 20,000 dalton fragment of the adenosine triphosphatase of sarcoplasmic reticulum. J. Biol. Chem. 251: 4147.

10. MacLennan, D.H. 1970. Purification and properties of an adenosine triphosphatase from sarcoplasmic reticulum. J. Biol. Chem. 245: 4508.

11. Lowry, O.H., Rosebrough, N.J., Farr, A.L., Randall, R.J. 1951. Protein measurement with the Folin phenol reagent. J. Biol. Chem. 193: 265.

12. Klip, A. and MacLennan, D.H. 1978. Zeroing in on the ionophoric site of Ca^{2+}+Mg^{2+}-ATPase. In "Frontiers of Biological Energetics: Electrons to Tissues". (A. Scarpa, F. Dutton, and J. Leigh, eds.) Academic Press, New York. In Press.

13. Shamoo, A.E. and Ryan, T.E. 1975. Isolation of ionophores from ion-transport systems. Ann. N.Y. Acad. Sci. 264: 83.

DISCUSSION

DR. ROTH: As somebody who is not familiar with this area very much, I would like to know whether the sarcoplasmic reticulum maintains the sodium potassium gradient as well as a calcium gradient.

DR. SHAMOO: There is no known sodium potassium ATPase activity in the sarcoplasmic reticulum.

DR. ROTH: I know that there are no sodium potassium ATPase activities but I thought that the Ca^{2+} pump activity was reponsible for maintaining the sodium potassium gradient.

DR. SHAMOO: There are no known data on the sodium concentrations except the recent X-ray microprobe analysis data. The details of counter transport are very debatable. Either it is potassium or magnesium, but I don't think there are many data to comment on that.

DR. BLUMENTHAL: The function of the calcium pump in the sarcoplasmic reticulum is to remove calcium from the muscle after muscle contraction and there is no teleological need for a sodium-potassium pump.

DR. ROTH: Is there a sodium-potassium gradient across these membranes?

DR. SHAMOO: There is no known evidence for that that I know of.

DR. HECTER: This is a beautiful hallucination, and I think it is congruent with the hallucinations of others. There are, however, some very specific statements you have made. You said, for example, that there is a highly selective divalent cation binding site, and that you have identified a piece of this peptide of about 13,000 dalton molecular weight.

DR. SHAMOO: Yes.

DR. HECHTER: And this selectively is maintained in liposomes or in black membranes?

DR. SHAMOO: Yes.

DR. HECHTER: Now, are you saying that in this channel that you have drawn, there is one point which looks like a gate, to which you assign the term "ionophore". Is it really a gate in a channel according to your hallucination?

DR. SHAMOO: Yes

DR. HECHTER: I just wanted to be clear about that. Then why do you use the term "ionophore"?

DR. SHAMOO: The question is why do I assign the 20,000 fragment as a selective gate ionophore, whereas the 45,000 fragment is really a channel within the artificial system we have measured.

Let me just give you an example of what is happening here. Take, for example the antibiotic channel former gramacidin and modify its selectivity by modifying the mouth of the gramicidin channel. These types of experiments are actually being done by Dr. Lauger's group.

You could do this by attaching the potassium selective ionophore valinomycin covalently to the mouth of the gramicidin. Now you have a potassium-selective "gated channel." If you cleave the valinomycin, of the gramicidin channel it will act as a "gated channel" ionophore in the reconstituted system. However, when it is in "situ", it is not really an ion carrier, as it was before.

DR. SINGER: You picture the single polypeptide chain as having all the channel properties. Do you have any evidence that the channel isn't formed by the conjunction of two chains, as it appears possible to be in the case of sodium potassium ATPase?

DR. SHAMOO: The only thing I could say, that we have no evidence for aggregation and that no more than a monomer is required. We make sure it is a monomer in all of our studies.

DR. SINGER: How do you know that it is a monomer in the black lipid membrane (BLM).

DR. SHAMOO: By succinylation the enzyme has become monomeric.

DR. SINGER: It is monomeric in solution, but how do you know, whether it remains monomeric when you put it into the bilayer.

DR. SHAMOO: The concentration - dependence of the conductance increase indicates that the enzyme acts as a monomeric channel in the BLM. The log of conductance in the BLM rises as a function of log C^n where n = 1.

DR. DE HAEN: Is the inhibition that you observe on reduction of disulfide bonds reversible?

DR. SHAMOO: Unfortunately not. We would like to have them reversible, but once the disulfide bond break, they could reform in a different fashion and with different adjacent sulfhydryls than before. However, we have run the obvious controls that we are possibly denaturing the protein and reducing its solubility. We found that the conductance inhibition appears to be specifically due to the reduction of the disulfide bonds.

THE NEGATIVE COOPERATIVITY OF THE INSULIN RECEPTORS: WHAT DO WE KNOW ABOUT THE MECHANISM?

PIERRE DEMEYTS
International Institute of Cellular and Molecular Pathology, 75 Avenue Hippocrate, B-1200 Brussels, Belgium

The first step in the action of insulin is its binding to a specific receptor located in the membrane of the target cell[1]. The interacting of insulin with its receptor does not appear to follow simply the law of mass action; when analyzed by the method of Scatchard or other graphical transformations, the binding isotherm appears complex and shows deviations consistent with either several classes of receptor sites or negative cooperativity, or both[2,3] (Figure 1B). In 1973, using a simple method to compare the dissociation rates of insulin from nearly empty receptors with its dissociation from insulin filled receptors, we showed that the dissociation rate of an insulin-receptor complex is markedly accelerated when neighboring receptor sites are occupied with insulin, consistent with interactions of the negative cooperative type[3].

The method that we described is as follows. Cultured human lymphocytes (IM-9) at high concentration in a single batch were reacted with radioiodinated hormone of high specific activity under conditions such that only a minority of receptor sites were occupied by the bound tracer. The cells were rinsed free of residual medium, resuspended in an identical volume, immediately divided into aliquots and diluted 100-fold in fresh medium; half were in hormone-free medium ("dilution only"), while half were in medium to which had been added an excess of unlabeled hormone ("dilution + unlabeled hormone"). Dissociation of the tracer from the receptor was monitored in both sets. If the sites were independent, the presence of unlabeled hormone, which at the concentration used rapidly fills the empty sites, should not affect the dissociation rate of the labeled hormone. Such was the case for growth hormone (Figure 1C), the binding of which gives a linear Scatchart plot (Figure 1A).

With insulin and its receptors a different result was obtained. A 100-fold dilution in the presence of unlabeled insulin (1.7×10^{-7} M) produced a much more rapid dissociation of the bound labeled insulin than did dilution alone; the half-time for dissociation of the bound hormone differed by 10-fold (Figure 1D). This suggests that insulin-filled sites lower the affinity of other sites for this hormone. Controls were built in the original set of experiments[3] to show the absence of rebinding of the dissociating tracer in the dilution used (100x), unless unstirred layers or retention effects are present[3,4] (see below).

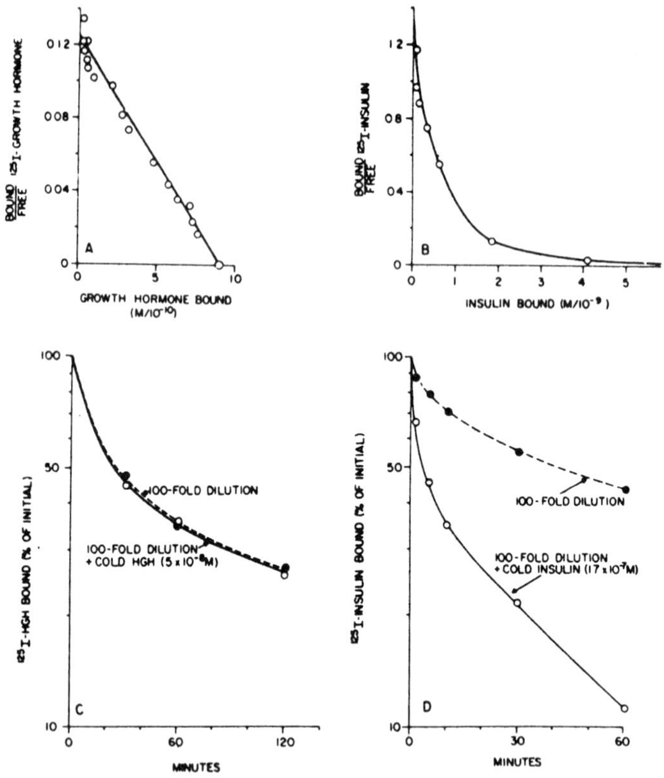

Fig. 1A and 1B. Binding of human growth hormone (hGH) or procine insulin to human lymphocytes. $[^{125}I]$hGH (2 x 10-11 M) at 30° or $[^{125}I]$ procine insulin (7 x 10^{-12} M) at 15° was incubated with 1.7 x 10^6 cultured lymphocytes (cell line IM-9) in the absence and presence of unlabeled hormone; at steady state, the cells were sedimented and radioactivity in cell pellet counted[5,6]. The bound/free of labeled hormone is plotted as a function of hormone bound to the cells; "nonspecific" binding, which constituted less than 5% of the total binding, has been subtracted[5,6]. Fig. 1C and D. Dissociation of $[^{125}I]$hGH or $[^{125}I]$ insulin from lymphocytes. $[^{125}I]$hGH (10^{-10}M) was incubated for 90 min at 30° and $[^{125}]$ insulin (5 x 10^{-11} M) for 30 min at 15° with 2.5 x 10^7 cells/ml, after which the cells were sedimented at 4°, the supernatant was replaced by an equal aliquot of chilled fresh medium, the cells were resuspended and aliquots (0.1 ml) were transferred to a series of tubes that contained 10 ml of medium in the presence and absence of unlabeled hormone (5 x 10^{-8} M, 30° for hGH; 1.7 x 10^{-7} M, 15°, insulin). At intervals, two tubes of each set were centrifuged, and the radioactivity in the cell pellet was counted. The radioactivity on the cells expressed as a percentage of the radioactivity present at t = 0, is plotted as a function of the time elapsed after the dilution of the system. Each point is a mean of duplicates; duplicates differed by less than 5%. The radioactivity at 5 = 0 was measured at the completion of the incubation, both before and after the first sedimentation step; both results were the same, indicating that trapped radioactivity and dissociation during the analyses were insignificant. At present we have no explanation why the dissociation of $[^{125}I]$hGH was not single order. (Reprinted from ref. 3 with permission.)

This experimental design has now become standard in the study of hormone receptors, and negative cooperativity has been found in several other surface receptors (for review, see 4) including for example those for nerve growth factor[5] and thrombin[6]. However, as we have stressed before[4], great caution should be exercised before interpreting the kinetic experiment as a straightforward indication of negative cooperativity. There are a number of models that can explain an acceleration of the dissociation rate of a labeled ligand by an excess of unlabeled ligand, some of which are listed in Table I.

TABLE I

SOME MODELS EXPLAINING AN ACCELERATION OF THE DISSOCIATION RATE OF LABELED LIGAND IN AN "INFINITE" DILUTION BY UNLABELED LIGAND

A. Non-Cooperative Models	
Interface effects:	Unstirred layer[7]
	Retention effect[8]
Ligand ligand interaction:	Concerted exchange[9]
B. Negative Cooperativity	
Direct ligand-ligand interaction:	ligand polymerization[10,11]
site-site interaction (= receptor-mediated ligand interaction):	
	- ligand-induced clustering of receptor sites[12,13]
	- ligand cross-linking of receptor sites[14-16]
	- flexible ligands, multi-contact binding, "extended" binding site[17-19]
	- dissociating or mobile receptor, two-step reactions[20-21]
	- ligand-induced subunit subunit interactions within an oligomer (including association or dissociation of receptor subunits)[22,23]

The models under A are qualified as non-cooperative because the modulation in the dissociation rate by increased free ligand concentration is a purely kinetic effect that does not effect the equilibrium level of binding. We qualify as negatively cooperative (B) models where the kinetic effect on the dissociation rate is responsible for an alteration in the thermodynamics of binding, that is a decrease in the apparent affinity of a receptor site for direct ligand interactions or indirect (receptor-mediated or "site-site") ligand interactions.

In the case of the insulin receptors, detailed studies[3,4,24,25] have led us to the conclusion that unstirred layers, retention effects or concerted exchange (Table I, A) are unlikely. Indeed, those models predict that any ligand that binds to the receptor would increase the xenocarbon rate in proportion to their affinity for the receptor. We have shown that in fact, insulin can be modified chemically in a way that retains some receptor binding but destroys the ability to accelerate the dissociation of labeled insulin from the receptors. Such analogs, like desalanine-desaparagine insulin and desoctapeptide insulin, were unable to accelerate the dissociation rate even at concentrations much greater than those which saturate the empty receptor sites[25] (Figure 2). The study of the localization of the initial chemical alterations allowed us to map a distinct binding domain ("cooperative site") in the surface of the insulin monomer that appears to be responsible for triggering the "negative cooperativity" (Figure 2).

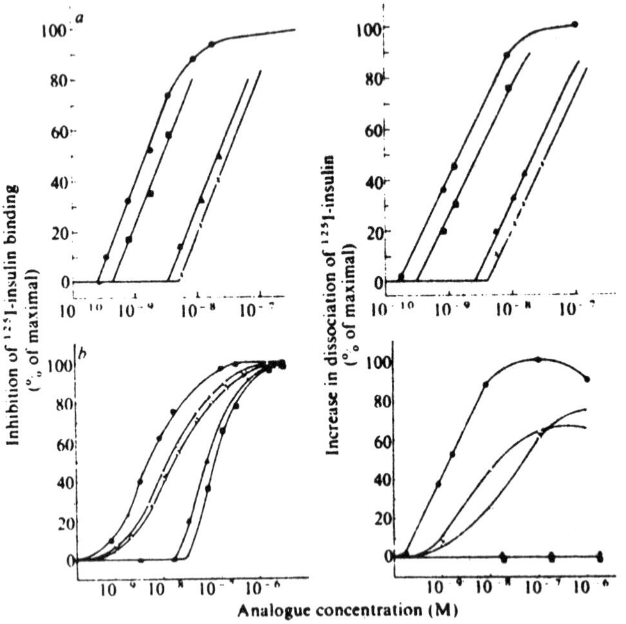

Fig. 2. Comparison between the ability of insulin analogues to inhibit ^{125}I-insulin binding to human cultured lymphocytes and their ability to induce negative cooperativity. a, Example of fully cooperative analogues; b, example of analogues with impaired or absent cooperativity. Left, binding assay. Cultured human lymphocytes are incubated in assay buffer at 15°C for 90 min with a tracer amount of ^{125}I-insulin alone or in the presence of increasing concentrations of unlabelled analogue. The inhibition of the binding of ^{125}I-insulin by the various concentrations of analogues is plotted in % of the maximal inhibition obtained, representing the extent of specific binding (usually 95% of the initial binding) as a function of analogue concentration. O, Insulin; , fish insulin; Δ, split proinsulin; χ, proinsulin. Kinetic assay for negative cooperativity (right). The assay is based on the property of cooperative analogues to accelerate the dissociation rate of a tracer amount of ^{125}I-insulin bound to the receptors. After binding of the ^{125}I-insulin to the lymphocytes has reached a steady state, an aliquot of the ^{125}I-insulin receptor complex is diluted 100-fold in the absence, or presence of increasing concentrations of unlabelled analogue in the diluting medium. The dissociation is allowed to proceed for 30 min, after which the tubes are centrifuged and the amount of radioactivity still bound is measured. The data are plotted as the radioactivity dissociated in the presence of a given concentration of the analogue, minus the dissociation in the absence of the analogue, in % of the maximal effect, as a function of the analogue concentration. (Reprinted from Ref. 25 with permission.)

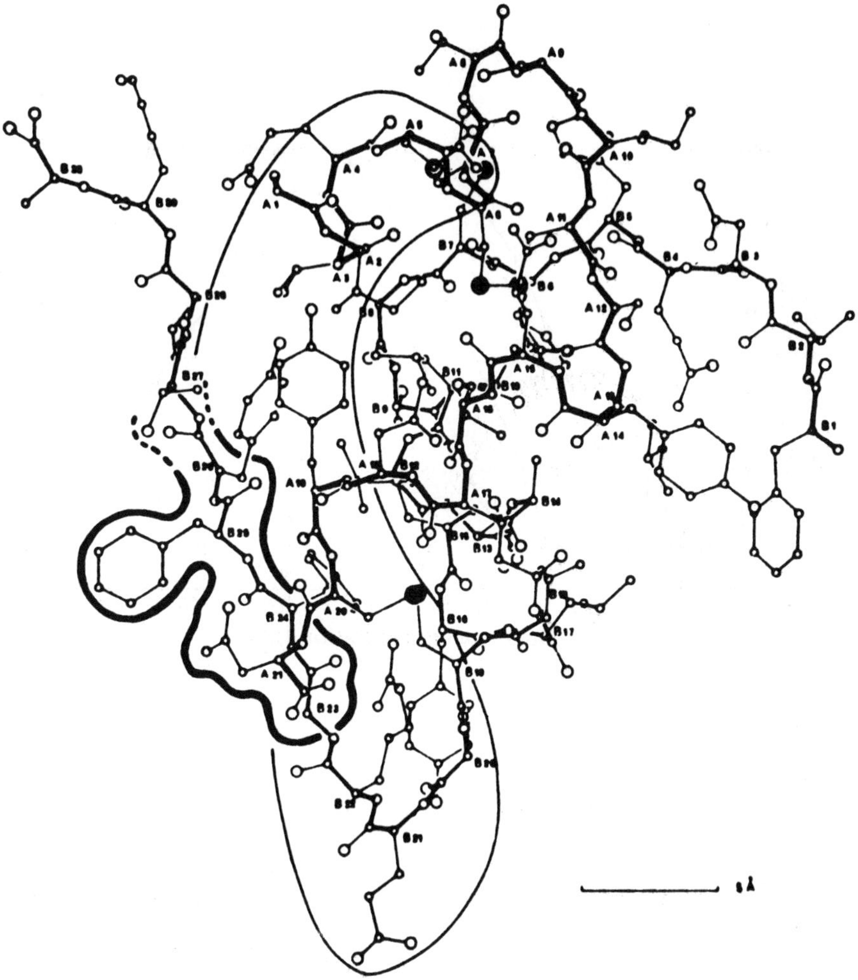

Fig. 3. Mapping of the cooperative site on the insulin monomer. The solid fine line represents the putative receptor-binding region ('bioactive site'); the thick line the cooperative site. (Figure due to the courtesy of T.L. Blundell.) (Reprinted from Ref. 25 with permission.)

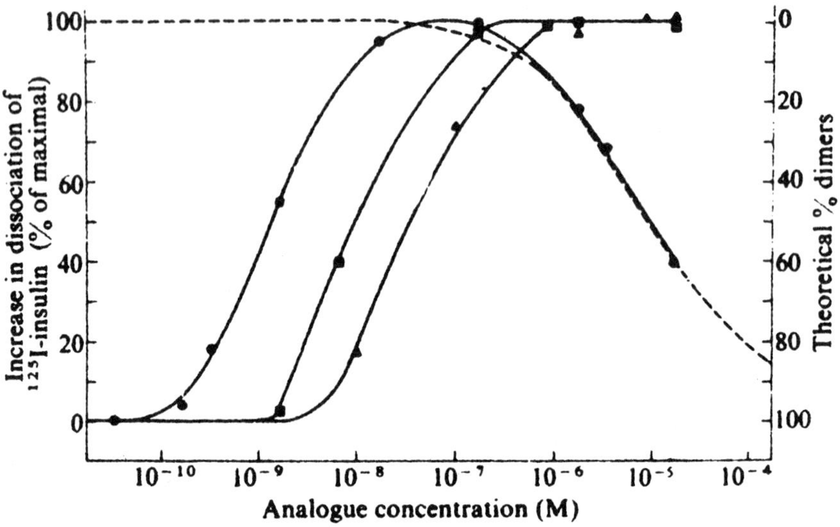

Fig. 4. Effect of insulin dmierisation on negative cooperativity. The ability of pork insulin and two non-dimerising insulins to induce negative cooperativity are measured in the same kinetic assay as Fig. 2, right. The assay includes high concentrations of the analogues, where dimerisation of porcine insulin becomes significant. The dashed line (note the inverted scale) represents the theoretical curve for dimerisation [mols dimer/mols (monomer + dimer)] x 100 calculated from a $K_{dimer} = 2.22 \times 10^5$ at neutral pH. 0, Pork insulin; , tetranitroinsulin; Δ, guinea-pig insulin. (Reprinted from Ref. 25 with permission.)

We have more recently labeled with ^{125}I one of the analogues that was unable to accelerate the dissociation rate of labeled insulin, desalanine-desaparagine (DAA) insulin. The ^{125}I DAA-insulin bound to the same number of receptor sites as ^{125}I-insulin, but yielded linear Scatchard plots[26] demonstrating directly that the kinetic effect was responsible for the alteration in equilibrium binding. This definitely excludes the model in Table I, A and favors the existence of a single class of points with negatively cooperative interactions. Furthermore, binding ^{125}I insulin to a small fraction of the receptor sites, and saturating the empty sites with unlabeled DAA-insulin prior to exposing the cells to an excess of unlabeled insulin prevented insulin from accelerating the dissociation rate (= antagonism for the negative-cooperativity)[26]. This is in favor of receptor-mediated interactions, since the same concentrations of free insulin is available for a "concerted exchange" even if DAA insulin, which does not by itself accelerate the off rate, has been added before.

A direct ligand-ligand interaction (dimerization of insulin on the receptor) was considered unlikely on the basis of the following arguments[4,24,25].

1. The acceleration in the dissociation rate of insulin is exerted at concentrations where insulin is monomeric in solution and in fact falls off at higher concentrations where insulin will dimerize in solution (Figure 4). We tentatively explained this by the fact that the "cooperative site" described above is covered in the natural dimer interface[25].

2. Non-dimerizing insulins (tetranitro insulin, guinea-pig insulin) are fully cooperative and the acceleration in the dissociating rate does not dissapear at high concentrations.

3. The ability of three dozen analogues to accelerate the dissociation is proportional to their affinity for the receptor site (except when modified in the "cooperative domain") and unrelated to their affinity for another insulin molecule (dimerization).

We are thus left with a variety of models that have loosely grouped under the label "site-site interactions" and which might be more properly labeled "receptor-mediated ligand interactions". I have no personal commitment to any of these models because I feel the physical chemical evidence that might allow to choose among them is still lacking. However, the conservation of negative cooperativity in solubilized and partially purified receptors[27-29] may be an argument for an intrinsic property of an oligomeric receptor rather than for long distance interactions linked to clustering, cross-linking or mobility in the membrane. It is quite easy, by computer curve-fitting using equations based on most of these models (Table I, B), to obtain a good description or even a good fit to our dissociation curves (dilutions alone and dilution + unlabeled insulin). However, the critical test of any model would be its ability to describe and fit simultaneously:

1. The two dissociation curves (dilution alone, dilution + unlabeled insulin)
2. The kinetics of association
3. The Scatchard plot
4. The dose-response curve for "cooperativity" with its fall off at high concentrations and, in the same time, account for the fact that desalanine-desasparagine and desoctapeptide insulin still bind to the receptor but have lost the ability to accelerate the dissociation.

Without giving to this description a too precise physical meaning, we interpret these data phenomenologically by saying that the insulin-receptor complex exists under two reversible states (we do not even say "conformations") a slow-dissociating state with a high affinity for insulin and a fast-dissociating state with a low affinity for insulin, and that increasing the concentration of bound insulin increases the fraction of complexes in the fast-dissociating state. Recently, Pollet et al.[9] have published experiments suggesting that it was not the amount of bound insulin but rather the free insulin, independently of receptor occupancy, that caused the increased dissociation rate ("concerted exchange"). This was based on the observation that preloading the receptors with increasing concentrations of insulin before diluting the system did not result in an accelerated dissociation rate by dilution alone, which one would expect with negative cooperativity since the occupancy has increased. We have not been able to reproduce these findings. We had published similar experiments in 1976[24] with results consistent with negative cooperativity and have repeated them since the publication of Pollet's paper with again the finding of a marked acceleration in the dissociation rate by dilution of "preloaded receptors" as compared with "empty" receptors. "Preloading" experiments have also been performed in 1975 for the thrombin receptors[6], which also exhibit curvilinear Scatchard plots and accelerated dissociation, with results and conclusions identical to ours. Olefsky, more recently[30], found also in agreement with us, marked acceleration in the dissociation rate of "preloaded" insulin receptors. He even concluded that it was too fast for being accounted for only by negative cooperativity, and postulated in addition some heterogeneity of binding sites. One should consider why in the same type of experiments, the off rate is too slow in Pollet, alright for us, and too fast for Olefsky. In fact, it should be observed that, once our original experimental design is performed, not with a tracer of ^{125}I insulin bound, but at a significant saturation of the sites, the method becomes much more subject to artifacts. Indeed, if the initial rate of dissociation accelerates, the wash step used to get rid of free hormone also gets rid of some of the bound tracer, leaving only the slow dissociating component on the cells. Thus, whatever the level of saturation, the tracer might dissociate slowly because the fast component has been missed. This is the only way we can reproduce Pollet's results. Also, since the fraction of ^{125}I-insulin bound decreases, the non-specific compartment becomes a larger proportion of total binding, and the correction for non-specific binding may introduce errors in the calculation of kinetic curves. Finally, rebinding of the dissociating tracer might be more significant at higher levels of occupancy than in tracer conditions. Due to these potential artifacts, kinetic data performed following our initial method but in non-tracer conditions should be taken with great care and are probably more qualitative than quantitative. In fact, Pollet's results seem to me internally inconsistent; since he also found a curvilinear Scatchard plot, and showed that the association kinetics was accounted for by a single rate constant, the two classes of binding sites must thus differ in affinity because of differing dissociation rate constants, which should then result in an accelerating dissociation rate with increased pre-loading, since the fraction of ligand bound to the low affinity component increases.

Another argument used by Pollet against negative cooperativity (as well as by Olefsky[30] in favor of additional sites heterogeneity) was that the dissociation rates are never first order even in limiting conditions where most of the sites are empty or most are filled. There is always a slow component which appears consistent with the "high affinity" component of two classes of sites and not with the initial affinity of a negatively cooperative model. We have always been aware of this problem[3], but we feel that it has nothing to do with the presence or absence of negative cooperativity. Indeed, the existence of a slow component in the dissociation curves is present even for receptors that yield linear Scatchard plots like growth hormone, as we had pointed out before[3], and gonadotrophins[31]. These two systems have linear Scatchard plots but show no acceleration in the dissociation rate by unlabeled ligand. In analyzing the binding of HCG to testis receptors, Ketelslegers[31] found that, although the Scatchard plot is perfectly linear, the dissociation process did not follow simple first-order kinetics, but a better fit was obtained with a double exponential model, one of the components having an extremely slow dissociation rate constant. Only the initial dissociation rate was consistent with the equilibrium data. The presence of this slow component cannot be thus simply taken as evidence against negative cooperativity or as evidence for a "high affinity" component, but rather as evidence that our concept of the hormone-receptor interaction is still too simple, even with built-in negative cooperativity. A scheme that would account for the insulin data must then comprise at least a second step, like an isomerization of the complex

$$H + R \rightleftharpoons HR \rightleftharpoons HR'$$

↑ modulation by insulin

or rather a complex reaction (e.g. internalization, cross-linking or simple effector binding), like in some two-step models proposed by Boeynaems[20]. Some of these models can explain negative cooperativity, but in our hands some of the simple ones were unable to fit all of the data simultaneously. Clearly much more work, both theoretical and experimental, is needed to unravel the intricacies of hormone-receptor interactions.

REFERENCES

1. Freychet, P., Roth, J. and Neville, D.M., Jr. (1971) Proc. Nat. Acad. Sci. USA 68, 1833-1837.
2. Kehn, C.R., Freychet, P., Roth, J. and Neville, D.M., Jr. (1974) J. Biol. Chem. 299, 2249-2257.
3. De Meyts, P., Roth, J., Neville, D.M., Jr., Garm, J.R. III, and Lesnack, M.A. (1973) Biochem. Biophys. Res. Commun. 55, 154-161.
4. De Meyts, P. (1976) J. Supram. Struct. 4, 241-258.
5. Bradshaw, R.A. and Frozier, W.A. (1977) Current topics in cellular regulation 12, 1.
6. Tollefsen, D.M. and Mayerus, P.W. (1975) Biochemistry 15, 2144.
7. Noyes, A.A. and Whitney, W.R. (1897) Z. Physik. Chem. 23, 687.
8. Silhavy, T.J., Szmelaman, S., Boos, W. and Schwartz, M. (1975) Proc. Natl. Acad. Sci. USA 72, 2120.
9. Pollet, J.R., Standaert, M.L. and Haase, B.A. (1977) J. Biol. Chem. 252, 5828.
10. Nichol, L.W., Jackson, W.J.H. and Winzor, D.J. (1967) Biochemistry 6, 2449.
11. Cuatrecasas, P. and Hollenberg, M. (1975) Biochem. Biophys. Res. Commun. 62, 31.
12. Singer, S.J. and Nicolson, G.L. (1972) Science 175, 720.
13. Levitzki, A. (1974) J. Theret. Biol. 44, 367.
14. Delisi, C. and Chabay, R. (1979) Cell. Biophysics, in press.
15. Kahn, C.R., Baird, K.L., Jarrett, D.B. and Flier, J.S. (1978) Proc. Natl. Acad. Sci. USA 75, 4209.
16. Schlessinger, J., this volume.
17. Schwartz, G. (1976) Biophysics of structure and mechanism 2, 7.
18. Sandvig, K., Olsenes, S. and Alexander, P. (1978) Europ. J. Biochem. 88, 307.
19. DeCean, A., Munson, P.J. and Robbard, D., Molecular Pharmacology, in press.
20. Boeynaesurs, J.M. and Dumont, J.E. (1975) J. Cyclic Nucleotide Res. 1, 123.
21. Jacobs, S. and Cuatrecasas, P. (1976) Biochem. Biophys. Acta. 433, 482.
22. Koshland, D.E., Jr., Heimethy, G. and Filmer, D. (1966) Biochemistry 5, 365-385.
23. Colosino, A., Brunoi, M. and Wyman, J. (1976) J. Mol. Biol. 100, 77.
24. De Meyts, P., Bianco, A.R. and Roth, J. (1976) J. Biol. Chem. 251, 1877.
25. De Meyts, P., van Obberghen, E., Roth, J., Wollmer, A. and Brandenberg, D. (1978) Nature 273, 504.
26. De Meyts, P. and Michaels-Place, M. (1978) Drabetologia 15, 227.
27. Harrison, L.C. Billington, T., East, I.J., Nichols, R.J. and Clarck, S. (1978) J. Endocrinology 102, 1485.
28. Ginsberg, B.H., Kahn, C.R., Roth, J. and De Meyts, P. (1976) Biochem. Biophys. Res. Commun. 73, 1068.
29. Ginsberg, B.H., Cohen, R.C., Kahn, C.E. and Roth, J. (1978) Biochem. Biophys. Acta 542, 88.
30. Olefsky, J. (1978) Diabetes 27, 946.
31. Ketelslegers, J.M., Knott, G.D. and Katt, K.J. (1975) Biochemistry 14, 3075.

DISCUSSION

DR. THAKUR: Pierre, you showed the Scatchard plot of the DAA interpreted as being linear. To me it seems to show positive cooperativity.

DR. DEMEYTS: Only the first point was below the straight line, but that may simply be an equilibrium problem where the tracer itself is not completely equilibrium, where the other points with higher concentrations are already. I don't think it indicates positive cooperativity, and certainly the competition curves are not suggestive of positive cooperativity. When translated into Scatchard coordinates, they fit a simple binding.

DR. CHANGEUX: An interesting result was that the dimer does not show negative cooperativity, whereas the monomer does. Do you think that the two sites which are interacting are close to each other, that there could be a steric effect due to the close relationship between the two sites? Actually, in the case of the acetyl choline receptor and alphatoxin binding, there are also some groups claiming that there is negative cooperativity, and the model there is a steric hindrance or some kind of steric interaction between the two molocules because they are so big that they interact at the level of the sites.

DR. DEMEYTS: I think the answer is no, because DAA is as big as insulin and does not do it. Also it seems to me that steric hindrance would explain how a first molecule sitting on a site prevents the next one from binding but not how the second one accelerates the off rate of the first one. The dimer data are easy to explain because the cooperative site is buried in the dimer, so since the structure-activity relationship data show that the area buried in the dimer is essential in triggering the effect, preventing the binding of that part by dimerisation lead to a fall in the cooperative effect. That is what Jossi Schlessinger and Alex Levitzki have described in the binding of NAD to glyceraldate phosphatic hydrogenase, where the negative cooperativity was triggered at the adenine subsite.

DR. LEVITSKI: I wanted just to make a comment concerning the competition experiments. You can do a very simple experiment, and that is, if you take the non-cooperative insulin binding and you do the binding experiment of insulin under conditions that the non-cooperative binding, then you are doing an insulin binding experiment where you can show very simply that a model which assumes a heterogenous population of sites will tell you that the shape of the curve will not change, but just the midpoint will change.

However, a non-cooperative ligand that will change the shape of the curve will convert the negative cooperativity either to more negative or less negative cooperativity, but never exceeding one actually, can be accommodated only by a kind of an induced type of model, and we have now done it with bis-3-phosphate dehydrogenase that can show that certain NAD analogs which are non-cooperative will convert the NAD binding to a non-cooperative binding altogether, and that can be only explained by a ligand-induced model, but not by pre-existing asymmetry type of model. That is the first point. The second point is that a recent paper in the European Journal of Biochemistry, where the binding of lectin to sepharose has been studied it has shown that although the binding is non-cooperative, there is an accelerated rate of dissociation when you do this occupancy study, as he has done in insulin, and in that case it is very clear that there is no negative cooperativity between receptors.

So they also claim that the only explanation could be that because there is an extended site for the lectin on the Sepharose, there is a moment where two lectins can bind to part of the two sites.

Now, I do not think that in the insulin case this argument has actually been dealt with properly. I mean, you do not really know. The fact that they have certain analogs that do not do it only means that analogs cannot bind simultaneously to the site of the negatively cooperative analog.

I think the approach that you have indicated concerning the competition experiment is a more rigorous approach, and that will actually tell you whether you are dealing with a really negatively cooperative system because you can really treat it in a thermodynamic fashion.

Now, the third point is that I do not follow very closely the arguments between you and Jesse's group and the other groups claim that --

DR. DEMEYTS: There is absolutely no argument between me and Jesse.

DR. LEVITZKI: No, no, no. I am talking about the alliance they made for Olefsky and the other groups in the western hemisphere who claim that there is no negative cooperativity for insulin binding, and sometimes it is very difficult for me to understand where is the difference, I mean, because you are doing the same experiments, and two sets of results are obtained, and this is somehow disturbing to me.

Now, the other point concerning this. The physiological range of insulin action -- does it cover the negatively cooperative region or only part of the binding curve?

DR. DEMEYTS: Let me try to answer the last question first. The physiological range of insulin action covers only part of the Scatchard plot because of spare receptors, and it could be explained either by the spare receptors or by negative cooperativity itself, which could cause non-linear coupling between binding and alteration.

DR. LEVITZKI: The previous question was why there is such a difference in the results between the groups.

DR. DEMEYTS: There is no big difference between the group. There has been a problem only in fat cells where, all groups reported negative cooperativity, except the group of Gliemann and Gammeltoft who worked at higher temperatures and for a while could not demonstrate it. More recently, with an improved binding technology, they have been able to demonstrate it, always to a lesser degree than in other cells.

There is more controversy in the interpretation of the data, especially the work of Pollet that I discussed above.

DR. LEVITZKI: And what about this point of the extended site and the experiment on the Sepharose-lectin interaction?

DR. DEMEYTS: That is one of the possible models which I have duly mentioned in Table I. As I have said I do not feel stuck to one given model. The point is well taken.

DR. DE HAEN: I am curious about your thoughts concerning the physiologic role of negative cooperativity because desoctapeptide insulin is a full agonist. The potency is slightly changed, but it is a full agonist, and even the biphasic response in antilipolysis has all the properties I am aware of that insulin has. So you do not need to imply that negative cooperativity has something to do with the basic function. What else could negative cooperativity have to do with?

DR. DEMEYTS: Well, the idea that DAA and DOP are full agonists came from the mistaken notion that the affinity for binding was the same as the biological potency. However, the affinity for binding was inferred from the distance between the competition curves. For example, the competition curve of DAA insulin is 50-fold apart from the curve obtained with insulin, at the top of the two curves, and DAA is also about 50 times less potent. However, we demonstrate now by direct binding studies that it is only 10 times less good in binding, so you would expect the biological effect to be also 10 times less good in binding, unless the negative cooperativity is playing some amplifying role in the ensuing response. We are presently retesting analogs like DAA and DOP in fat cell bioassays to see if they are really full agonists or if they may be partial antagonists that have not been recognized previously.

When you measure directly the binding of some analogs like DAA, the relative potencies in a biological assay do not appear to reflect any more truly the affinities for binding, which might indicate that the negative cooperativity plays a role in the biological potency of insulin by amplifying it. It is very speculative, but that is an interesting possibility to explore. Otherwise, the role that is normally postulated is a buffering of the response, but Alex Levitzki has also shown in a "clustering" model of negative cooperativity that if the response of proportion to the fraction of sites turned off, you get an amplification and not a buffering.

So there is a physiological implication, It can be studied precisely now, especially with these non-cooperative analogs.

As I said, we have now good reasons to think that hormone-receptor interactions are indeed not simple bimolecular reactions in irreversible equilibrium, and that the reaction scheme is probably more complex. We probably have at least a two-step reaction which resolves the thermodynamic "paradox" that you were mentioning and which is well taken. We shown in fact, that in competition curves, insulin, which does accelerate the dissociation rate of labeled insulin, also converts labeled DAA binding from a non-cooperative one to a negatively cooperative one, since labeled DAA is complete for by unlabeled DAA in a non-cooperative way (2 log units) and by labeled insulin in a negatively cooperative way (4 log units); conversely, the binding of labeled insulin is competed for by unlabeled DAA in a non-cooperative way and by unlabeled insulin in a cooperative way. I think this answers Alex's point.

Dr. Rodbard: First, I would like to congratulate you on a very elegant series of studies with the "non-cooperative" analogs of insulin, which considerably strengthen the case for the role of negative cooperativity. You have observed that insulin binding affects the kinetics of DAA-insulin dissociation, whereas DAA-insulin binding does not influence the kinetics of insulin dissociation. Many persons would be likely to interpret this as implying that insulin affects the equilibrium constant(s) of association for DAA-insulin, although the equlibrium constants and the association kinetics were not actually measured for each ligand in the presence of the other. Likewise, many persons might interpret the lack of effect of DAA-insulin on insulin dissociation kinetics as indicating the absence of an effect on the equilibrium constant for insulin. However, such interpretations (and I am not attributing them to you) would be premature and would be likely to be erroneous. Just as one cannot extrapolate from equlibrium data to kinetic mechanism (as noted by you when you first proposed the possibility that the non-linear Scatchard plots for insulin binding might be due to negative cooperativity), one cannot extrapolate from kinetic data to infer equilibrium properties.

I would like to present a simple model which may facilitate analysis of the problem. Figure 1A shows the case of a bivalent receptor which may be filled with one ligand (L_1, e.g. insulin) or another (L_2, e.g. DAA-insulin). All possible equilibrium pools are shown. This model is the equivalent of the stepwise or "Adair" model for two ligands with exclusive binding, and provides a completely general approach to equilibrium binding, as noted by Fletcher in the case of a single ligand. It may readily be extended to consider a receptor with four binding sites (Fig. 1B), with any desired valency, or with non-exclusive binding. According to the "principle of detailed balance", one must satisfy the requirement that the product of the stepwise equilibrium constants around different paths in a loop must be the same, e.g. $K_1 x K_2 = K_3 x K_4$. Thus, at equilibrium, it would be impossible for one ligand to exert a cooperative effect on a second ligand, without having the second ligand exert a cooperative or allosteric effect on the first.

Accordingly, in the insulin--DAA-insulin system, the association rate constants must have changed (although such changes may be too small to measure accurately), or else the system is not at equilibrium. A similar conclusion would still apply if the model were to include several isomerizing steps (e.g. if each of the pool classes in Fig. 1 were to exist in an equilibrium between two or more conformational states). However, the principle of detailed balance - and the constraint which it imposes - would no longer be applicable if any of the steps in Figure 1 were irreversible or energy requiring. Thus, if accurate measurements of the association rate constants subsequently indicate no change in k-on corresponding to or compensating for the observed changes in k-off, then one would have to seriously consider the possibility that the system is not a reversible equilibrium, but rather one that involves energy input or an irreversible step.

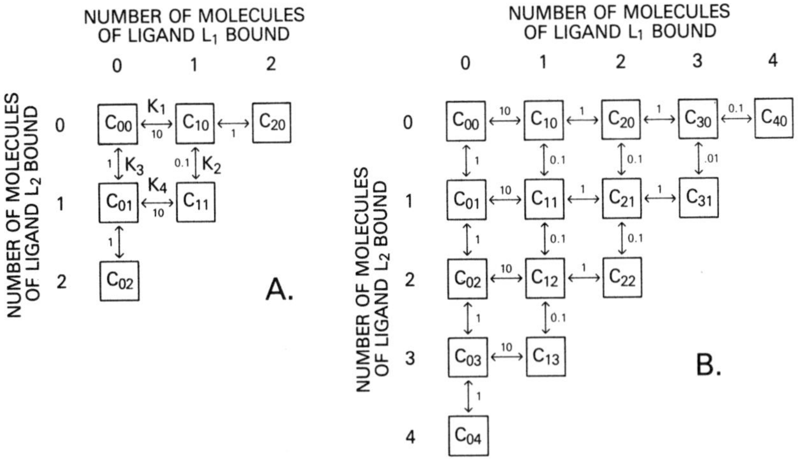

Figure 1A. Schematic model of all possible "pools" (states of receptor) when a bivalent ligand interacts with two ligands which bind at the same site. By the "principle of detailed balance", K1xK2 = K3xK4. Thus <u>at equilibrium</u> it would be impossible for binding of ligand L_1 to affect ligand L_2 and not have ligand L_2 also alter the binding of ligand L_1. The numerical values for K shown here illustrate the hypothetical case where binding of ligand L_1 decreases the stepwise affinities for either L_1 or L_2 by a factor of 10, whereas binding of L_2 does not affect the stepwise K values for either ligand. Such a case would be impossible, irrespective of the valence of the receptor.

II
MOBILITY OF CELL SURFACE RECEPTORS

CHAIRMAN'S REMARKS

"Mobility" is an ambiguous word. For example it appears in titles as lateral mobility, translational mobility. The word begs the question of just how things are moving, what drives them, and this afternoon I think the papers are going to cover a range of kinds of mobility, as you will see as the session unfolds.

I thought I would begin by trying to define a few terms that I think are generally either misused or neglected, and pointing out that one person's mobility needn't be another's.

The first term that I wanted to discuss is "fluidity." That word too appears in lots of paper titles. Though fluidity has a specific physical definition, reciprocal viscosity, too often when it appears in a text or a paper title, it is used just in the sense of "Yes, we can see something moving in the membrane."

I want to make clear -- you can infer fluidity of membrane lipids from the way things move in them, but you are not directly measuring fluidity especially if you are looking at proteins rearranging themselves after addition of ligands or in similar situations.

Mobility, as I said, is not formally defined, as is fluidity but it may be useful to make some comparisons, first of all, of scales of motion. There are people who experiment on molecular mobilities in cell membranes and look over Angstrom scales, and I think we are going to hear a fair amount about that scale this afternoon (and indeed at other places in this session when you come to talk about molecular associations).

Then there are experiments on molecular mobility and scales that run up to micrometers. These usually measure motion in some kind of dynamic way. The associations (Angstrom scale) of molecules undergoing the long (micron scale) treks are not clear. Nevertheless, it is mainly these micron scale measurements, at least for lateral diffusion, that give you constants that can be plugged into theories about shorter range, smaller scale molecular interactions.

So there are roughly, operationally anyway, two scales of mobility and distance that are worked on, and of course, then, it falls out that time scales vary, too. There are people looking at very fast motions, nanosecond, microsecond motion, and others dealing with time scales that can go out to days.

Another pair of distinctions -- people talk about mobility and it is not clear what the driving force is, and perhaps this is the central place where the discussion is muddled and needs to be clarified. In experiments in which ligands are added and configurations at cell surfaces change, one suspects that you are dealing largely with chemically driven processes. Something is pulling something somewhere else, whatever happened early on in the process. That is, whether or not patching is diffusionally driven, capping certainly is not.

On the other hand, there are mobility measurements over about the same time and distance scales as capping which appear to be measurements of thermally driven mobility, of diffusion, and these two are often lumped together. They connect in the end. I think they definitely connect, if only because it seems that true lateral diffusion is somehow constrained by elements that also participate in capping process.

I just will close with speculation that perhaps all these scale differences that we come to because we are doing different kinds of experiments have some biological meaning. It appears the small scale motions, whether they are diffusion-driven or driven by contractile elements in the cytoskeleton, can have some biological usefulness.

They seem to be important when membrane molecules interact to transmit signals through membranes, to reconfigure surfaces in various ways.

On the other hand, if you simply consider the topography of cells in tissues, the very nice way in which so many cells, in particular epithelial cells, lay out their surface enzymes in distinct spatial patterns, you begin to wonder if while small scale motions help you work with the cell, so-to-speak, the cells are fighting large scale motions. A cell has to invest something in keeping a membrane arranged, organized, so that for example in intestinal epithelium, the hydrolases are at one end at the microvilli, and the pumps are at the other end, and the two sets of enzymes don't get intermixed.

So there are some contrasts, if you like. We can group them all under the term "mobility" in the membrane.

RECEPTOR AGGREGATION AS A MECHANISM FOR TRANSMEMBRANE SIGNALLING: MODELS FOR HORMONE ACTION

JOSEPH SCHLESSINGER

Immunology Branch, National Cancer Institute, National Institutes of Health, Bethesda, Maryland 20014 (USA) and Department of Immunology, The Weizmann Institute of Science, Rehovot (Israel)

ABSTRACT

In this paper I propose a receptor aggregation model for transmembrane signalling. This model is based on a variety of experiment evidence which demonstrates that receptor motion and clustering (cross linking) in the plane of the membrane plays an important role in the molecular events which lead to hormone action, degranulation of mast cells and other important physiological processes.

I postulate that clustering of mobile receptors (local aggregation) causes the formation of a molecular product. This putative molecule can activate various transport systems across the membrane. It also forms an ion channel or triggers an active transport system of ions across the plasma membrane. The elevated intracellular ions (Ca^{++}?) could, in turn, activate intracellular enzymes and thereby trigger the various biological responses to the hormone. The local aggregation of receptors is a rapid and efficient process and it is not a rate determining step for the action of insulin, EGF and other peptide hormones.

I propose the following steps to describe the action, regulation and metabolism of insulin and EGF.

(1) The unoccupied receptors are mobile and homogenously distributed over the cell membrane.

(2) The hormone binds to diffuse receptors.

(3) Hormone binding induced local aggregation of hormone receptor complexes. This process provides the "unit signal" for triggering the various transport systems which are activated by hormone binding to their surface receptors.

(4) The mobile microaggregates cluster at coated regions of the plasma membrane and become immobile (global aggregation). The global aggregation process is related to receptor down regulation and precedes the internalization and degradation of the hormones.

(5) The clustered hormone receptor complexes and the coated pits pinch off and form coated vesicles. Later the coated vesicles interact with other intracellular organelles.

INTRODUCTION

In this paper, I summarize experimental evidence which demonstrates that clustering (crosslinking) of hormone receptors plays an important role in the molecular events which lead to hormone action. I propose a simple aggregation model which is based on the assumption that receptor clustering in the plane of the membrane can produce (or activate) a molecular product to which the biological response to the hormone is proportional. I am particularly interested in understanding the mode of action of insulin, epidermal growth factor (EGF) and other peptide hormones. However, this simple model is also applicable to describe the initial membrane events which lead to other physiological processes such as mast cell degranulation[1] and activation of lymphocytes[2,3].

Several authors have already proposed that the lateral motion of hormone receptors plays a role in the first steps of hormone action[4]. Lateral motion of the hormone receptor in the plasma membrane would enable it to interact with various effector molecules, some of which might be mobile, and others located at fixed sites. The existence of a mobile receptor is consistent with the fact that hormones that bind to different receptors can apparently activate the same adenylate cyclase molecule. Furthermore, β-adrenergic receptors from turkey erythrocytes can activate the adenylate cyclase of other cell types after the two are fused by Sendai virus[5].

In the first part of the introduction, I will summarize the current knowledge about the mechanism of action of insulin and EGF. In the second part, I will summarize some questions of mechanism which were obtained from studies with insulin.

Insulin is a potent hormone which can influence a wide variety of metabolic processes in various tissues[6-8]. Insulin is known to stimulate the membrane transport of glucose, amino acids, and certain ions. It can affect the activity of intracellular enzymes, increase the storage of glycogen, and stimulate the synthesis of proteins, RNA and DNA[6-8].

Despite many years of research concerning the mechanism of action of insulin, it is still not clear how the hormone regulates these diverse effects on target cells. Most investigators believe that insulin acts directly on the plasma membrane of target cells and thereby alters the permeability and transport mechanisms of the membrane. The concept of insulin acting on the plasma membrane was supported by experiments which have shown that insulin can bind to specific receptors on isolated plasma membrane preparations and promote biological response. Recently the receptor for insulin from various cell types[9-11] has been partially purified. With the discovery that cAMP is the second messenger for glucagon and other hormones, a few reports have demonstrated effects of insulin on the level of cAMP. It was shown that under certain circumstances insulin can inhibit adenylate cyclase[12], and under other conditions it can elevate cellular levels of cAMP[13]. Several other substances have been proposed as second messengers of insulin, among them cGMP, certain ions, ATPase, phosphodiesterase, phosphorylation-dephosphorylation reactions, oxidation of sulfhydryl groups, internalized insulin or internalized insulin fragment and alteration in membrane electrical potential[14]. At present, however, there is insufficient evidence to support the notion that any of these proposed second messengers can mediate the dive biological responses of insulin.

There is some evidence that insulin enters target cells. Stein and Gross have shown[15] by binding and autoradiographic techniques that insulin can enter the cytoplasm of muscle and liver cells. Terris and Steiner have demonstrated that insulin can be degraded after binding to isolated hepatocytes [16-17]. They concluded that insulin must enter the hepatocyte in order to be degraded. Cellular uptake of insulin has been demonstrated in cultured lymphocytes where 10% of the uptake is localized to the nuclear fraction[18]. Goldfine et al have shown that the nuclear binding is saturable, that it displays a slower time course than surface membrane uptake[18], and that the nuclear receptors and membrane receptors are immunologically distinct. It was recently proposed that the delayed effect on the synthesis of DNA and RNA may require the entry of either insulin or its fragments into the interior of the cell [16-19]. At present the physiological role of the intracellular insulin is unknown.

Epidermal growth factor (EGF) is a small polypeptide hormone which is isolated from the submaxillary glands of mice. EGF is a potent mitogen of epidermal and epithelial tissues both in vivo and in vitro[20-21]. It can stimulate the proliferation of various cultured cells, among them human fibroblasts[22]. glial cells[23] and mouse 3T3 cells[24]. Human fibroblasts exhibit approximately 100,000 receptors for EGF with an apparent dissociation constant of 3×10^{-10}M. No other peptide hormone competes for the binding sites of EGF. Carpenter et al have shown that EGF binding to human fibroblasts is followed by internalization and degradation[25-26]. The degradation of EGF is blocked by chloroquine, inhibitors of metabolic energy, local anesthetics and ammonium ions. Carpenter and Cohen have proposed that the occupied EGF receptors are internalized and that EGF is degraded by lysosomal enzymes. The hormone-mediated receptor loss is termed "down regulation" of receptors. Anderson et al pointed out that the kinetics of entry of human EGF into human fibroblasts are strikingly similar to those observed for low density lipoprotein. Therefore, they proposed that EGF may also enter via clustered receptors at coated regions[27]. In a recent report Aharonov et al suggested that the initial EGF internalization and receptor down regulation are not sufficient for EGF mitogenesis[28]. They have shown that the mitogenic response of the cells to EGF requires the continued presence of EGF over a 3-4 day period; in contrast, down regulation is completed within four hours. Therefore continued occupancy of a portion of the remaining and restored EGF surface receptors is essential for mitogenesis. They suggest that the major role of the down regulation of EGF receptors is to adjust the cell's sensitivity to EGF. On the other hand, Das and Fox propose[29] that internalization and lysosomal processing of the EGF-receptor complex yields a species which is the second messenger or has the capabilities of producing one. They suggest that this putative second messenger can induce DNA synthesis. In addition to the mitogenic response, EGF can stimulate the transport of sugar[30], convert ribosomes into polysomes[31], induce a transient increase in the activity of ornithine decarboxylase[32] and specifically increase the binding of Swiss-3T3 cells to Con A-coated nylon fibers[33]. EGF has also been reported to stimulate the synthesis of a major cell surface glycoprotein fibronectin[34]. It has been shown recently that mutant cells defective in the glycosylation of cell surface proteins have diminished amounts of EGF receptors on their surface and it seems possible that cellular transformation by certain sarcoma viruses, which can also decrease the amount of EGF receptors, has a similar effect[35,36].

Questions of Mechanism.

In the next paragraph I will summarize some questions of mechanism which are related to the mode of action of peptide hormones. I will describe in detail some studies with insulin because of the considerable background literature on this hormone. Many of the mechanistic aspects mentioned in this paragraph are either similar or have not yet been studied in detail in other hormonal systems.

a. Studies with fluorescent hormone analogues. Shechter et al have prepared various types of fluorescent analogues of insulin and EGF. The fluorescent analogues retained considerable binding affinity[37,38] and biological potency[38]. Schlessinger et al have used these analogues to study the lateral motion and distribution of insulin and EGF on cultured fibroblasts. The cells labeled with the fluorescent analogues were visualized with a sensitive video intensified camera[39] and the quantitative lateral diffusion coefficient were measured by the fluorescent photobleaching recovery (FPR)[40-43] method. Schlessinger et al have shown that insulin and EGF are initially mobile[38] in the plane of the membrane with $D \sim 5 \times 10^{-10}$ cm^2/sec (see table 1) and homogenously distributed over the cell surface[39]. Within a few minutes at 23°C or 37°C, the hormone receptor complexes aggregate into patches on the cell surface. The hormone receptor complexes which were initially mobile in the plane of the membrane become immobilized later as a consequence of hormone mediated receptor aggregation[38,39]. Within 30 minutes at 37°C, much of the aggregated hormone (patches) become internalized into endocytic vesicles[39]. The aggregation process does not require metabolic energy but internalization of hormones requires metabolic energy. Using fluorescein labeled α_2 macroglobulin and rhodamine labeled insulin or EGF, it was shown that all three polypeptides collect at the same patches on

TABLE 1

LATERAL DIFFUSION COEFFICIENTS OF THE HORMONE RECEPTOR COMPLEXES OF INSULIN AND EGF ON CELL MEMBRANE

Hormone and binding conditions	Diffusion coefficients (cm^2/sec)	Mobile fraction (% R)
Fluorescent insulin for 20 minutes at 23°C	$(4\pm0.8)\times10^{-10}$	40-80
Fluorescent insulin for 20 minutes at 37°C	10^{-11} -- 10^{-10}	<10
Fluorescent EGF for 20 minutes at 23°C	$(3.4\pm0.5)\times10^{-10}$	50-85
Fluorescent EGF for 20 minutes at 23°C, incubated for 30 minutes more at 37°C	10^{-11} -- 10^{-10}	<10

the cell surface and are internalized within the same vesicles[44]. Since α_2-macroglobulin was shown to patch over coated regions and to be internalized by coated vesicles[45], we concluded that insulin, EGF and α_2-macroglobulin indeed enter the cell via clustered receptors at coated regions as proposed by Anderson et al[27].

b. Studies with antibodies against insulin receptors. Recently, Kahn, Flier and their colleagues have shown that sera of some patients with insulin-resistant diabetes contain autoantibodies to insulin receptor. These antibodies bind to insulin receptors[46], block insulin binding[47] and initiate many of the biological effects of insulin[48]. These include stimulation of glucose and amino acid transport, glucose oxidation, incorporation of glucose into glycogen and alteration in the activity of two cytoplasmic enzymes: inhibition of phosphorylase and activation of glycogen synthase[49]. It was also shown that these antibodies are approximately as potent as insulin on a molar basis. In a recent publication, Kahn et al[49] have reported that both monovalent Fab' and divalent $(Fab')_2$ fragment of the anti-receptor antibody are fully active in inhibiting the binding of ^{125}I-insulin. Bivalent $(Fab')_2$ also retains its insulin-like biological potency. In contrast, the monovalent Fab' loses almost all ability to stimulate glucose oxidation, and acts as a competitive antagonist of insulin-stimulated glucose oxidation. Addition of anti-Fab' antibodies which cause the cross-linking of the Fab' receptor complexes results in a restoration of the insulin-like activity of the antibody. Furthermore, Kahn et al have shown that the exposure of cells to submaximal doses of insulin followed by addition of low amounts of anti-insulin antibodies enhances the biological activity of insulin[49].

These results together with studies on the mobility[38] and aggregation[39] of insulin and EGF on the cell membrane suggest that receptor aggregation might be the first membrane alteration which leads to the biological response to the hormone. The studies with anti-receptor antibodies give further insight into the mechanism of insulin action. They demonstrate, as pointed out by Kahn et al[49], that receptor occupancy by ligand alone is not sufficient for the generation of the biological response to the hormone, and they suggest that the receptor molecule probably contains all the biological information necessary to initiate hormone action.

Recently Jacobs and Cuatrecasas have shown that antibodies raised against purified insulin receptor can also initiate some of the biological effects of insulin. However, these antibodies do not compete with insulin for the same sites[50]. Similar results were obtained when fat cells were sensitized with anti-idiotype antibodies directed against antibodies against insulin[51].

c. Negative cooperativity of insulin binding. DeMeyts et al[32] have suggested that insulin binding to cell membrane receptors exhibits negative cooperativity. The major evidence for negative cooperativity was the increased rate of dissociation of ^{125}I-labeled insulin from lymphocytes after addition of unlabeled insulin and the non-linear Scatchard plots obtained for insulin binding to various cells. The basis for the negatively cooperative binding has been interpreted to be due to site-site interaction among insulin receptors. DeMeyts[52] proposes that the biological function of this phenomenon is to be a regulatory feedback step which prevents the effects of hormone excess.

Gliemann et al[53] failed to observe negative cooperativity of insulin binding to fat cells and recently Pollet et al[54] suggested that the lymphocyte insulin receptors behave kinetically and at equilibrium as a homogeneous class of independent high affinity binding sites up to insulin concentrations well above the physiological range.

Evidently more effort is necessary in order to determine the significance of this interesting possibility.

d. Hormone effect on membrane potential. Insulin acts in several ways on the electrical properties of the plasma membrane of target cells[55]. It is now well established that insulin hyperpolarizes the membrane of its target tissue[55]. A related observation is the effect of insulin on fluxes and intracellular concentration of Na and K ions[11,55]. The functional significance of the electrical change is not clear: it may be affected directly by insulin or may be byproducts of its actions. Most of the intracellular recording was performed by classical electrophysiological techniques using glass microelectrodes. The tissues used were rat skeletal muscle and white adipose tissue. Recently Petrozzo and Zierler performed (57) similar studies on isolated fat cells. They monitored the membrane potential by measuring the fluorescent intensity of a carbocyanine dye which was embedded in the plasma membrane of the fat cells. They have observed a similar hyperpolarization effect as a consequence of insulin binding to the plasma membrane of an isolated fat cell. Changes in the intracellular concentration of CA^{+2} ions appear to be related to some of the cellular changes induced by insulin. There is evidence for a role of intracellular Ca^{++} in insulin-mediated glucose transport in various tissues[14,56].

Although it appears possible that certain selective metabolic events could be affected by electrical alterations in the plasma membrane, there is no direct evidence to suggest that this change is a primary and general effect in hormone action.

e. Down regulation of receptors. Cells of obese insulin resistant patients and obese rodents which have increased level of circulating insulin have abnormally few insulin receptors[58]. In both, the number of binding sites increase toward normal when serum insulin is lowered by prolonged restriction of food intake[58], suggesting that insulin induces a specific loss of insulin-binding sites from the cell surface. A similar phenomenon can be demonstrated *in vitro*. Gavin et al found a time-dependent loss[59] of insulin receptors when human IM-9 lymphocytes were cultured in media containing insulin. With each insulin concentration, a particular rate of loss and steady-state level was achieved. The ability of insulin to "down regulate" its receptors is dependent on hormone concentration, time and temperature. When the cells are returned to a medium free of insulin, there is a return of receptors to normal levels within 16 hours. Both the down regulation of receptors and the return are blocked by cycloheximide.

The biological significance of down regulation of receptors is not clear. Many investigators believe that the function of receptor down regulation is to adjust the cell's sensitivity to the hormone. Raff proposed[60] that the function of receptor down regulation is to regulate the level of the cell surface receptors to show maximal biological response at fractional ligand occupancy.

LOCAL AGGREGATION AND GLOBAL AGGREGATION

I postulate that hormone induced receptor aggregation is related to the hormone responses described earlier in this paper. These include: the effect of antibodies against insulin receptor, hormone induced patching and hormone induced change in the electrical properties of the plasma membrane. I speculate that insulin mediated membrane hyperpolarization and other membrane electrical changes are coupled to the observed hormone receptor aggregation process. I postulate that the aggregated hormone receptor complex forms an ion channel or triggers some active transport of ions across the membrane. This, in turn, could activate some cytoplasmic or membrane enzymes and thereby trigger the various biological responses to insulin.

I have divided the hormone induced aggregation of receptors into two processes: local aggregation and global aggregation.

Local aggregation of receptors is defined as the process in which occupied receptors encounter each other and move a distance which is in the dimensions of 50-200 nm and the aggregation involves only a few receptor molecules. The time scale of such a process will be in the range of 50 msec up to 1 sec.

Global aggregation of occupied hormone receptor complexes is defined as the process in which the microaggregates of hormone receptor complexes move a distance of several microns to form visible patches which contains hundreds of receptor molecules. The time scale of such a process would be in the time dimension of a few seconds up to 10 minutes.

My major hypotheses are that local receptor aggregation is related to the rapid membrane responses to insulin (and to some extent EGF) while the global aggregation is related to receptor down regulation and preceeds the internalization and degradation of insulin and EGF.

a. Local aggregation of receptor as a model for insulin action. I have divided the molecular events in our model into two chemical reactions. For simplicity, I assume a minimum model and our "unit signal" is composed of two occupied receptors.

The time scale of a random encounter between two receptors can be calculated simply assuming that the cell possess approximately 50,000 receptors which correspond to 50 receptors per $(\mu m)^2$. The mean distance between two molecules in two dimensions equals $0,5\sqrt{1/c}$ (61) where c is the concentration of the receptors. For $c = 50$ receptor/$(\mu m)^2$ the mean distance r equals 0.07 μm. For $D = 5x10^{-10}$ cm^2/sec the encounter time $r^2/4D = (0.07\ 10^{-4})^2/(4 \cdot 5\ 10^{-10}) = 0.03$ sec. This corresponds to 30 collisions per second.

Figure 1 presents the proposed scheme. Reaction I describes a fast reversible binding of hormone (H) to its membrane receptor (R). Reaction II describes the interaction between two occupied receptors (HR) which can diffuse in the plasma membrane and form a product (C) through an intermediate encounter complex (HRRH). Reaction II is described by the theory of Smoluchowsky[62] with some modifications by Eigen[62] and Bell[64]. Finally, we assume that the biological response is proportional to (C).

$$\text{I. } R + H \underset{k_{-1}}{\overset{k_1}{\rightleftharpoons}} RH$$

$$\text{II. } RH + RH \underset{k_{-2}}{\overset{k_2}{\rightleftharpoons}} HRRH \underset{k_{-3}}{\overset{k_3}{\rightleftharpoons}} C$$

$$\text{III. } C \Longrightarrow \text{Biological Response}$$

Fig. 1: Local aggregation of receptors as a model for hormone action. Reaction I describes a fast reversible binding of hormone (H) to its membrane receptor (R). Reaction II shows that two occupied receptors (HR) which are free to diffuse in the plasma membrane form a product (C) through an intermediate encounter complex (HRRH). I assume that the biological response is proportional to (C) (reaction III).

The mass law for Reaction I gives

$$[RH] = K_1[R][H];\ K_1 = k_1/k_{-1} \tag{1}$$

As previously shown, k_2 and k_{-2} can be calculated as for a diffusion limited reaction follows[57,58]:

$$k_2 = 2\pi(2D) = 4\pi D \tag{2}$$

$$k_{-2} = 2\pi(2D)\ d^{-2} = 4D/d^2 \tag{3}$$

where D is the lateral diffusion coefficient of RH in the plane of the membrane, and d is the encounter distance (see figure 2).

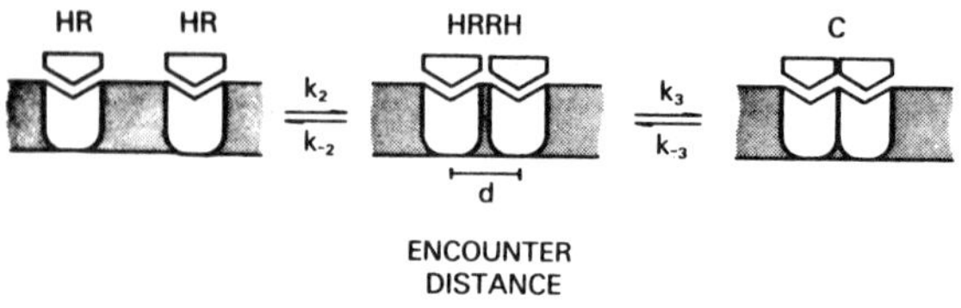

Fig. 2: The formation of an active product C ("unit signal") from two occupied receptors through an intermediate complex HRRH. The intermediate complex is formed when the distance between two diffusing HR molecules is shorter than the encounter distance d.

For a reaction in equilibrium, Eigen has shown[57] that:

$$\vec{k} = \frac{k_2 \cdot k_3}{k_{-2}+k_3} \; ; \; \overleftarrow{k} = \frac{k_{-2} \cdot k_{-3}}{k_{-2}+k_3} \tag{4}$$

where $\vec{k}$ and $\overleftarrow{k}$ are the overall forward and background rate constants for reaction II. The equilibrium constant of reaction II for very small [HRRH] equals:

$$K = \frac{\vec{k}}{\overleftarrow{k}} = \frac{k_2 . k_3}{k_{-2} . k_{-3}} = K_2 . K_3 \tag{5}$$

from equations (2) and (3)

$$K_2 = \frac{k_2}{k_{-2}} = \frac{4\pi D}{4D} d^2 = \pi d^2 \tag{6}$$

and

$$K = d^2 K_3 \tag{7}$$

Assuming quasiequilibrium conditions for reaction II we get the following:

$$\begin{aligned}[C] &= K[RH]^2 = \pi K_3 d^2 [RH]^2 = \\ &= \pi K_3 K^2 d^2 [R]^2 [H]^2\end{aligned} \tag{8}$$

The conclusion from equation (8) is that the overall <u>quasi equibrium constant</u> does not depend on the diffusion coefficient of the receptor. Since we assume that the biological response is proportional to [C], this means that the amplitude of the biological response does not depend on the rate of diffusion of the receptors.

The rate equation of reaction II is as follows:

$$\frac{d[C]}{dt} = \vec{k} [RH]^2 - \overleftarrow{k} [C] \tag{9}$$

Let us define [RH] at t=o as a and [C] at any given time t as x then

$$\frac{dx}{d+} = \vec{k} (a-2x)^2 - \overleftarrow{k} x \tag{10}$$

$$\frac{dx}{dt} = 4\vec{k}x^2 - (4a\vec{k}+\overleftarrow{k})x + \vec{k}a^2 \qquad (11)$$

If we define $A = 4\vec{k}$; $B = -(4a\vec{k} + \overleftarrow{k})$ and $C = \vec{k}\,a^2$ then it is necessary to solve the following integral:

$$\int \frac{dx}{Ax^2 + Bx + C} = dt \qquad (12)$$

The complete solution of this problem is presented in the appendix.

If we assume that $k_{-3} << k_3$, which means that the second step of reaction II is irreversible, then the complicated expression which is described in the appendix can be reduced to the following simple solution:

$$[C] = \frac{a^2\pi 4\ Dt}{1+\ 8\pi aDt} \qquad (13)$$

The same expression is obtained if we assume that $\overleftarrow{k} = 0$ and solve the correspondingly simplified equation (10).

At $t << 1/8\pi aD$, [C] is approximately equal to $4a^2Dt$. At $t >> 1/8\pi aD$, [C] reaches a plateau which equals to $a/2$. The time constant of this reaction is $1/8\pi aD$.

Table 2 presents the time constants of the reaction for various levels of a. The surprising conclusion is that this process is very rapid even at very low levels of receptor occupancy. Even at the level of 1 occupied receptor per $(\mu m)^2$ (which corresponds to 2500 receptors per cell) the time constant of this process equals 0.8 sec. The major conclusion from this calculation is that receptor diffusion and formation of dimers do not seem to be the limiting step in the cascade of events which lead to the biological response to the hormone.

TABLE 2

CALCULATED TIME CONSTANT OF THE RATE OF APPEARANCE OF C FOR VARIOUS LEVELS OF OCCUPIED RECEPTORS(a)

Number of receptors per $(\mu M)^2$	Time constant (msec)
1	800
5	160
50	16
500	1.6

b. Global aggregation of hormone receptors. Global aggregation of occupied hormone receptor complexes is the process in which a few hundred of occupied receptors move in the plane of the membrane and form visible clusters; patches or caps. Goldstein and Brown[65] have demonstrated that coated regions of the plasma membrane of human fibroblasts are involved in the process of clustering and endocytosis[66] of low density lipoprotein (LDL)[65]. On the basis of these results, Anderson et al[27] proposed a mechanism for the regulation of the receptors for low density lipoprotein on fibroblasts. They postulate that newly synthesized receptors for LDL are inserted at random sites along the plasma membrane and that they cluster together independent of the binding of LDL in the regions of the membrane containing "coated pits", which internalize to form coated vesicles. They propose that the same mechanism might exist for the internalization of EGF on fibroblasts[27].

There is one difference between the results on the patching and internalization of insulin and EGF and the mechanism proposed by Anderson et al[27]. Schlessinger et al find that insulin and EGF receptors are initially diffuse before the hormones are bound and that patching and endocytosis are triggered by hormone binding[39]. Anderson et al propose that LDL binds to preexisting clusters of receptors[27].

It is possible that LDL-receptors cluster without the binding of LDL. However, the binding of ligands to clustered receptors is a less efficient process than binding to diffuse mobile receptors which cluster after becoming occupied. Berg and Purcell[67] have developed a theory which allows us to compare these two (and other) possibilities. The following equation can be used for our purpose:

$$J = J_{max}\, NS/(NS+ A) \qquad (17)$$

where J is the diffusional flux of a ligand which can bind to a membrane receptor; N is the total number of receptors per cell; S and A are the radius of the receptor and the cell respectively.

I have calculated J for three different geometries. In all three examples, the total number of the receptors is 10,000 the radius of the receptor is 10A° and the radius of the cell is 5 μm. In the first example, the receptors are diffuse over the cell membrane. In the second example, the receptors are clustered within 100 randomly distributed patches each containing 100 receptors. In the third example, all the receptors are clustered in a single cap. Table 3 summarizes the results. The diffusional flux to clustered receptors (patches) is 15% of the flux to diffuse receptors and the flux to a single cap is only 1.6% of the flux to the diffusely distributed receptors.

TABLE 3
THE EFFECT OF RECEPTOR DISTRIBUTION ON THE DIFFUSIONAL FLOW OF THEIR LIGAND

Receptor Distribution	Relative Diffusional Flow
Diffuse	1
Patches	0.15
Cap	0.016

This calculation demonstrates that simple topological considerations can drastically affect the binding property of a ligand to the cell membrane. It demonstrates that diffuse distribution of mobile receptors followed by patching is more efficient for selectively adsorbing solutes than binding to preclustered receptors. Since the level of LDL in the serum is high (1-3 mg/ml) it is possible that the cell prefers to use less efficient processes when the solute is largely available. This does not seem to be the case for insulin, EGF and other peptide hormones which are available at very low concentrations (1 ng/ml).

Recently Schlessinger et al have shown that the distribution of rhodamine-insulin on human cultured lymphocytes (IM-9) is different than on fibroblasts. When IM-9 cells were incubated at 37°C in the presence of R-insulin, the hormone receptor complexes migrated and formed a single cap on one pole of the cell. Sodium azide prevented the capping process but did not effect the patching[68]. The fraction of capped cells has correlated well with the extent of receptor down regulation suggesting that receptor capping might be related to insulin mediated receptor loss[68].

DISCUSSION

The most convincing biological system in which a relationship between receptor motion, receptor aggregation and biological response was demonstrated is the process of degranulation of mast cells and basophils. (For a theoretical treatment see paper by C. DeLisi in this volume.)

In a set of elegant experiments performed primarily by Metzger and coworkers, it was shown that the Fc receptors are monovalent toward the IgE molecule[69,70], and that the IgE:Fc receptors complex can move laterally in the plane of the membran with $D \sim 3 \times 10^{-10}$ cm^2/sec[70]. The degranulation of mast cells and basophils can be accomplished by each of the following treatments:

a) The binding of IgE to the membrane receptor followed by crosslinking with anti-IgE antibodies[1].

b) Binding of chemically cross-linked IgE dimers[1].

c) Binding of antibodies against IgE receptors[71]. Fab' fragments bind to the IgE receptor but do not cause degranulation; $(Fab')_2$ binds to the membrane receptors and induces degranulation[71]. The presence of Ca ions is necessary in order to obtain degranulation in each of these treatments.

The major conclusions of these studies are that aggregation of IgE receptors provides the signal for triggering degranulation[71] and that the "unit signal" is a dimer of IgE receptors[1].

Similar experiments were performed with autoantibodies against insulin receptors[46-49] (summarized in the Introduction), anti idiotype antibodies against anti-insulin[56], and antibodies prepared against purified insulin-receptor[50]. It is not known yet what is the "unit signal"[1] (minimal number of clustered receptors which produces the biological response) for triggering the insulin-like response of insulin by these various antibodies.

The local aggregation model proposed in this paper is adequate for describing the first events which lead to mast cell or basophil degranulation. Using equation 13, we can calculate the time constant of this process given that D (Fc receptors) $= 3 \times 10^{-10}$ cm^2/sec (70) and that the number of Fc receptors equals 2000 per $(\mu m)^2$. For this values, the time constant equals 0.6 msec. When 1% of the receptors are occupied, the time constant equals 60 msec.

The fact that antibodies against Fc receptors can trigger histamine release[71] indicates that the information for generating the signal which leads to degranulation exists in the receptor molecules and that the IgE molecules serves only as a linkage between the antigen and the receptor[72]. Therefore it is puzzling why randomly colliding empty Fc receptors (at a frequency of few hundred of collisions per second) do not provide random "unit signals" for degranulation. Several possible explantions were proposed by Metzger, Isersky and their colleagues[72] in order to explain this discrepancy:

1) that external crosslinking provides the additional energy to overcome the activation energy barrier of this process;

2) that crosslinking changes the conformation of the receptor increasing their affinity toward each other;

3) that crosslinking increases the period in which the receptors are within the encounter distance d;

4) that crosslinking orients the receptors in a necessary orientation.

At present it is not known which of these possibilities is relevant. For insulin, EGF and other peptide hormones, the local aggregation model serves only as a minimal model. More adequate models could be formalized after the valence of the receptor toward the hormone and the "unit signal" are elucidated. At present, we are examining the effect of formation of higher aggregates of $(HR)_n$ $(n>2)$ on the kinetics of the formation of C[73].

Our simple model has several predictions which can be tested experimentally. It shows that the absolute value of the biological response should not depend on the diffusion coefficient of the receptor while the kinetics of the biological response is proportional to $4\pi a^2 Dt$ at $t<<1/8\pi aD$ and it reaches a plateau at $[C] = a/2$. Therefore changing the diffusion coefficient of the membrane receptor should in principle cause a decrease in the rate of the biological response but should not affect the absolute value of the hormone effect. Similarly, if the hormone induced membrane electrical change originates from the local aggregation process, than the kinetics of the hyperpolarization (on muscle or fat cells) should be affected by addition of cholesterol to the membrane (which decrease the diffusion coefficient) while the amplitude of the hyperpolarization should remain unchanged. Addition of cholesterol, which is known to be a membrane "stiffener", should decrease D (receptor) and thereby attenuate the rate of the local aggregation process and therefore decrease the kinetics of the hormone induced hyperpolarization. Similar experiments can be done with mast cells and basophils. Another major conclusion from the local aggregation model is that the formation of a "unit signal" is a very rapid process even at receptor occupancy of 1% or less. This means that the aggregation process is not a limiting step in the cascade of events which lead to the metabolic effects of the hormone. Ciarldi and Olefsky have recently demonstrated the existance of a time lag in the coupling between insulin receptors and the glucose transport system and that the duration of this time lag is dependent on temperature[74]. Olefsky have previously proposed that the lateral notion of insulin receptors is not an essential feature of insulin action since insulin mainly causes an increase in already available glucose carriers rather than enhancing intrinsic activity of already functioning carriers[75]. The local aggregation is clearly consistent with the notion that the lateral diffusion of insulin receptors within the plasma membrane is not a rate-determining step in insulin action. Therefore it is very likely that the time lag in the coupling between insulin binding to its receptor and the glucose transport system is due to some other temperature sensitive processes. It does not mean however that the lateral motion of insulin receptor does not play a physiological role[75]. On the contrary, it could provide a rapid, efficient means for producing a molecular aggregate which triggers a biological signal.

There are at least two more biological systems in which local aggregation of receptors (and their lateral motion) seem to play a physiological role. It was recently demonstrated that intact (divalent) antibodies or $(Fab)_2$ fragments of antibodies against acetycholine receptor (isolated from patients with mysthenia gravis) induce receptor degradation (76), while their Fab fragments bind to the receptor but do not trigger degradation of receptors (77). It was also shown that antibodies against TSH-receptors (isolated from patients with Grave's disease) can mimic TSH and activate the adenylate cyclase of thyroid cells (78,79).

I predict that our simple local aggregation model can be applied to other biological systems, and as more membrane receptors are purified and antibodies against them prepared more examples will appear.

On the basis of the studies summarized in this article, I propose the following model to describe the various steps in the action, regulation and metabolism of insulin, EGF and perhaps other peptide hormones. Figure 3 summarizes the various steps involved in our proposed model.

(I) The unoccupied receptors are mobile (D 5×10^{-10} cm^2/sec) and homogeneously distributed over the cell membrane.

(II) The hormone binds to diffuse receptors. At 9°C the hormone receptor complexes remain diffuse for at least 90 minutes.

(III) Hormone binding (at 37°C) induces local aggregation of hormone receptor complexes The local aggregation process provides the "unit signal" for triggering the membrane effects of insulin and EGF. I speculate that this aggregation process activates the various transport systems (glucose transport, amino acid transport, etc.). I propose that the hormone induced change in the electrical properties of the plasma membrane is also induced by the local aggregation of receptors. The following mechanism can cause the electrical changes: (a) The molecular product c which is formed from two HR moieties (see Fig. 2) has the properties of an ion channel. It is possible, for example, that each Fc receptor on mast cells is "half" of Ca^{++} channel. Therefore a dimer of Fc receptors forms a Ca^{++} channel which provides a "unit signal" for further products. (b) The molecular product c activates a preexisting ion channel. (c) The molecular product c triggers an active transport system.

Each of these possibilities can cause the influx of Ca^{++} ions (following efflux of potassium ions) and thereby cause the observed hyperpolarization of the plasma membrane. The increase in the intracellular Ca^{++} will cause the activation of various cytoplasmic enzymes which are known to be activated by insulin, EGF or other growth factors. It was recently shown by Kyte[80] that a $(Na^{+}+K^{+})$ ATPase which is responsible for the active transport of sodium and potassium across the plasma membrane of animal cells is a dimer of two identical subunits. Kyte provided experimental evidence which demonstrate that the ion channel passes between the subunits through the symmetry axis of the dimer[80].

(IV) The mobile hormone receptor complexes (microaggregates) cluster at coated region of the plasma membrane and become immobile. I call this step the global aggregation process. According to this hypothesis, the formation of microaggregates (step III) causes elevation of intracellular Ca^{++}. The elevated Ca^{++} could activate, among other things, an enzyme which anchores the hormone receptors to clathrin which is a major constituent of coated vesicles from calf brain[81]. Other possible structures which could be involved in the process of clustering and immobilization of receptors over coated pits are contractile proteins of the cytoskeleton of the cell. Such or similar processes could account for the clustering and immobilization of hormone-receptor complexes over coated pits. It was previously postulated by Edelman[82] and Singer[83] that cytoskeletal elements participate in the process of patching and capping of cell surface receptors.

Recently, Schlessinger et al have shown that surface aggregation and immobilization does not necessarily imply clustering over coated pits. It was shown that artificial lipopolysacharide[84] inserted into the plasma membrane and cross linked by antibodies form immobile[85] patches on areas which do not involve coated regions[86]. Similar results were obtained when H-2 antigens are cross linked by anti H-2 antibodies[86].

(V) The clustered hormone receptor complexes and the coated pits pinch off and form coated vesicles. We postulate that this step is the mechanism of receptor mediated endocytosis[66] and hormone mediated receptor loss (down regulation)[58] of insulin, EGF, α_2-macroglobulin, LDL and other polypeptides. Both phenomena depend on time, temperature and hormone concentration.

Recently, I have shown that when IM-9 lymphocytes were incubated with R-insulin the hormone receptor complexes form a single cap on one pole of the cell[66]. Since the fraction of capped cells correlate well with the extent of receptor down regulation, it is possible the observed insulin mediated receptor loss on these cells is caused by insulin induced capping of hormone receptors.

The generality of these proposals for other cell types and hormones has to be established by further studies.

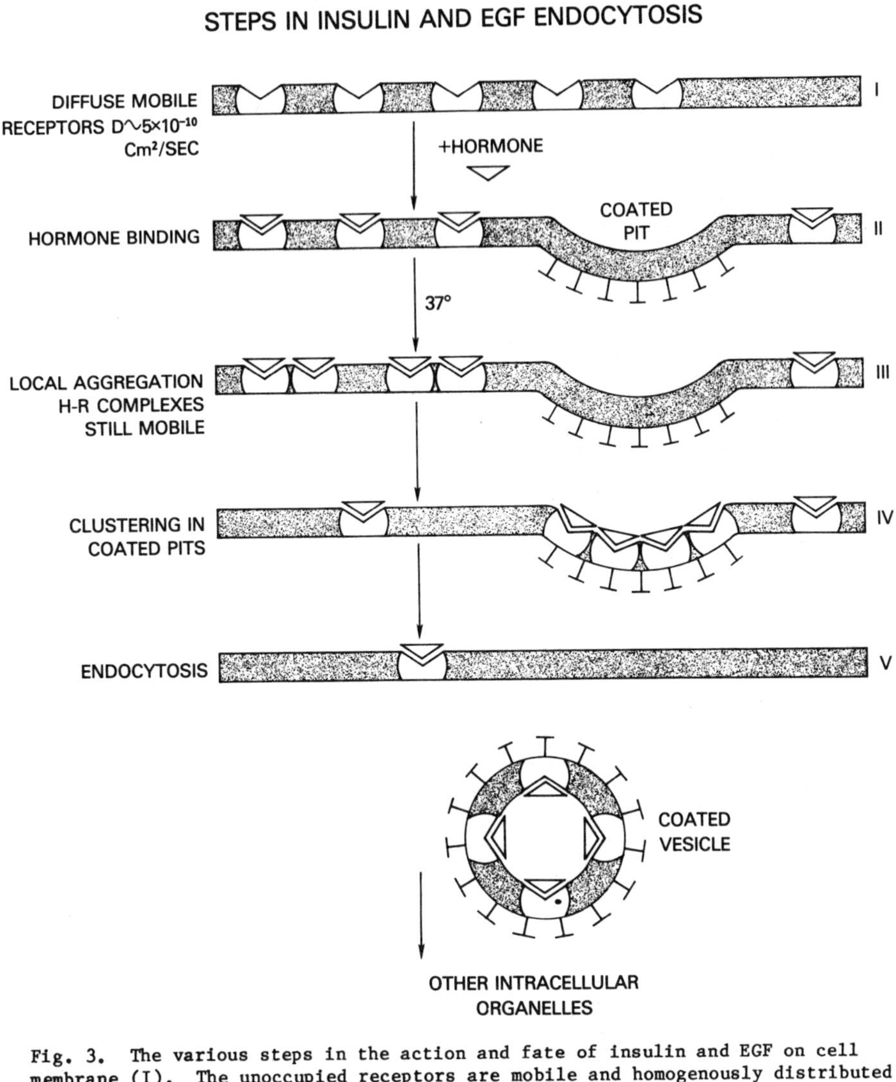

Fig. 3. The various steps in the action and fate of insulin and EGF on cell membrane (I). The unoccupied receptors are mobile and homogenously distributed over the cell membrane. (II). The hormone binds to diffuse receptors. (III). Hormone binding induces local aggregation of hormone receptor complexes. I postulate that this process provides the "unit signal" for triggering the various transport systems which are activated by hormone binding. (IV) The mobile hormone receptor complexes cluster at coated regions of the plasma membrane and become immobile. I call this step as the global aggregation process. (V) The clustered hormone receptor complexes and the coated pits pinch off and form a coated vesicle. I postulate that this step is the mechanism of receptor mediated endocytosis[66] and hormone mediated receptor loss (down regulation)[58] of EGF, insulin, LDL, α_2-macroglobulin and other polypeptide). Later the coated vesicle interact with other intracellular organelles.

APPENDIX

The solution of the integral

$$\int \frac{dx}{AX^2 + BX + C} = dt \tag{1}$$

equals to

$$\Gamma + t = \frac{1}{\sqrt{B^2 - 4AC}} \cdot \log \frac{2AX + B - \sqrt{B^2 - 4AC}}{2AX + B + \sqrt{B^2 - 4AC}} \tag{2}$$

when t = 0, x = a and therefore

$$\Gamma = \frac{1}{\sqrt{B^2 - 4AC}} \cdot \log \frac{B - \sqrt{B^2 - 4AC}}{B + \sqrt{B^2 - 4AC}} \tag{3}$$

after simple substitution it can be shown that

$$X = \frac{B - \sqrt{B^2 - 4AC} - (B + \sqrt{B^2 - 4AC}\; e^{\sqrt{B^2 - 4AC}\,(\Gamma + t)})}{2A(e^{(\sqrt{B^2 - 4AC}\,(\Gamma + 1)} - 1)} \tag{4}$$

The final expression is:

$$X = \frac{B - \sqrt{B^2 - 4AC} - (B + \sqrt{B^2 - 4AC})\; e^{\left[\log \frac{B - \sqrt{B^2 - 4AC}}{B + \sqrt{B^2 - 4AC}} + \sqrt{B^2 - 4AC}\; t\right]}}{2A\left[e^{\operatorname{Log} \frac{B - \sqrt{B^2 - 4AC}}{B + \sqrt{B^2 - 4AC}} + \sqrt{B^2 - 4AC}\,.\,t)} + 1\right]} \tag{5}$$

where $A = 16\ D$; $B = -16\pi aD - \overleftarrow{k}$ and $C = 4\pi Da^2$

ACKNOWLEDGMENTS

I would like to thank my friends and colleagues Y. Shechter, P. Dragsten, E.L. Elson, P. Henkart, K. Yamada, R. Blumenthal, E. Van Oberghen, E. Ralston, H. Metzger, and F. Maxfield for their very stimulating discussions and collaborations.

REFERENCES

1. Segal, D. et al. (1974) Proc. Natl. Acad. Sci. USA, 74, 2293-2997.
2. Yahara, I. and Edelman, G.M. (1972) Proc. Natl. Acad. Sci. USA, 69, 608-612.
3. Dintzis, H.M. et al. (1976) Proc. Natl. Sci. USA, 73, 3621-3675.
4. Cuatrecasas, P., (1974) Ann. Rev. Biochem., 43, 169-214.
5. Orly, J. and Schramm, M. (1976) Proc. Natl. Acad. Sci. USA, 73, 4410-4416.
6. Krahl, M.E. (1961) The Action of Insulin on Cells, New York, Academic Press.
7. Fain, J.N. (1974) in MTP International Review of Science, Biochemistry Series 1, Vol. 8, Rickenberg, H.V. ed., Park Press, Baltimore, Md. pp. 1-23.
8. Park, C.R. and Pilkis, S.J. (1974) Ann. Rev. of Pharmacology, 14, 368-385.
9. Cuatrecasas, P. (1972) Proc. Natl. Acad. Sci. USA, 69, 318-322.
10. Cuatrecasas, P. and Tell, G.P.F. (1973) Proc. Natl. Acad. Sci. USA, 70, 485.
11. Jacobs, S. et al. (1977) Biochem. Biophys. Res. Commun., 77, 981-988.
12. Hepp, K.D. (1971) FEBS Lett., 12, 263-266.
13. Kane, T. and Barham, F.W. (1973) J. Biol. Chem., 248, 7417-7426.
14. Czech, M.P. (1977) Ann. Rev. Biochemistry, 46, 359-384.
15. Stein, O. and Gross, J. (1959) Endocrinology, 65, 707-724.
16. Terris, S. and Steiner, D.F. (1975) J. Biol. Chem., 250, 8389-8398.
17. Steiner, D.F. (1977) Diabetes, 26, 322-340.
18. Goldfine, I.D. et al. (1977) Proc. Natl. Acad. Sci. USA, 74, 1368-1372.
19. Goldfine, I.D. et al. (1978) Nature, in press.
20. Cohen, S. and Taylor, J.M. (1974) Rec. Prog. in Hormone Res., 30, 533-550.
21. Cohen, S. and Savage, C.R. (1974) Rec. Prog. in Hormone, 30, 551-574.
22. Hollenberg, M.D. and Cuatrecasas, P. (1973) Proc. Natl. Acad. Sci. USA, 70, 2964-2968.
23. Westermark, B. (1976) Biophys. Res. Commun., 69, 304-310.
24. Rose, S.P. et al. (1975) J. Cell Physiol., 86, 593-598.
25. Carpenter, G. et al. (1975) J. Biol. Chem., 250, 4297-4304.
26. Carpenter, G. and Cohen, S. (1976) J. Cell Biol., 71, 159-171.
27. Anderson, R.G.W. et al. (1977) Nature, 270, 695-699.
28. Aharonov, A. et al. (1978) J. Biol. Chem., 253, 3970-3977.
29. Das, M. and Fox, C.F. (1978) Proc. Natl. Acad. Sci. USA, 75, 2644-2648.
30. Barnes, D. and Colowick, S.P. (1976) J. Cell Physiol., 89, 633-640.
31. Hoober, J.K. and Cohen, S. (1967) Biochim. Biophys. Acta, 138, 347-356.
32. Stasny, M. (1970) Biochim. Biophys. Acta, (1970) 578-589.
33. Aharonov, A. et al. (1978) J. Cell Physiol., 95, 195-202.

34.Chen, L.B. et al. (1977) Science, 187, 776-778.
35.Pratt, R.M. and Pastan, I. (1978) Nature, 272, 68-70.
36.Shields, R. (1978) Nature, 272, 670-671.
37.Schecter, Y. et al. (1978) Proc. Natl. Acad. Sci. USA, 75, 2135-2139.
38.Schlessinger, J. et al. (1978) Proc. Natl. Acad. Sci. USA, in press.
39.Schlessinger, J. et al. (1978) Proc. Natl. Acad. Sci. USA, 75, 2659-2663.
40.Koppel, D.F. et al. (1976) Biophys. J., 16, 1315-1329.
41.Schlessinger, J. (1976) Proc. Natl. Acad. Sci. USA, 73, 2409-2413.
42.Jacobson, K. et al. (1976) Biochim. Biophys. Acta, 433, 215-222.
43.Edidin, M. et al. (1976) Science, 196, 466-468.
44.Maxfield, F. et al. (1978) Cell, 14, 805-810.
45.Willingham, M.C. et al. (1978) Cell, submitted.
46.Jarrett, D.B. et al. (1976) Proc. Natl. Acad. Sci. USA, 73, 4115-4119.
47.Flier, J.S. et al. (1976) J. Clin. Invest., 58, 1442-1449.
48.Kahn, C.R. et al. (1977) J. Clin. Invest., 60, 1094-1106.
49.Kahn, C.R. et al. (1978) Proc. Natl. Acad. Sci. USA, 75, 4209-4213.
50.Jacobs, S. et al. (1978) Science, 200, 1283-1284.
51.Sege, K. and Peterson, P.A. (1978) Proc. Natl. Acad. Sci. USA, 75, 2443-2448.
52.DeMeyts, P. et al. (1976) J. Biol. Chem., 251, 1877-1888.
53.Gliemann, J. et al. (1975) J. Biol. Chem., 250, 3368-3374.
54.Pollet, R.J. et al. (1977) J. Biol. Chem., 252, 5828-5834.
55.Zierler, K.L. (1972) Handbook Physiol. Sect. 7 Endocrinol., 347-368.
56.Rasmussen, H. and Goodman, D.B.P. (1977) Physiol. Rev., 57, 421-509.
57.Petrozzo, P. and Zierler, K. (1976) Fed. Proc. 33 (Abst. No. 2178), 602.
58.See review by C.R. Kahn (1976) J. Cell Biol., 70, 261-286.
59.Gavin, J.R. et al. (1974) Proc. Natl. Acad. Sci. USA, 71, 84-88.
60.Raff, M. (1976) Nature, 259, 265-266.
61.Chandraselchar, S. (1943) Rev. Med. Phys., 15-1-143.
62.Smoluchowsky, M. and Physik, Z. (1976) Chem. 92, 557 and 585.
63.Eigen, M. (1974) in Quantum Statistical Mechanics in the Natural Sciences, Mintz, S.L. and Wiedermayer, S.M., eds., Plenum, New York.
64.Bell, G.I. (1978) Science, 200, 618-627.
65.Goldstein, J.L. and Brown, M.S. (1977) Ann. Rev. Biochem., 46, 897-930.
66.Silverstein S.C. et al. (1977) Ann. Rev. Biochem., 46, 669-722.
67.Berg, H.C. and Purcell, E.M. (1977) Biophysical J., 20, 193-219.
68.Schlessinger, J. et al. (1978) in preparation.
69.Mendoza, G. and Metzger, H. (1976) Nature, 264, 548-549.
70.Schlessinger, J. et al. (1976) Nature, 264, 550-551.
71.Isersky, C. et al. (1978) J. Immunol., in press.

72.Metzger, H. and Isersky. Private communication.

73.Buas, M. (1977) Ph.D. Thesis.

74.Ciaraldi, T. and Olefsky, J. (1978) Proc. of 38th Ann. Meeting of the American Diabetes Assoc., Diabetes Abs., 87, p. 452.

75.Olefsky, J.M. (1978) Biochem. J., 172, 137-145.

76.Stanley, E.F. and Drachman, D.B. (1978) Science, 200, 1285-1287.

77.Drachman, D.B. et al. (1978) New Engl. J. Med., 298, 1116-1122.

78.Smith, B.R. and Hall, R. (1974) Lancet ii, 427-430.

79.Mehdi, S.A. and Nussey, S.S. (1975) Biochem. J., 145, 105-111.

80.Kyte, J.E. (1978) Biophys. J., 24, 153a.

81.Pearse, B.M.F. (1976) Proc. Natl. Acad. Sci. USA, 73, 1255-1259.

82.Edeleman, G.M. (1976) Science, 192, 218-226.

83.Ash, J.F. et al. (1977) Proc. Natl. Acad. Sci. USA, 74, 5584-5588.

84.Wolf, D. et al. **(1977) Biochemistry**, 16, 3476-3483.

85.Wolf, D. et al. (1978) in preparation.

86.Schlessinger, J. (1978) in preparation.

DISCUSSION

DR. SINGER: I am a little confused about where certain fluorescence patterns that you have seen may correspond with the kinds of pictures I have been talking about. For example, you showed early in your talk an experiment in which you allow the insulin to sit on the cell surface at 37 degrees for a certain time, and then according to your photobleaching experiment, you get no mobility. Then, when you look at the cell surface at that point, you see a fairly small number of fairly large patches distributed over the cell surface.

DR. SCHLESSINGER: Three hundred.

DR. SINGER: Three hundred. But you are talking about 100,000 or 500,000 receptors. The question I have in part is whether you can be sure that the patching you see is relevant to the immobilization that you measured by the laser-induced photobleaching, whether in fact the patching you see has anything to do with the physiological responsive receptors, and whether it is some kind of artifact, a sideline of what happens to some of the receptors.

What I am not clear about is how you can be certain that what you are seeing is related to the physiology of the biological response, whether you might not have several kinds of phenomena in which you are only able to observe one; namely, fluorescence patches that are large enough to be observed.

DR. SCHLESSINGER: I think I can clarify the problem by providing some additional information. The size of the laser beam is about one micron. The size of a patch is approximately .1 micron. Since the patches are beyond the resolution of light microscopy they look like round bright spots. The size of the spot doesn't tell us anything.

So in the area which is illuminated by the laser beam we have approximately 10 patches or even more. This is because they are randomly distributed. Under these conditions the fluorescence photobleaching experiments show that the hormone receptor complexes are immobile.

The second thing is that under the condition in which we have diffuse (4°C) or partial aggregation of receptors we can measure the mobility of the receptors but when the temperature is raised to 32°C the receptors become immobile and visual patches can be seen.

DR. SINGER: Now, one other point about that: Do you envision that, say, the EGF receptors are present in coated pits to begin with, and the coated pits are then clustered together? Because that is a very different situation from Joe Goldstein's with the LDL receptor. There all the receptors in a given situation are in the coated pits to begin with, which are very small structures, and the coated pits are not clustered together in any way. When the LDL binds to the receptor in the coated pits, there is an immediate process of invagination of the coated pits. There is not a state in the process in which the bound receptors are clustered together, as I take it you are suggesting.

So the relationship between what you are getting with the EGF receptor and what Joe Goldstein gets with the LDL is not obviously one to one.

DR. SCHLESSINGER: This is clearly true. What we suggest for insulin and for EGF is that the hormones induce clustering of receptors which eventually migrate to the coated pits. The coated pits pinch off and form coated vesicles. I think maybe Joe Goldstein would like to comment. Goldstein and Brown suggest that the receptors for LDL preexist in the coated pits, which pinch off and form these coated vesicles.

I would like to emphasize the hormone induced aggregation of receptors before they are localized at the coated pits. This very rapid process of local microaggregation is presumably related to the studies of Kahn with the anti-insulin receptor antibodies. We have calculated from the measured diffusion coefficient that this microaggregation process would be very rapid (250 msec).

Recently we have some new results which are consistent with that. Yoram Shechter has prepared a derivative of EGF which binds to its membrane receptors but does not produce a biological response. We found that if we cross-link this analogue with anti-EGF antibodies the biological response is restored. The fluorescent analogue of this derivative is diffuse on the cell surface, but when we added anti-EGF antibodies they also patch.

We now know that at the dose of antibody in which the biological response is restored we cannot see visual patches. Under the conditions in which we see these patches, the biological response is already starting to decline.

This is very similar to the system of the mast cells, and I think Henry Metzger can comment on that. Conditions in which you see patching on mast cells are inhibitory for histamine release. In both biological systems micro-aggregation of receptors is a key process which triggers biological response.

So the story about coated pits is nice and it seems to be related to the processes of down regulation or perhaps desensitization of receptors but I think that the biological response is triggered much earlier.

DR. ISERSKY: Especially on the receptor you never see.

DR. SCHLESSINGER: Chaviva, I think, is going to say that with anti-receptor, in fact, you never see patching, and ~~it~~ can mimic the biological response. Right?

DR. ISERSKY: Yes

DR. SCHLESSINGER: Okay. So I can predict.

DR. HECHTER: This very striking paper went so very fast with so very many features. I thought -- it was like in a movie, where it was speeded up three or four times, and some of the questions that came, before I could formulate them, disappeared.

I would now like to recapitulate very slowly. We had in the first slide the diverse metabolic effects of insulin, and the statement that these effects depend on variable hormone occupancy of receptors. When we speak about insulin there are many target cells. One could think of muscle. One could think of fat cells. One can think of liver cells, in addition to culture cells and so on. The term, biological response, has to be operationally then defined in terms of what specific cell type and what biological response. We already know that in the fat cell the biological response of insulin for glucose oxidation and for certain other things -- it is so old that maybe some people don't know it -- can be activated by a variety of proteolytic enzymes, nucleotides; a whole group of things can do interesting things in fat cells.

It is so old that it might need to be rediscovered. Is this the biological response? The effects on other parameters are much later.

The second question was the diagram with the hypothetical second messenger. If you have a unitary receptor and you have events at the membrane, you have events inside, how are they coupled? And of course, one way to do it is through a second messenger. But once you bring, as you have already done, hormone inside the cell, why do you need a second messenger? Why can't you use that same receptor, much as steroid hormones are going to be utilized to activate various parts of protein synthesis, whether through one mechanism or another. That is a detail. I could ask three or four, but just these two would be enough.

DR. SCHLESSINGER: Unfortunately, I am not sure that I understand those two. What I described in those two slides is a summary of the current knowledge about the mechanism of action of insulin. I am not talking about second messengers, up to now many second messengers were proposed and many disproved. I wanted to give a brief description of what is known about the time scale of the various responses of insulin and this was just some kind of background slide.

DR. HECHTER: What about these antibodies that make the response?

DR. SCHLESSINGER: I think Ron Kahn would be the one to answer this question I think it is known that the antibodies can mimic many of the responses of insulin, and Ron is the man who did much of that work. He should talk about it.

DR. KAHN: We will comment some more on this tomorrow, but basically, the antibody to the insulin receptor appears to be able to mimic both the long and short-term responses, membrane transport effects, activation and inactivation of various cellular enzymes.

So presumably I think your point is a good point. Maybe the concept that we need a classic second messenger for insulin action is not necessary here. We don't really know. I mean, that is one thing that has always been proposed by analogy to other peptides.

DR. HECHTER: What cell type?

DR. KAHN: In fat cells, muscle and liver. Therefore, with respect to your comment and with respect to what Yossi showed, I think one plausible mechanism that Yossi and Jesse and I have sort of toyed around with is that the internalization is actually bringing the hormone in and possibly the receptor in. We do not know whether this by itself is the second messenger whether it is still generating yet another signal which could be a second messenger.

And for insulin, unfortunately, compared to things which act through cyclic nucleotides, we really do not know the answer to that question, but this is the way we are thinking about the problem.

DR. SCHLESSINGER: Ron, did you test the long-term response of mitogenesis by these antibodies?

DR. KAHN: Not mitogenesis, but protein synthesis required for lipoprotein-lipase stimulation has been tested.

DR. SCHLESSINGER: So you don't need internalization to trigger this process.

DR. KAHN: Right.

DR. RAFF: Could I just clarify a point that has already come up or ask for clarification on the need for cross-linking? With antibodies, I think it is true to say in no instance has anyone ever demonstrated a monovalent antibody binding to a cell surface antigen that will patch and cap. You can get ingestion, but you don't see visible patches and you don't see caps.

Now, is it known in the case of EGF, insulin or alpha-macroglobulin that the binding to the receptor is divalent or multivalent, or do you think it is monovalent?

DR. SCHLESSINGER: First of all it is not known, but we can suggest a few speculations. We talked a lot about the possibility of insulin molecules aggregating on the surface and thereby cross-linking adjacent receptors. There are now a lot of reasons to rule out this possibility and a more plausible mechanism would be that the hormone binds, changes something in the receptor that is really not clear, (presumably a conformational change) which causes it to be more sticky. This I think is the most plausible mechanism for triggering hormone receptor clustering.

DR. RAFF: But then you are going to say that Fab' antibody binding won't do that?

DR. SCHLESSINGER: Well, we find that Fab fragments do not cause patching.

DR. RAFF: Therefore, they don't patch, they don't signal, so insulin is behaving differently than the monovalent antibody.

DR. SCHLESSINGER: I agree.

DR. RAFF: Right, so isn't it more likely that the insulin is in fact multivalent, just like it was assumed initially that cholera toxin capped GM-1, that it was a monovalent interaction and therefore you didn't need cross-linking? It turns out it is multivalent, and I would predict that all of these things that redistribute and signal are in fact cross-linking the receptor.

DR. SCHLESSINGER: Well, Jesse wants to answer right?

DR. ROTH: I think, Marty, you are left with a circular argument; namely, if they cause biological effects, they must be divalent, and therefore, the ligands must cross-link. If you postulate that insulin cross-links in the membrane, that there are insulin-insulin interactions in the membrane, then they would have to be of a unique kind that have not been described for insulin previously. In fact, if you take a whole series of biologically active insulins, their ability to form dimers or not to form dimers appears to be poorly related to their ability to cause biological effects. So if you take pork insulin or pork proinsulin or guinea pig insulin or some of the dealaninated desasparagine or deoxypeptide insulins, these insulins have a predictable kind of biological response. They are all full agonists, and yet some of them can't form dimers at all in solution, even at very high concentrations, where others do. Now, this doesn't rule out your circular argument, but certainly doesn't strengthen it.

DR. RAFF: I am not suggesting insulin is aggregating with insulin in cross-linking. I am suggesting that insulin has more than one binding site for the receptor. We already have been told that is true. It has a negative cooperative binding site. It has what we will call an inductive site, so at least there are two. There could be more, and why not have one negative site found in one receptor and the inductive site in another receptor, and you cross-link?

DR. SCHLESSINGER: My intuition would say that this aggregation can cause positive cooperativity, but maybe I am wrong.

DR. DEMEYTS: That is not right. It can cause negative cooperativety.

DR. EDIDIN: Jesse, are you going to be the lead spokesman for this? One more comment?

DR. ROTH: No. I may need reinforcements. The argument against the negative cooperative site being involved, at least in the major role in biological activation of the cell, is the fact that the analogs that Pierre showed this morning, like DAA insulin and DOP insulin, which are unable to bind to the negative cooperative site, or the naturally occurring dimers of insulin, in fact, are biologically active full agonists. So at least if you needed a second site, you have got to create one, but not one that is already known. The known insulin-insulin reactions and the known insulin-receptor reactions don't account for this. Now, if you want to still have dimerization, you have got to create a new dimer site. Fine. But just recognize that there is no evidence for it.

DR. SCHLESSINGER: I would just like to add something more. We have prepared recently an analog of EGF that does not patch and is not biologically active. I already said that, if you patch this analogue with antibodies, the biological activity is restored. So there must be a domain on the EGF molecule which can activate receptor clustering.

DR. RAFF: It is monovalent.

DR. SINGER: Marty's point is something that I don't get the logic of. You are suggesting that since antibodies produce cross-linking by being multivalent and Fab fragments don't, that therefore, if a molecule like insulin induces an aggregation, it must be doing it as a multivalent ligand. But it is equally plausible, at least, that insulin is doing it as a monovalent ligand, but is doing something to its receptor that an Fab fragment doesn't do to its receptor, and that it is especially adapted to do because it is a hormone and that is what the nature of a hormone-receptor interaction is like.

DR. RAFF: It is just that there is no precedent for that.

DR. SINGER: No, the precedent is not the logical impasse.

DR. RAFF: We have logical precedents now for cross-linking, so you have a bet.

DR. SINGER: Again, I don't see the logic to that. The only precedent you have is work that has been done with lectins and antibodies. It has nothing to do with hormone activity, and it is equally plausible, if you want to talk about precedents, we have many precedents where the binding of a small ligand to a protein molecule induces a conformational change.

DR. RAFF: Not aggregation

DR. SINGER: I don't think that is exactly relevant to the precedent. The point is that it produces a conformational change. If you induce a conformational change, you may well induce one that produces an aggregation or stickiness.

DR. LEVITZKI: I have a question and a comment. First the question: What receptor occupancy do you need in order to see the patching that you are watching?

DR. SCHLESSINGER: This is a technical question. In order to see patching --we need at least 30 percent occupancy of receptors.

DR. LEVITZKI: Thirty percent, so that is much above the range of saturation of physiological response?

DR. SCHLESSINGER: Well, what is the physiological response? For example, in order to get maximum amino acid uptake, you need 100 percent. In order to get glucose uptake you need three percent, so what is the definition of saturation of physiological response? At different doses, it has maximal response for different effects.

DR. LEVITZKI: The reason I am asking is because one should look at the minimal dose of fluorescent agonists that you have to get the maximum response, calculate what is the amount of dimer that you can form within a certain amount of time and see whether it makes any sense biolgically.

DR. SCHLESSINGER: I did that calculation. If you assume that 1 receptor per $(\mu\ m)^2$ diffuses with $D \cong 5X10^{-10}\ cm^2/sec$ you would get cross-linking within less than a second. So even at very low occupancy with this diffusion coefficient, the receptors move very rapidly and encounter each other.

DR. LEVITZKI: My question is whether under these conditions you really have dimers. I mean, if you really need, let's say, 10-fold or 50-fold less hormone in order to generate the maximal response, is it still true that you still need RH-RH complex in order to generate the response which you measure by amino acid uptake or whatever you want to use?

DR. SCHLESSINGER: I don't know how to answer that question because we don't know the dissociation constant of RH complexes in the cell membrane. We speculate that they form microaggregates on the basis of the knowledge that at higher concentration of hormone they form patches and on the basis of the studies with the anti insulin receptor antibodies.

DR. LEVITZKI: Then the comment I wanted to make is that I don't think that the analogy between the antibodies and the insulin is really something that we can take for granted, because obviously the antibodies to the insulin receptor are doing something else. So maybe in order to get the right "Kvech" (sqeeze) on the insulin receptor to get the biological response, you need aggregation when you use antibodies, but don't need it when you use insulin. So the fact that you really show that you need aggregation for the response using the antibody is not an argument that this is the way insulin works.

DR. SCHLESSINGER: Alex, maybe you did not hear, but we have a recent derivative for EGF which does not aggregate, and when you artificially cause aggregation with antibodies against EGF -- the biological activity is restored.

DR. GORDON: One point about precedent. There was no precedent for internalization of insulin. In fact, the phenomenon was shown, and now the reasons for it are going to have to be figured out later.

The other point is that I would just like to comment. You have emphasized that autoradiography is very tedious, but very patient people can do tedious things.

We have actually studied binding and internalization using autoradiography. If it does not mean anything else, at least it is a model for what we study kinetically in vitro, and that is all I will say. We found that physiologic amounts --, that is, concentrations of insulin present in portal vein, can be used to study the binding interaction of insulin with isolated rat hepatocytes, and that there is essentially a one-to-one correlation between binding and internalization from inital interaction, where it is all on the surface, to steady-state, which is reached at 10 to 20 minutes at 37 degrees. The steady-state is essentially maintained as long as you can carry out this kind of a study.

So the fact is it can be done. Now, we can obviously not show in our system that we are getting dynamic aggregation that you have shown so beautifully. We think we do show perhaps in a better way, the internalization phenomenon and where the insulin is actually going. It appears to be going directly into structures that look like lysosomes, secondary lysosomes, which is totally analogous to what has been seen by Brown and Goldstein for the lipoprotein systems. So I think that we do have some kind of a general phenomenology going on.

THE ROLE OF RECEPTOR AGGREGATION IN INSULIN ACTION

C. RONALD KAHN and KATHLEEN L. BAIRD
Section on Cellular and Molecular Physiology
Diabetes Branch, National Institute of Health
Bethesda, Maryland 20014

THE PROBLEM OF INSULIN ACTION

Despite more than 50 years of study, the mechanism of insulin action remains a mystery. Like other peptide hormones, insulin binds to a receptor on the plasma membrane of the cell[1]. How this interaction is transformed into a message which is carried across the membrane and eventually results in stimulation of intracellular metabolism, however, is unknown.

Two general models of insulin action have been considered (Figure 1). The first would be a mechanism similar to that for the peptide hormones which activate adenylate cyclase; namely, insulin would bind to some surface membrane receptor and this interaction would result in an activation of another, perhaps independent, membrane-associated enzyme or transport protein which in turn would generate an intra-cellular messenger for hormone action[1,2,3]. The other possiblity is that insulin itself is its second messenger[3,4]. Certainly, there is more and more evidence that insulin gets into the cell[5,6,7], and it is possible that this entry, or internalization, of insulin is important for generation of the signal required for insulin action - either because insulin is degraded to some active fragment, such as happens with the LDL[8], or perhaps the insulin itself can bind to some intracellular receptors and generate its biological effect[1,5]. At present, there are insufficient data to choose between these two models.

RECEPTOR AGGREGATION IN INSULIN ACTION

Several bits of evidence suggest that insulin receptors are mobile in the plane of the membrane[10] or even oligomeric[9,11]. The possibility that receptor aggregation may play a role in insulin action has come from some recent work in our laboratory using antibodies to the insulin receptor. These are naturally occurring antibodies found in the serum of some patients with a form of insulin-resistant diabetes [1,2,13]. These are polyclonal IgG, so they are bivalent. They can be shown to bind to cells in proportion to the number of insulin receptors. They inhibit insulin binding, and likewise insulin inhibits their binding, so in a sense they are behaving as true competitive antagonists of one

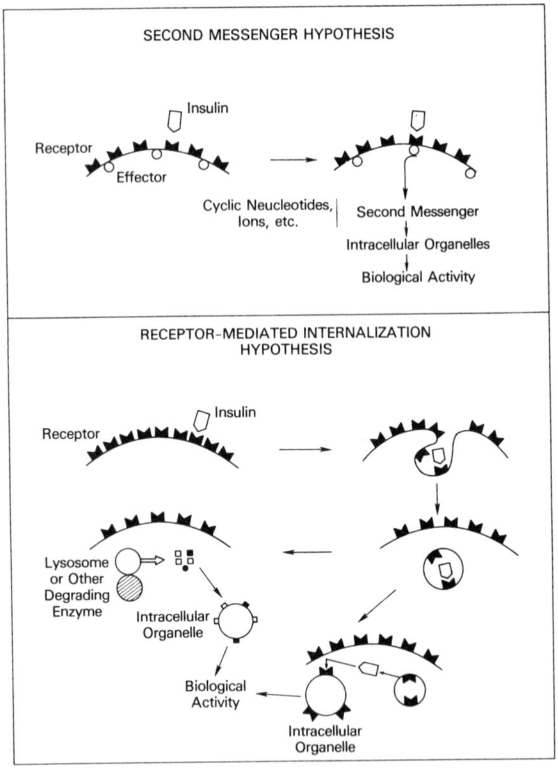

Figure 1

TWO POSSIBLE MECHANISMS OF INSULIN ACTION

another[14]. This inhibition of insulin binding occurs across a wide variety of species and tissue lines, suggesting that the insulin receptor is highly conserved through evolution[12]. These anti-receptor antibodies are also able to immunoprecipitate solubilized receptors quantitatively[15]. And finally, these antibodies can initiate many, and perhaps all, of insulin's biological effects. For example, in the fat cell, they have been observed to stimulate glucose transport and glucose metabolism, inhibit lipolysis, stimulate leucine incorporation into protein, activate glycogen synthase (either in the presence or the absence of glucose) and inhibit phosphorylase[16-18]. In some recent studies in our laboratory using 3T3-L1 cells, an adipocyte-like cell line in tissue culture, we have shown that these antibodies also activate the enzyme lipoprotein lipase, an enzyme induction which requires new protein synthesis (Van Obberghen, E., Spooner, P., Grunfeld, C. and Kahn, C. R., unpublished observation).

The insulin-like activity of these antibodies, in contrast to their ability to simply bind to the receptor, appears to depend on bivalency (Figure 2)[20]. Thus, the bivalent IgG, the bivalent $F(ab')_2$, and the monovalent Fab' all inhibit binding (the monovalent Fab' being only slightly less potent than the bivalent derivatives), whereas only the bivalent insulin receptor antibodies can activate the cell. The small amount of biological activity which is associated with the monovalent fragment can be accounted for simply by contamination of this preparation with a small amount of bivalent antibody and is totally eliminated in more purified preparations. If one calculates the activity of these antibodies on a molar basis, assuming that about 1-2% of the total IgG is receptor antibody, then the antibody is essentially equipotent with insulin on a molar basis for this type of biological response.

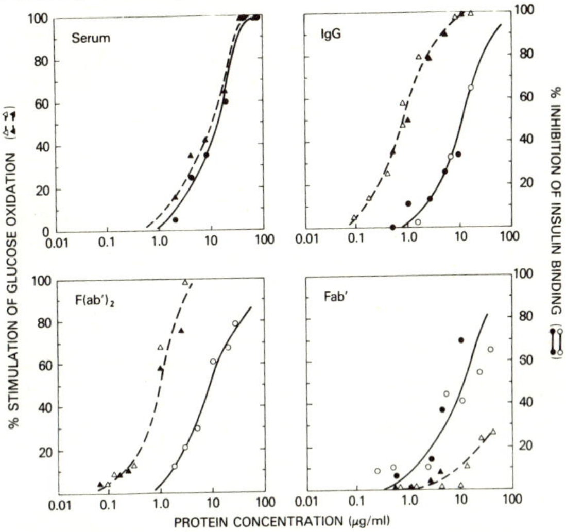

Fig. 2 Comparison of the effects of whole serum, IgG $F(ab')_2$ and Fab' from patient B-2 (Ref 13) on insulin binding and glucose oxidation by isolated adipocytes The inhibitory effect on binding and the stimulatory effect on glucose oxidation were calculated as a percentage of the maximal effect produced by insulin in each system.

The fact that bivalency is required and that the loss of activity is not simply the result of chemical modification is based on two further observations. First, the activity of the monovalent antibody can be restored substantially if cells treated with monovalent Fab' are subsequently exposed to a second antibody, such as an anti-human Fab' serum or anti-human IgG, which cross-links the Fab'-receptor complexes. In addition, similar results are obtained with monovalent Fab fragments prepared by papain digestion without alkylation.

Thus, with respect to the antibody activity, we could conclude the following (Figure 3). The bivalent antibody can bind to the receptor, block insulin binding, and induce some form of receptor aggregation, and this combination of events generates a biological response. The monovalent antibody can still bind to the receptor, as evidenced by the fact that it inhibits insulin binding equally well as the bivalent antibody, but this binding does not generate a biological response, although it can be made to generate a biological response if this antibody is cross-linked by the addition of a second antibody. Now,

Figure 3

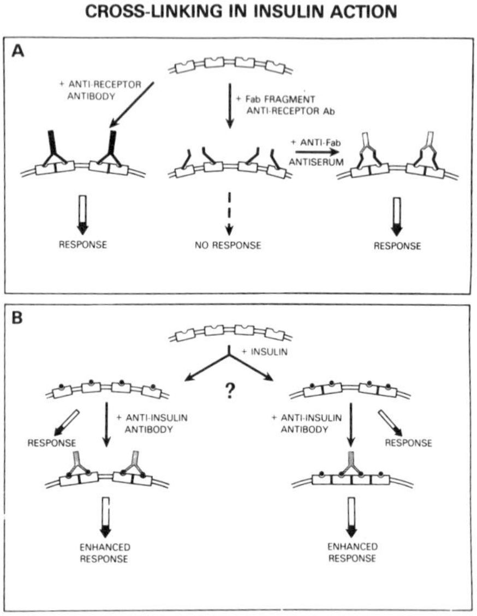

the real question is: Is this really relevant to the question of insulin action, or is this only something that has to do with the antibody itself?

To attempt to look at this question, we decided to investigate the effects of anti-insulin antibodies on insulin action. While this is a well-known method to inhibit insulin action, our approach was to use a much more dilute antibody solution in hopes that we could actually cross-link some of the insulin on the surface of the cell. The results are shown in Figure 4. Note that by adding an anti-insulin antibody in very high dilution the first effect is to potentiate slightly insulin's biological effect. At higher concentrations of antibody the effect is primarily inhibitory. These data suggest that in fact a small amount of the anti-insulin antibody may be able to bind to insulin on the cell, cross-link the insulin, and potentiate its effect.

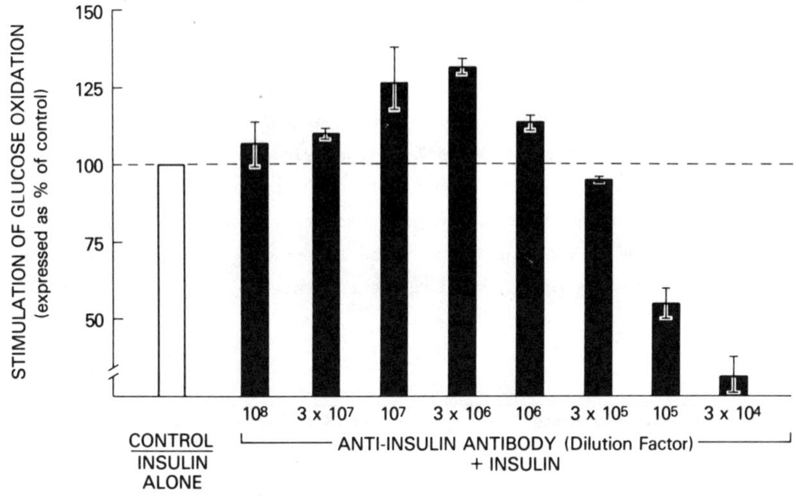

Fig. 4 Effect of anti-insulin antibodies on insulin stimulated glucose oxidation. Control cells were incubated with 0.3 ng of insulin/ml at 22° for 30 min. Anti-insulin serum was then added at the indicated dilutions and glucose oxidation measured for the next 60 min. (From Ref. 20).

In terms of our previous model (Figure 3), two possible mechanisms exist. It is possible that insulin, which is certainly monomeric in solution at the concentrations where it is biologically active, binds as a monovalent ligand and does not induce any receptor aggregation, but still is able to generate a biological response. The enhanced activity in the presence of anti-insulin antibodies would then represent some effect of receptor aggregation in addition to the intrinsic activity. Alternatively, insulin is somehow able to induce receptor aggregation, which is required for biological response, independent of the external cross-linkage[21]; the enhancement of activity produced by anti-insulin antibody in this case would simply be due to enhanced aggregation.

So far we have been unable to identify any component of the cytoskeleton, which is important in this aggregation process. Thus, the standard antimicrotubular and antimicrofilament agents such as colchicine, vincristine, vinblastine, and the cytochalasins do not block insulin or antibody stimulation of glucose oxidation (Figure 5). Perhaps the receptors exist in preformed clusters or aggregates. Dr. Dietrich Brandenburg and his colleagues at Aachen have shown that insulins cross-linked with a suberoyl (six carbon chain) cross-linking reagent are more active than the substituted monomers suggesting that both of the molecules in the dimer can occupy receptors simultaneously[22]. This means that at some time in the history of these receptors, they are close enough together to be bound by two insulins which are separated by only 11 Å.

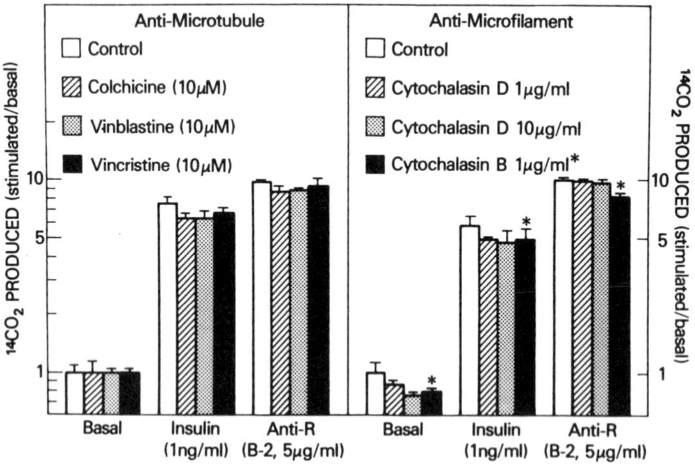

Fig. 5 Effect of anti-microtubular and anti-microfiloment agents on glucose oxidation by isolated adipocytes. Cytochalasin B has a direct inhibitory effect on glucose transport and therefore the scale for cytochalasin B has been expanded x 10.

SUMMARY AND CONCLUSIONS

In summary, these studies demonstrate three points with respect to the mechanism of insulin action. First, many, and perhaps all, of insulin's actions can be initiated by interaction of ligands other than insulin with the insulin receptor. This would suggest that, as alluded to earlier in this conference by Dr. Roth, the critical information for activation of the cell is probably not in the insulin molecule, but in the receptor itself. Certainly, these data would be against the hypothesis that the internalization and degradation of insulin to produce an active fragment is important for biological activity, since it would be unlikely that insulin and receptor antibodies would generate the same fragment, even if both are internalized. It is possible that both ligands aid in the internalization of the receptor and that some degradation fragment of the receptor plays a role in activation of the cell.

The second point is that occupancy of the insulin receptor is not sufficient for signal generation. Thus, monovalent antibody will occupy the receptor, at least as judged by its ability to block insulin binding, but not activate the cell. This is unique among substances which bind to the receptor, since thus far no competitive antagonist of insulin at the receptor level has been discovered. Such a competitive antagonist should allow us to ask new questions about insulin action, such as: Does the unoccupied insulin receptor exert any tonic effect on cellular metabolism which would be altered when occupied by an inactive analogue? Are the acute metabolic effects and the long term growth effects of insulin mediated through the same or different receptors? Such questions are currently under study in our laboratory.

Finally, it appears from these data that aggregation or cross-linking of the receptors is important for insulin action. This interaction is probably occurring at a molecular level and should not be confused with formation of microscopically visible patches or caps discussed elsewhere in this symposium[10]. Whether this requires some action on the part of cytoskeleton elements or is intrinsic to the insulin-receptor complex remains to be determined.

ACKNOWLEDGEMENTS

The authors wish to thank Mrs. Beverly Knight for her excellent secretarial assistance.

REFERENCES

1. Kahn, C. R. (1976) J. Cell Biol. 70, 261-286.
2. Czech, M. P. (1977) Annu. Rev. Biochem. 46, 359-384.
3. Goldfine, I. D. (1977) Diabetes 26, 148-155.
4. Steiner, D. F. (1977) Diabetes 26, 322-340.
5. Goldfine, I. D., Smith, G. J., Wong, K. Y. & Jones A. L. (1977) Proc. Natl. Acad. Sci. USA 74, 1368-1372.
6. Carpentier, J. L., Gorden, P., Amherdt, M., Van Obberghen, E., Kahn, C. R. & Orci, L. (1978) J. Clin. Inv. 61, 1057-1070.
7. Kahn, C. R. & Baird, K. L. (1978) J. Biol. Chem. 253, 4900-4906.
8. Brown, M. S. and Goldstein, J. L. (1976) N. Engl. J. Med. 294, 1386-1390.
9. Jarrett, L. & Smith, R. M. (1974) J. Biol. Chem. 249, 7024-7031.
10. Schlessinger, J. This volume.
11. Ginsberg, B. H., Kahn, C. R., Roth, J. and De Meyts, P. (1976) Biochem. Biophys. Res. Commun. 73, 1068-1074.
12. Flier, J. S., Kahn, C. R., Roth, J. & Bar, R. S. (1975) Science 190, 63-65.
13. Kahn, C. R., Flier, J. S., Bar, R. S., Archer, J. A., Gorden, P., Martin, M. M. & Roth, J. (1976) N. Engl. J. Med. 294, 739-745.
14. Jarrett, D. B., Roth, J., Kahn, C. R. & Flier, J. S. (1976) Proc. Natl. Acad. Sci. USA 73, 4115-4119.
15. Harrison, L. C., Flier, J. S., Roth, J., Karlsson, F. A. and Kahn, C. R. (1978) J. Clin. Endo. Metab., in press.
16. Kahn, C. R., Baird, K. L., Flier, J. S. & Jarrett, D. B. (1977) J. Clin, Inv. 60, 1094-1106.
17. Kasuga, M., Akanuma, Y., Tsushima, T., Suzuki, K., Kosaka, K. & Kibata, M. (1978) J. Clin. Endocrinol. Metab. 47, 66-77.
18. Lawrence, J. C., Jr., Larner, J., Kahn, C. R. and Roth, J. (1978) Mol. Cell. Biochem., in press.
19. Van Obberghen, E., Karlsson, F. A., Grunfeld, C., Kahn, C. R. and Roth, J. (1978) Diabetologia 14, 278.
20. Kahn, C. R., Baird, K. L., Jarrett, D. B. and Flier, J. S. (1978) Proc. Natl. Acad. Sci., USA, 75, 4209-4213.
21. Singer, J. S., Bradshaw, R. A., Frazier, W. A., Merrell, R. C., Gottlieb, and Hogue-Angelletti, R. A., ed., (1976) in Surface Membrane Receptors, Plenum Publ. Corp., New York, pp. 1-24.
22. Willey, K. P., Tatnell, M. A., Khalili-Shirazi, A. Jones, R. H., Thomas, S. H. L., Schüttler, A. and Brandenburg, D. (1978) Diabetologia 14, 281.

DISCUSSION

DR. DE HAEN: I would like to share with you some thoughts that have evolved from some experiments in our lab concerning a new candidate for second messenger of insulin. It is of course only a candidate.

As you probably are aware, a bewildering number of substances mimic insulin, e.g. polyamines and cystein as well as a whole bunch of other compounds. Some of these have now been shown to require Cohn fraction V bovine serum albumin for their action. Contaminant proteins use the insulin mimicking agents as substrates to generate H_2O_2. Michael Czech[1] first demonstrated that in fat cells exogenous H_2O_2 increases glucose transport and increases glucose C-1 oxidation. We observed that H_2O_2 is anti-lipolytic and stimulates glucose incorporation into lipid. Smith-Johannsen et al.[2] described an effect of M_2O_2 on amino acid transport, and Lawrence Larner[3] as well as my collegues showed stimulation of glycogen synthesis by H_2O_2 (unpublished observation).

We thus became interested in the question of whether H_2O_2 could be produced by the fat cell under the effect of insulin. Intracellular H_2O_2 production was estimated via formate oxidation by the endogenous catalase. Formate plus H_2O_2 gives CO_2 plus water.

We could indeed demonstrate that insulin stimulates formate oxidation by H_2O_2 with a dose response curve that paralleled the other effects, and also with the complex bell-shaped glucose concentration dependence characteristic of glucose C-1 oxidation and glucose H-3 incorporation into lipids.[4]

While we were investigating this system, Mukherjee and Lynn [5] published a paper in which they showed that membranes from fat cells pre-exposed to insulin showed 6 to 7-fold increased NADPH oxidase activity associated with stoichiometric H_2O_2 production. Thus there seems to exist an insulin-sensitive H_2O_2 generating enzyme.

This, in my judgment, puts H_2O_2 in the fore front as a candidate for second messenger of insulin. I think we should take a closer look at that possibility.

REFERENCES

1. Czech, M.P., Lawrence, J.C., Jr., and Lynn, W.S. (1974) Proc. Nat. Acad. Sci., USA 71, 4173-4177.

2. Smith-Johannsen, H., Perdue, J.F., Ramjecsingh, M., and Kahlersberg, A. (1977) J. Supramolec. Struct. 7, 37-48.

3. Lawrence, J.D., Jr., and Larner, J. (1978) J. Biol. Chem. 253, 2104-2113.

4. May, J.M. and De Haen, C. (1979) J. Biol. Chem. (in press).

5. Mukherjec, S.P., and Lynn, W.S. (1977) Arch. Biochem. Biophys. 184, 69-76.

MOBILITY IN MEMBRANES IN TWO DIMENSIONS

LEONARD FINEGOLD

Department of Physics, Drexel University, Philadelphia, Pennsylvania 19104 (U.S.A.)

ABSTRACT

The fluid mosaic model of membranes has been simulated with a computer by causing "intra-membrane particles", initially placed randomly on a two-dimensional "membrane" with fluid properties, to diffuse in the plane of the membrane and to interact with one another. The resulting particle aggregates are similar to those observed of freeze-fracture particles.

This simulation can be used as a uniquely defined test of any proposed measure of aggregation.

Particles in real membranes may be packed at a density (of particles per unit area of membrane) which approaches the maximum possible for non-overlapping discs on a surface (the "car-parking problem" in two dimensions). This high density then has implications for models of membrane biogenesis.

INTRODUCTION

The fluid mosaic model predicts that cell membrane entities will move in the plane of the membrane. The particles seen in freeze-fracture electron microscopy will aggregate under given conditions. Hence one immediately asks if simulated freeze-fracture particles would indeed produce "pictures" like those seen under the electron microscope. Now, for particles to aggregate from an initially unaggregated state, they must first move and then attach to one another. One may have an intuitive picture of how the particles might aggregate, derived from one's everyday macroscopic experience (e.g. of corks on water), but one's conventional intuition can break down in the microscopic realm where, for example, viscous forces are enormous relative to inertial forces. This work describes a model[1] which predicts the morphology of particle aggregates in a membrane, and compares the results with those seen in freeze-fracture micrographs of membranes -- see Figures 1 and 2.

For the analysis of freeze-fracture micrographs, it would be valuable to the practicing electron microscopist to have a measure of aggregation. To compare proposed measures of aggregation, one needs a set of pictures of gradually aggregating particles. How is one to specify that the set of pictures is indeed graduated in aggregation? The only certain way is to use the results of the model previously described, in which the particles are inexorably forced to aggregate.

The particles, being treated as discs on a two-dimensional surface, can be at such high densities in a real membrane that they approach the maximum mathematical packing of placing circular discs on a surface. . . the "car-parking problem". (We have a first result for this problem for discs.) This density is so high that it enables

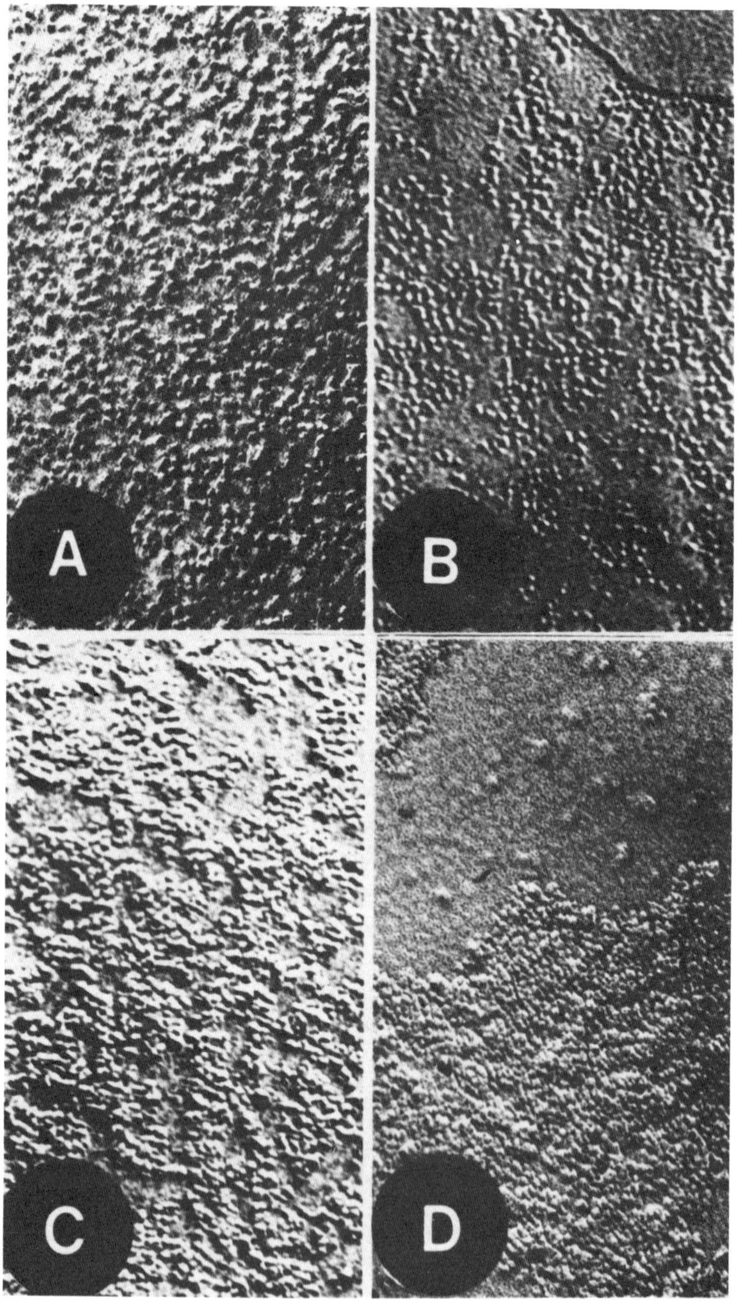

Fig. 1. Freeze-fracture electron micrograph of human (BFS) erythrocytes, at differ - ent degrees of temperature-induced agregation. The circles around the letters are 200 nm diameter. (Courtesy of Dr. B.F. Schall of this laboratory.)

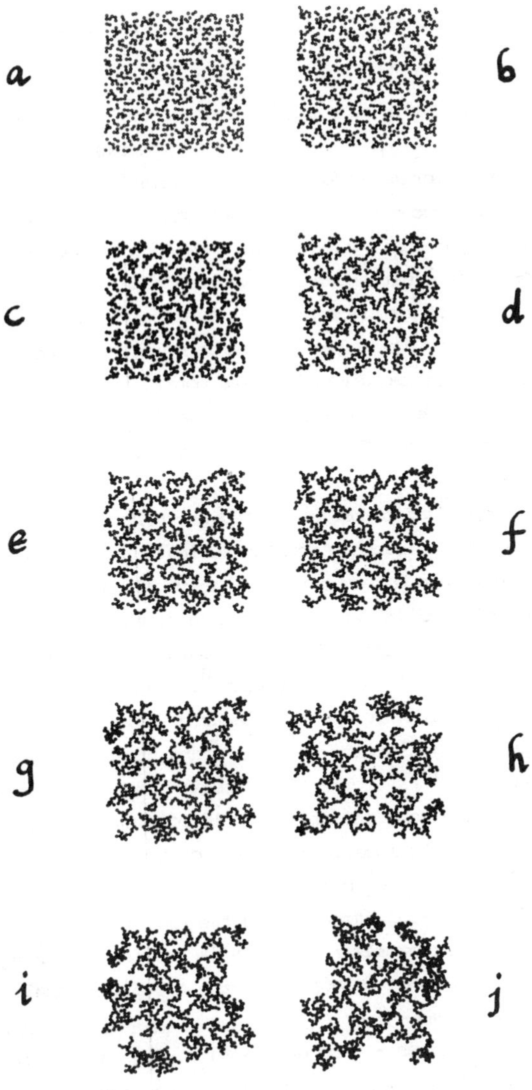

Fig. 2. Particles Aggregation simulated by computer.

statements to be made about membrane fluidity during biogenesis.

BACKGROUND

The elucidation of the structure and function of cell membranes remains one of the most important problems in biology. The field of membranes is one of the most active fields; indeed present membrane workers are colloquially called "The Membrane Generation". It is now widely accepted that a common phenomenon in cell membranes is the mobility of surface components. Our present knowledge is due to the following advances of recent years: (a) great advances have been made in our understanding of the biochemical architecture of membrane proteins and lipids, and even amino-acid sequencing has begun; (b) we now have electron microscopical (EM) methods for visualizing large areas of membranes in two dimensions (as contrasted with previous painstaking sectioning methods), e.g. ferritin-conjugated labelling and freeze-fracture techniques; (c) a wide variety of physical techniques has successfully complemented many aspects of the preceeding.

The fluid mosaic model[2] of membranes, based on thermodynamics[3], is strongly supported by a wide range of evidence, from the seminal experiments demonstrating macroscopic movement of antigenic determinants in heterokaryons[4], antigen-induced "capping" in lymphocytes[5] and cell fusion[6] to translocation of particles observed in electron microscopy by the freeze-fracture technique in chloroplast and other membranes[7] and the diffusion of rhodospin molecules in retinal[8]. Membrane components can diffuse under the influence of externally-applied electric fields[9,10]. There are several excellent reviews emphasizing fluidity[11,12,13,14] and its connection with trans-membrane effects in nucleated cells[15,16]. The fluid mosaic model stresses that protein-associated elements of the membrane are not held completely rigidly in particular locations, but may be free to move in the plane of the membrane. The reader is also referred to other papers in these Proceedings.

Membrane fluidity is clearly important to the cell physiologically, for cells appear to maintain a constant membrane fluidity in the face of changes in temperature[17] or nutrition[18,19]. It is very likely that the two-dimensionality of membranes[20] leads to a much higher frequency of interaction of membrane components - which may include membrane-bound enzymes - than does a three-dimensional system, provided of course that the membrane is "fluid". However, at low component densities (and/or high membrane viscosity) the components will but rarely interact. Conversely, if the component density be too high (and/or the membrane viscosity be too low), the components interact only with their close neighbors. Hence the actual physiological fluidity observed in membranes may be, and provide, a happy medium for membrane component interaction.

Cytoplasmic involvement of the membrane[15,21] is of course extremely important in cell function, but is probably only indirect and minimal at the scale of freeze-fracture particles[22]. The "fluidity" or "mobility" of cell membranes is expressed over a large range of time and distance scales[23], from the ruffling of the leading edge of a motile cell seen by time-lapse optical microscopy (minutes and micrometers) down to the motion detected by added spin probes (less than nanoseconds and nanometres)[11]. Fluidity or mobility of membrane components is quantified by the diffusion constant D.

To measure the diffusion constant D of entities in an intact membrane, in principle one need only (i) label them (for identification), (ii) start them off in an identifiable pattern, and (iii) measure a distance x moved with time; then $\overline{x^2} = 4Dt$, where the bar denotes an average in time. Unfortunately, this measurement is feasible under physiological conditions only in a limited number of cases[11]. Many measurements have been done using optical techniques with fluorescent labels[25]: these are very valuable but are perforce limited by the wavelength of light, and by the difficulty of establishing the exact membrane component(s) to which the label is (preferably covalently) attached. The particles observed in practically all membranes[26, 27,28] including myelin[29] - called "PPPPP" for Peripatetic Presumably Predominantly Proteinaceous Particles[1], "IMP" for Intra-Membranous Particles or "MAP" for Membrane Associated Particles - are naturally occuring, observable without biochemical labelling, express mobility by aggregating, and are of close-to-molecular dimensions (10 nm). They are probably predominantly proteinaceous[30] and may well be (the sites of) enzymes. To estimate their fluidity is a challenge which we are engaging.

THE MEMBRANE MODEL

A simple[31] model is chosen. The model assumes that particles of 8 nm diameter are to be found initially at random in two dimensions, at a packing density of 4,300 particles/μm^2 in an <u>effective</u> viscosity of 1 poise[11]. These values (6.2×10^5 particles per cell[32] of 145 μm^2) in area [33] are typical for mammalian (human) erythrocyte membranes at least. The Brownian motion is mimicked by moving each particle in turn separately and randomly in distance and direction. The random translations are given by Einstein's equation and particle hydrodynamics. When all the particles have been randomly moved once, then the distance between all of them is measured in turn to see if any two are close enough to collide. Since, at present, we do not know the interaction parameters between particles, we have simply assumed that when two particles approach to within the interparticle distance observed in electron micrographs of the aggregated particles, they bind. A resonable interparticle distance is 12.5 nm. This general technique is known as the Monte Carlo method, since part of it uses processes very intensively studied in the European town of that name[36]. A particular power here is that it gives morphological results; most statistical-mechanical methods deliberately average over space and hence lose spatial information. A power of our approach is that in contradistinction to other approaches[37,38] - the particle motions or interactions are not constrained to lie within certain directions, nor on a lattice.

After every particle and cluster has been translated and rotated, again the computer tests for which particle and/or clusters are close enough to combine, and the whole process is repeated cyclically until practically (due to cost limitations) every particle has combined into a cluster, and the clusters have aggregated. After any desired cycle the positions of each of the particles is drawn. This completes one machine "cycle", which may be thought of as one picture frame in a cinematograph of the particles on the membrane. The frequency with which pictures are "taken", i.e. frequency of machine cycles, controls the simulated time rate of aggregation. Succeeding machine cycles proceed as before, but each cluster of two or more particles is translated and rotated randomly as a whole about its common centre of mass. Larger clusters both translate (drift) and rotate more slowly[35]. The boundaries of the "computer membrane area" are treated with cyclic spatial conditions, i.e. as a toroid, to diminish artificial effects.

The following refinements have been made to the original program (Finegold) with Mr. J. T. Donnell. The random number generator has been improved, tested, and shown to be good[39]. The hydrodynamic motion of the particles is represented more realistically[35]. Much of the computer time is spent in checking if collisions have occurred: this part of the program now runs considerably faster. If, as the result of a collision, two particles would overlap then they are backed-off until they no longer infringe each other's territory, i.e. the particles are hard-cored. The collision-checking of (groups of) particles across the (toroidal boundaries is also improved. Incidentally, the overall problem of clustering is also met in ecology[40].

The simple assumptions made in the model — basically that of membrane fluidity and aggregation - are sufficient to duplicate the morphology of particle patterns seen in electron microscopy[41], and the refined computer model confirms the earlier work[1]. Hence the assumptions are quite consistent with, and strongly support, the fluid mosaic model[2,12,13]. Moreover, because the pH-induced aggregation experiments[41] are not done kinetically as frames in a time sequence, it appears that the effect of pH is to release the membrane particles (which are relatively immobile in the human erythrocyte ghost membrane under normal conditions) from their constraints. These constraints involve spectrin [42] via calcium [43].

MEASURES OF AGGREGATION

The electron microscopist is adept at visually recognizing and comparing degrees of aggregation. (e.g. [41]). It is most desirable to have a quantitative, objective measure of aggregation which is also simple and does not require special equipment. These criteria cannot be simultaneously fully satisfied. With Mr. J. T. Donnell, we have used the results of our computer program, which provides a unique set of inexorably-aggregating particle pictures to test simple measures of aggregation . If one uses electron micrographs as test fields, then one has no independent criteria - other than visual - for distinguishing closely-aggregated states[44-48]. In principle, the information in a picture (of aggregated particles) is contained in a two dimensional Fourier transform. One obtains a visual impression of a truncated Fourier transform by gradually squinting at a field of particles - "patterns" appear. However, a full

Fourier transform may require (lengthy) exact measurement of particle positions and much computer time. (Optical Fourier transform techniques would be useful if particle contrast were high.) We have investigated these measures of degree of aggregation: mean cluster size, radial distribution function, nearest neighbor separation, and a "boxing" technique, using as "micrograph" particles in various degrees of aggregation generated by our computer model. A pictorial example is shown in figure 3.

Fig.3. Simulated Freeze-Fracture Electron Micrograph.

The last measure appears to be the method of choice: A coarse, but effective measurement of particle positions is made by placing a transparent sheet bearing a simple rectangular grid over the electron micrograph, and manually recording which of the grid boxes contain a particle. A computer program then examines groups (2 x 2, 3 x 3, 4 x 4 etc.) of boxes in turn and records the fluctuations of particles per group as a standard deviation. The results are normalized by comparison with the positions of random non-aggregating particles at the same density (random data is available for a wide variety of densities). The normalized standard deviation then increases monotonically with time. For a typical picture, the manual recording of the data takes 45 minutes and the program 7 seconds on an IBM 370/168 computer. We are in process of using this technique to analyze actual freeze-fracture micrographs, taken in our laboratory by Dr. B. F. Schall.

To simplify the analysis of aggregation of micrographs (of freeze-fracture particles, antibodies or lectins visualized by conjugated ferritin, or metal labels such as collodial iron hydroxide, etc.) we intend to use image processing: A micrograph is placed in front of a television-type camera, and is immediately scanned and analyzed by a dedicated computer. Human "manual" intervention will (at least initially) be necessary to avoid confronting elaborate image-recognition problems, and also problems of picture contrast. Our ultimate goal is to be able to help other workers by processing their micrographs - which would be mailed to us - with a turn-around time of less than a week.

PARTICLES AND CAR-PARKING

The problem of finding the maximum fractional area covered by randomly sequentially placed squares, with the same orientation, is known as the car-parking problem in two dimensions and has not been solved in closed form[49]. Our recent computer result [Finegold and Donnell, unpublished] for the maximum fractional area covered by randomly sequentially placed circular discs, is 0.503 ± 0.002.

During membrane biogenesis[52,53] proteins can be inserted sequentially into the lipid bilayer at many places. Since this process occurs at the molecular level, the mechanism should be relatively simple. Identifying membrane particles as proteins then implies, from the spatial distribution of freeze-fracture particles, that the process is also spatially random. Initially there would be many possible places to place a particle but as the membrane becomes filled there would be fewer and fewer possible vacant sites, and very many more "attempts" would be needed by the process to achieve the observed particle density -- a situation which is unfavorable to the cell both in time and in energy.

These time and energy costs are avoided, yet a high particle density is still achieved, if the membrane particles are permitted some lateral motion, say of the order of a few of their diameters, during biogenesis. Hence the observed high particle density is consistent with short-range fluidity in membranes at a molecular scale[11,13].

Particles in the mature intact human erythrocyte are held immobile via spectrin[43], so the above discussion relates to the immature cell.

ACKNOWLEDGEMENTS

I thank Dr. B. F. Schall for discussions and micrographs, Mr. J. T. Donnell for a computer-generated photograph, and the National Science Foundation for grant PCM 75003542. Computer funds were kindly supplied by the Graduate School. The camera-ready manuscript was typed by Ms. Cindy Lee.

REFERENCES

1. Finegold, L. (1976) Biochim. Biophys. Acta 448: 393-398.
2. Singer, S.J., Nicolson, G.L. (1972) Science 175: 720-731.
3. Singer, S.J. (1971) Structure and Function of Biological Membranes, ed. L. Rothfield, 145-222, New York: Academic.
4. Frye, L.D., Edidin, M. (1970) J. Cell. Sci. 7: 319-335.
5. de Petris, S., Raff, M.C. (1973) Nature (N.B.) 241: 257 - 259.
6. Zakai, N., Kulka, R.G., Loyter, A. (1977) Proc. Nat. Acad. Sci. USA. 74: 2417-21.
7. Ojakian, G.K., Satir, P. (1974) Proc. Nat. Acad. Sci. USA. 71: 2052-2056.
8. Poo, M-m., Cone, R.A. (1974) Nature 247: 438-41.
9. Poo, M-m., Robinson, K.R. (1977) Nature 265: 602-605.
10. Orida, N., and Poo, M-m. (1978) Nature 275: 31-35.
11. Edidin, M. (1974) Annu. Rev. Biophys. Bioengin. 3: 179-201.
12. Singer, S.J. (1974) Annu. Rev. Biochem. 43: 805-833.
13. Singer, S.J. (1974) Adv. Immunol. 19: 1-66.
14. Cherry, R.J. (1976) Biological Membranes. ed. D. Chapman, D.F.H. Wallach, 3: 47-102, New York: Academic.
15. Edelman, G.M. (1976) Science 192: 218-226.
16. Nicolson, G.L. (1976) Biochim. Biophys. Acta 457: 57 - 108; ibid.
17. Melchior, D.L., Steim, J.M. (1976) Annu. Rev. Biophys. Acta 472: 55-92.
18. Thompson, G.A., Nozawa, Y. (1977) Biochim. Biophys. Acta 472: 55-92.
   van Megan, W., Snook, I. (1976) Nature 262: 571-572.
19. Vessey, D.A., Zakim, D. (1974) Horizons Biochem. Biophys. 1: 138-174.
   Weinstein, R.S. (1976) Cancer Res. 36: 2518-2524.
20. Adam, G., Delbrück, M. (1968) Structural Chemistry and Molecular Biology, ed. A. Rich, N. Davidson, pp. 198-215 San Francisco: Freeman.
21. Ash, J.F., Vogt, P.K., Singer, S.J. (1976) Proc. Nat. Acad. Sci. USA 73: 3603-3607.
22. Loor, F. (1976) Nature 264: 272—73.
23. Heraclitus. 530 BCE. παντα ρει De Caelo 3.I.18 ed. C.H. Weise (1834) Lipsiae.
24. Albrecht-Buehler, G. (1977) J. Cell Biol. 72: 595 - 603.
25. Schlessinger, J., Axelrod, D., Koppel, D.E., Webb, W.W., Elson, E.L. (1977) Science 195: 307-309.

26. Branton, D. and Kirchanski, S. J. Microsc. 111, 117 - 124 (1977).

27. Bullivant, S., J. Microsc. 111, 101 - 116 (1977).

28. Sleytr, U.B. and Robards, A.W. J. Microsc., 111, 77 - 100 (1977).

29. Miller, R.G., Torreyson, P. (1977) Biochim. Biophys. Act. 466: 325-335.

30. Guidotti, G. (1972) Annu. Rev. Biochem. 41: 731-752.

31. Occam, W. (1343) Works (1963) Manchester: University Press.

32. Branton, D., Deamer, D. (1972) Protoplasmatologia II/E/1: 1-70.

33. Evans, E., Fung, Y.C. (1972) Microvascular Res. 4: 335-347.

34. Jost, W. (1952) Diffusion. New York: Academic.

35. Saffman, P.G. and Delbrück, M. (1975) Proc. Natl. Acad. Sci. U.S.A. 72: 3111-3113.

36. Fluendy, M. (1970) Markov Chains and Monte Carlo Calculations in Polymer Science. ed. Lowry, G.G. pp. 45-90. New York: Dekker.

37. Maruyama, M., Oosawa, F. (1975) J. Theor. Biol. 49: 249-262.

38. Antonelli, R., McLaren, D.I., Rogers, T.D., Lathrop, M. and Willard, M.A. (1975) J. Theor. Biol. 49: 385-400.

39. Yakowitz, S.J. (1977) Computational Probability and Simulation. Reading, Mass.: Addison-Wesley. Chap. 1.

40. Pielou, E.C. (1977) Mathematical Ecology, New York: Wiley.

41. Elgsaeter, A., Branton, D. (1974) J. Cell. Biol. 63: 1018-1030.

42. Nicolson, G.L., Painter, R.G. (1973) J. Cell. Biol. 59: 395-405.

43. Bennett, V., Branton, D. (1977) J. Biol. Chem. 252: 2753-2763.

44. Abbas, A.K., Ault, K.A., Karnovsky, M., Unanue, E.R. (1975) J. Immunol. 114: 1197-1203.

45. Irimura, T., Nakajima, M., Hirano, H., Osawa, T. (1975) Biochim. Biophys. Acta 413: 192-201.

46. Copps, T.P., Chelack, W.S., Petkau, A. (1976) J. Ultrastruct. Res. 55: 1-3.

47. Melhorn, R.J., Packer, L. (1976) Biophys. J. 16: 613-625.

48. Weinstein, R.S. (1976) Cancer Res. 36: 2518-2524.

49. Solomon, H., Proc. Fifth Berkeley Symposium Math. Statist. Probability, 3, 119-134 (1967).

50. Akeda, Y. and Hori, M. Biometrica 63, 361 - 366 (1976).

51. Palasti, I. Magy. tudom. Akad. Mat. Kut. Intéz. Közl 5, 353 - 60 (1960).

52. Rothman, J.E. and Lenard, J. Science 195, 743 - 753 (1977).

53. Singer, S.J. J. Supramolecular Struct. 6, 313 - 323 (1977).

DISCUSSION

DR. SINGER: I think one of the things that is very interesting from the first part of your presentation, Len, and I didn't quite appreciate this before, was that from your simulation, what I think you are suggesting is a little surprising to me, and that is that when you arrive at a stage where every particle is attached to another particle -- I mean, none of them is free -- that the overall area occupied by the aggregate is not very much less than the area these particles started out in. In other words, that when you have achieved a state of complete aggregation, you haven't compressed the particles into a much smaller area than they started out in.

Now, that is very interesting from a number of points of view. It means that what you are saying is that when you do this particular kind of simulation, you wind up with an aggregate which has a very open structure, practically as open as you started with when the particles were all free.

Of course, that may be related to one of the things that Marty was asking before, that you may have a trapping mechanism involved in an aggregate if the structure of the aggregate is very open. It is possible that there is some connection there, but I think that the assumption that goes into your model is that any place on the circumference of the particle is allowed as an interaction site for another particle, and that is why you get this open structure.

But if you had a particle with binding sites that were geometrically related, as you probably have in something like bacteria rhodopsin, now you are allowing aggregation to occur only in specified directions with, say, a hexagonal lattice or something like that, and now you get a compression of an aggregate into an area that is much more confined.

So I think what I would like to ask you is: How do these simulations look as a function of the initial particle density? That is, what you could perhaps define is the ratio of an area you start with, which is filled by the particles which are free, to the envelope of the area occupied by the aggregate -- how does that ratio vary with the starting particle density?

It is like a problem in gas liquefaction, and the question is: If you start with a very high density, then when you do this aggregating, you are going to wind up with a net area which is very similar to the one you started with, but at low density you must collect into a much smaller area. So I think you should ask: How does this picture you have gotten change as you change the starting particles?

DR. FINEGOLD: Jon raised several points. The first one that I should mention is that these particles are taken to be incompressible, and so if you were to measure the area of each of these particles when you begin, it will be the same area as when you finish. So the same fractional area is covered by the particles at the beginning as at the end of the simulation. However, visually one does have just this feeling that you mentioned about particle density: I guess it is that the size of each cluster is important.

What one is doing mentally is drawing a little boundary around certain clusters and saying, "Here, the clusters look as though they are a particular size, and it turns out that this is a very difficult thing to measure by computer. I understand that some automated cytology depends upon the measurement of cluster size and it has recently been possible to measure cluster size from boundaries for certain simple cells. But the problem here is a very difficult one, and even with our controlled system we don't have any simple measures to use for boundaries of clusters.

You see, we can do all the statistics we like because we have got the system completely under our control, but the one you mentioned is a very difficult one to compute.

You mentioned the sweeping-up aspects, Marty Raff's idea where as some particles move, could one have another set of particles which the first set also sweeps up? I have wondered about this, but without doing the simulations for specific problems, and without more information about the sweeping and swept particles, it is hard to say. I can think of pros and cons for both sides.

You asked, are all places of attachment allowed? The answer is yes. In this model the particles can stick together anywhere. However, one must remember that the maximum number of attached particles you can have is six; you can geometrically get only six circles around another one. (We would like to plan and do a model of this for antibodies).

You mentioned a hexagonal lattice. It turns out that you do not need to have hexagonal symmetry to get a hexagonal lattice. If you think of bubbles on a surface which is a well known thing in (should we call it) bubble physics, you can make lots of pretty rafts of bubbles, and the bubbles have circular symmetry. There is no hexagonal symmetry, and yet the bubbles will form beautiful hexagonal arrays.

DR. SINGER: They are freely mobile.

DR. FINEGOLD: They are freely mobile.

DR. SINGER: And they are free to pack.

DR. FINEGOLD: They are free to pack.

DR. SINGER: What you are saying is that things are not free to pack. Once they make the contact, they are stuck.

DR. FINEGOLD: Yes, right. In my simulation I do have a different system from free bubbles. Concerning your question about the particle density -- we have looked at doing simulations at different particle densities. In today's simulation there is a simple scaling. I could show you pictures.

DR. RODBARD: I think there may be another very simple way to measure the degree of clustering. If we wish to measure the degree of clustering, we want to know the average size of a typical particle which we would know if we had the surface to volume ratio.

DR. FINEGOLD: Did you mean of a typical cluster?

DR. RODBARD: Yes. We want to know the mean size of a typical cluster, and you can know this if you measure its surface to volume ratio, and this can be measured very easily by the techniques of quantitative histomorphy developed by the light microscopists such as Weibel, and by simply putting in a random grid of points and lines on the photograph on the microscope, you can measure the number of intersections of the points and the number of intersections of the lines and get the surface to volume ratio and the volume fraction or the mean length of the circumference of the cluster.

This is a very simple technique, easy to use, which has been around, of course, for 20 or 30 years in the hands of the light microscopists. That may be an easier way to quantitate.

DR. FINEGOLD: We have, of course, looked at it as one of our starting points, and all that it does is that it gives you the very first parameter, which is very easy to measure. That is very easy, but it is not sophisticated enough for this work. I think the simple answer is that when a human looks at an electron micrograph, (s)he interprets much more subtle things than is given just by a simple measurement of clustering.

DR. DE HAEN: I would like to mention to you something that occurred to us by looking at insulin receptors stained on fat cell membranes observed _en face_, so we have a flat sheet in front of us and many insulin receptors are on it, and the question was: How do we determine quantitatively the amount of clustering that was there?

We started to put, just as suggested by David Rodbard, we put some arbitrary grid on it and counted how many receptors were in each cell. I took two extremes. First consider a very large grid that encompasses hundreds of receptors. Every grid square will have exactly the same number on the average.

Now take the opposite extreme. I am going to make a grid that has only the size of one ferritin molecule, and on the average, most of the time I will have either one or none in my grid, in my individual square.

The conclusion is: In the two extremes, you can't find clustering, and in between you get a degree of clustering that depends on the arbitrary grid you put down. Now this is a problem that has been addressed in astronomy before namely, when you ask the question of clustering of stars and galaxies, and in ecology it is the question of distribution of males in a female environment of gazelles and things like that.

Anyhow, it is really indeed a very difficult problem, and I suggest that a recent book that has come out and that is fantastic on this subject, called "Fractals, Form, Chance and Dimension," should be considered in those contexts because I think the clustering of receptors is a fractal problem.

DR. FINEGOLD: I spent a happy day talking with Benoit Mandelbrot, who is the author of "Fractals" about this, because he thought that some of our data might be just what he wanted for his work. It turned out that we are at opposite extremes of density, and that to run our programs, the density that he needs would be very expensive, even though he is at IBM.

DR. ROTH: You have raised some very fascinating considerations, but there you are considering a homogeneous set of particles interacting randomly with an infinite number of contacts -- in other words, they can contact on any part of their surface.

DR. FINEGOLD: Yes, but that is actually not too confining because these particles, whatever they are, will be rotating. Incidentally, we have put in that they rotate i.e. we have included rotational diffusion.

DR. ROTH: Now, among the other more interesting kinds of things that happen on cell surfaces are where proteins have a distinct preference for a like protein, and might then have affinities that are multi-fold or are orders of magnitude greater, so that we talk about receptor aggregations or ligand-induced aggregations. We are talking about aggregations usually of like proteins or of a very limited number of species of proteins.

I am curious then, given the extraordinary insights that you have had, what happens when you introduce the idea that there are self-associations -- say, one percent of the proteins have an extraordinary affinity for one another, whereas they have a relatively low affinity for others. What are the kinds of constraints that you get?

DR. FINEGOLD: One of the fun parts about this branch of Monte Carlo work is that you really are doing experiments by computer. I try of course, always to predict what results I am going to get -- and I have often been surprised.

So I would simply not like to stick my neck out and even guess the answer to your kind of problem because it is distinct enough from what I have done already that I don't have enough intuition built up. I do have intuition for problems that are close to my simulation, your problem is something that could be done given NIH funds for computing.

DR. PERELSON: Since this topic is brought up, I have looked at clustering problems a bit and I actually have a few slides showing some indices of clustering that one can use, plus other methods of computing them, and I thought if we had a few minutes I would display them. If not, anybody who was interested, I could show them after the lecture.

DR. FINEGOLD: Yes, I would like to see them.

DR. PERELSON: There is a paper in Experimental Cell Research (112, 309, 1978) on using radial distribution functions to characterize surface IgE distribtuions. I have mentioned some of the indices that are used in ecology that can be applied to describing these.

DR. FINEGOLD: Yes. One can use these, but they can require a fantastic amount of computing. However, the average electron microscopist doesn't have a lot of big computers nor does he want to spend time playing with them. This is the way we have viewed many of the possible indices.

DR. PERELSON: Just two comments really, the first on using a grid to analyze degrees of clumping, since it was brought up in astronomy. This type of analogy can be used in which you actually just take a grid and overlay it onto some sort of pattern and count the number of particles within each grid point, and as you point out, depending upon the size, the grid, you get different results.

This problem has been analyzed quite heavily by statisticians, and there is sort of a rule of thumb that you have to make your grid size sufficiently large that you get on the order of five or six particles, and then you can show that in that range, you are sort of in a statistically relevant type of range.

This type of analysis, for example, was applied by Ken Ault and Emile Unanue's group to analyze surface Ig patterns, and what you can do to ask if the pattern is random or not is just calculate the mean number of particles in each box and, for example, see if it fits a poisson distribution.

The characteristic of a Poisson is that the mean number of particles per box will equal the variance, so if you measure those two parameters and they are equal or relatively equal, you know that the distribution is random.

It turns out that if the variance is greater than the mean, you can show the particles are clustered, and this has been used by a number of statisticians to come up with a single number to try and describe the degree of aggregation of particles. David and Moore in 1954 developed something which he called the index of clumping, in which you just take the variance over the mean and subtract one from it, so if the variance is greater than the mean, you will get a number that is larger than one, indicating that it is clustering.

This index has a strange property, that if points are removed randomly from your area -- say, if animals are killed in your population, if you make the density lower just by random, then that index will decrease, and so statisticians developed some other indices that don't have that property.

One is Lloyd's index of mean crowding, and what you do is to compute the number of points that share a box with a given point, and then you take the average that way. You can take that number then and divide it by the mean, and that will give you an index of clumping that is independent of the density, and I just bring these up because there are a lot of such indices.

Let me just mention one other method which comes out as statistical-mechanical analysis of, say, structures of fluids, and that is called the radial distribution function. What you do here is that you have a set of particles, and say you center on one random one, and you go out a given distance, R, and calculate the number of particles that lie in that distance, say, in an annulus.

If you then divide that by the area, because as you go out further and further, you will obviously get more and more particles -- if you take the number of particles at a given distance and divide it by the area, what it is really giving you is the local density of particles since it is number per unit area. That can then be used to characterize whether the particles are clustered or, say, randomly distributed.

If they are randomly distributed, then that local density should be the same as the overall density, and this number would just be called the radial distribution function, and when you do that and you average over all particles, it should be approximately one.

DR. RODBARD: You mean clustered around that point?

DR. PERELSON: No. You do the analysis over that one point, and you repeat it over every other point and average.

DR. RODBARD: Every other existing figure? Is that right?

DR. PERELSON: Yes. You do that around point one and you get the radial distribution function around point one, and then you do the same thing over every other point in the picture and average that.

DR. RODBARD: But there are a lot of galaxies. That is pretty tedious.

Dr. PERELSON: You can do it for sample points, too, in terms of computation. There are some other clever ways of doing it, because you don't have to calculate the distance between every pair of points in the picture.

DR. RODBARD: It is an optimal way to do it.

DR. PERELSON: Yes. Just to show you what these things look like, for a random pattern it should be one. If you do it on a finite grid, you get something that goes around one.

It is used in statistical physics, for example, to analyze, say a crystal. Say you take a real lattice and have a real size annulus and you do it and you find a bunch of spikes. The width of a spike has to do with the width of the annulus that you are using in a computation.

You can compute these things in statistical mechanics for liquids and gases, and there what you find is that instead of having a real crystal structure, that the molecules can move a bit, and therefore, you see sort of peaks now that correspond to nearest neighbor distances, but instead of seeing many of them, you may only see a first or second peak for first and second nearest neighbors.

The reason I bring it up is because we did this analysis on surface Ig molecules that were labeled first with a monovalent Fab anti-immunoglobulin and that was labeled with fluorescein so you could pick up B cells, and then we used a monovalent anti-fluorescein that was tagged with ferritin. This was work that came out of Emile Unanue's lab and was done by Adil Addabus, Unanue and Ken Ault, who is here.

I looked at some of their patterns. The only interesting point is that by doing this type of analysis, you get a radial distribution function that would show for immunoglobulin receptors that were not cross-linked, that you get something that sort of resembles a typical liquid type of radial distribution function with points being clustered. One can argue whether this is an artifact of the freeze-fracture procedure that we use, or the fixing.

But there are mathematical methods of analyzing this type of clustering distribution which I just wanted to bring up.

DR. EDIDIN: I will take two comments on that.

DR. FINEGOLD: Mine is very brief. I would thank Alan for showing these because these are just the techniques that we have quite independently used, and they are natural ones to use for our simulation using the data from our computer modeling where particles aggregate. Unfortunately, these techniques aren't always as sensitive to the aggregation as one might wish.

However, many of these techniques are certainly easy to use, (except for the radial distribution function, which does take a lot of computing).

DR. GERSHON: I would like just to add to the work of Alan Perelson. We have developed, when I was on sabbatical with the group of Mike Edidin at Hopkins, a similar method. We developed it independently. We have applied it to virus-infected cells on the GP-70 particles, unlabeled ones, and also on gap junctions, and we find also different amounts of clustering by using an analogous method.

A MODEL FOR THE REVERSIBLE BINDING OF BIVALENT ANTIGEN TO CELLS*

ALAN S. PERELSON

Theoretical Division, University of California, Los Alamos Scientific Laboratory, Los Alamos, NM 87545, and Division of Biology and Medicine, Brown University, Providence, RI 02912

ABSTRACT

A simple model for the reversible binding of bivalent antigen to cell surfaces is developed. The model is used to quantitatively predict the dynamics of receptor cross-linking and the amount of cross-linking that occurs at chemical equilibrium. To illustrate the usefulness of the theory, it is applied to the analysis of histamine release by basophils.

INTRODUCTION

Bell[1] has developed a model for the binding of multivalent antigen to cells which has proved useful in analyzing the response of B lymphocytes to multivalent antigens[1,2] and in understanding the phenomena of "patching" of membrane receptors[3]. The model, because it assumes the antigen is bound irreversibly to the cell once it is attached to more than one receptor, may not be appropriate for bivalent antigen. Here I develop a generalization of Bell's model for the specific case of the binding of bivalent molecules to cells, and as an example show how it can be used to qualitatively analyze the relationship between histamine release and the cross-linking of IgE molecules on the surfaces of mast cells or basophils. This example was motivated by the work of Dembo, Goldstein, Lichtenstein

*Work performed under the auspices of the U.S. Department of Energy.

and Sobotka[4-7] in which this quantitative relationship was first described. The model should also prove useful in analyzing the binding of other molecules, such as hormones, to cellular receptors.

MODEL

Consider a system containing bivalent antigen with two identical determinants and cells to which the antigen can bind. Let C, m and M be the concentrations of free, singly bound and doubly bound antigen, respectively. Assume that the concentration of antigen in the medium surrounding the cells is sufficiently large that the binding of antigen to cells does not appreciably change C, i.e., assume C is constant. Further, let S and S_0 be the free and total receptor site surface concentrations, respectively. For concreteness, assume the receptors are bivalent immunoglobulin molecules, so S_0 is twice the receptor density. An analogous treatment holds for monovalent or f-valent receptors.

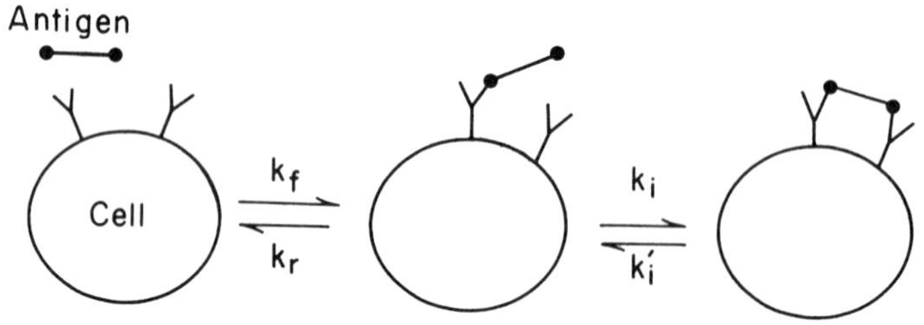

Fig. 1. Model for the reversible binding of bivalent antigen to cellular receptors.

The binding model is depicted in Fig. 1. Free antigen is assumed to bind to a single receptor site with forward rate constant k_f and to dissociate from the receptor with rate constant k_r. Once the antigen is singly bound it can cross-link two receptors with rate constant k_i and become doubly bound. This reaction is assumed to be reversible with k_i' being the rate constant for the transition from doubly bound to singly bound antigen.

The equations describing this model are:

$$S = S_0 - m - 2M, \quad (1)$$

$$\frac{dm}{dt} = k_f CS - k_r m - k_i Sm + 2k_i' M, \quad (2)$$

$$\frac{dM}{dt} = k_i Sm - 2k_i' M. \quad (3)$$

At $t = 0$ all the antigen is assumed to be free. Thus $m(0) = 0 = M(0)$, and $S(0) = S_0$. This kinetic model assumes that all antigen sites are equivalent and that all antibody sites are equivalent with regard to their chemical reactivity. Consequently the same rate constant, k_i, describes the rate at which an antigenic determinant attached to any site on an antigen-receptor aggregate combines with a free receptor site on another antigen-receptor aggregate, irrespective of its size. This assumption, however, restricts the use of the model to systems in which one can neglect reactions between a receptor site and an antigenic determinant on the <u>same</u> aggregate, that form rings or cyclic aggregates. Cycle formation is an intramolecular change which requires a rate constant different from the bimolecular rate constant k_i.[4]

Although the most novel aspects of this model relate to its ability to predict the dynamics of antigen binding and receptor cross-linking, the dynamic results can best be understood in light of the equilibrium properties of this model. Dembo and Goldstein[4] have previously analyzed the equilibrium binding of a symmetric bivalent antigen to IgE on the surface of basophils. By using an infinite system of equations which described all possible aggregate states, they were able to explicitly include the possibility of ring formation and dependence of the reaction rate constants on aggregate size. Wofsy et al[7] extended this treatment to the case of asymmetric bivalent antigens. DeLisi and

Perelson[3] and DeLisi and Thakur[8] considered both the equilibrium and dynamic properties of receptor cross-linking by f-valent antigens, $f \geq 2$, using a modification of Bell's model[1] which still assumed irreversible attachment of cross-linked antigen to the cell surface.

EQUILIBRIUM PROPERTIES OF THE MODEL

Let us examine the equilibrium properties of the model described by (1) - (3). At equilibrium, dm/dt = dM/dt = 0. Therefore,

$$K \equiv \frac{k_f}{k_r} = \frac{m}{CS} \tag{4}$$

and

$$K_i \equiv \frac{k_i}{k_i'} \equiv \frac{2M}{mS}, \tag{5}$$

where K is the equilibrium constant describing the binding of the whole antigen to a single receptor site, and K_i is the equilibrium constant for the cross-linking reaction. Substituting (1) into (4) and (5) and solving the resulting system of equations gives[9]

$$\bar{m} = \frac{-1 + \sqrt{1 + 4\beta(1-\beta)K_i S_0}}{2(1-\beta)K_i} \tag{6}$$

and

$$\bar{M} = \frac{1 + 2\beta(1-\beta)K_i S_0 - \sqrt{1 + 4\beta(1-\beta)K_i S_0}}{4K_i\beta(1-\beta)} \tag{7}$$

where the overbar denotes an equilibrium value and

$$\beta \equiv \frac{KC}{1+KC} \ . \tag{8}$$

Notice β can never equal 1 and thus (6) and (7) are valid for all values of KC. However, as KC $\to \infty$, $\beta \to 1$ and one can show $\bar{m} \to S_0$ and $\bar{M} \to 0$. This is exactly what one would expect. At high antigen concentration all receptor sites will bind a separate antigen ($\bar{m} = S_0$), making cross-linking impossible ($\bar{M} = 0$).

The equilibrium concentration of cross-links on the cell surface is $\bar{M}$. From equation (7) one can now show[9]: i) $\bar{M} = 0$ when $\beta = 0$ or $\beta = 1$, ii) $\bar{M}$ reaches a single maximum at $\beta = 1/2$ (KC = 1), iii) a plot of $\bar{M}$ vs. β is symmetric about $\beta = 1/2$, and iv) a plot of $\bar{M}$ vs. log KC is symmetric about the origin. These properties, illustrated in Figs. 2 and 3 have more general validity than indicated by this model. Dembo and Goldstein[4] established equivalent results for a model including ring formation and cross-linking rates which depend on aggregate size.

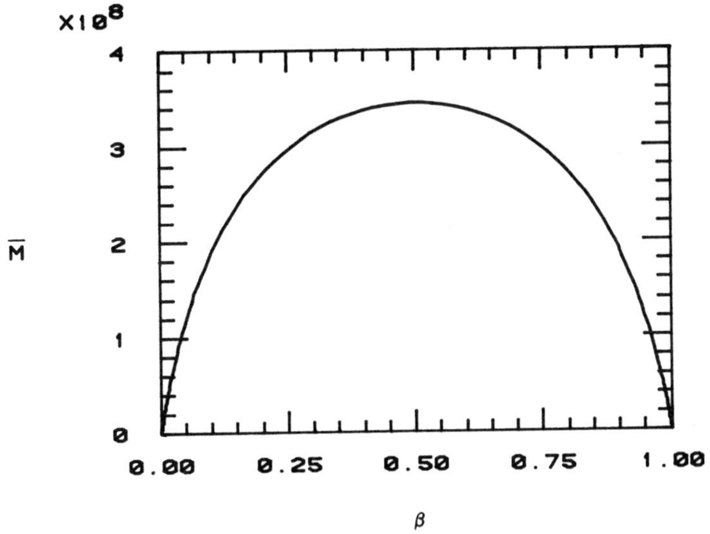

Fig. 2. The equilibrium concentration of doubly bound antigen $\bar{M}$ (cross-links) versus $\beta \equiv KC/(1 + KC)$, with $S_0 = 2 \times 10^9$ molecules cm^{-2} and $K_i = 1.6 \times 10^{-9} cm^2$ molecule^{-1}.

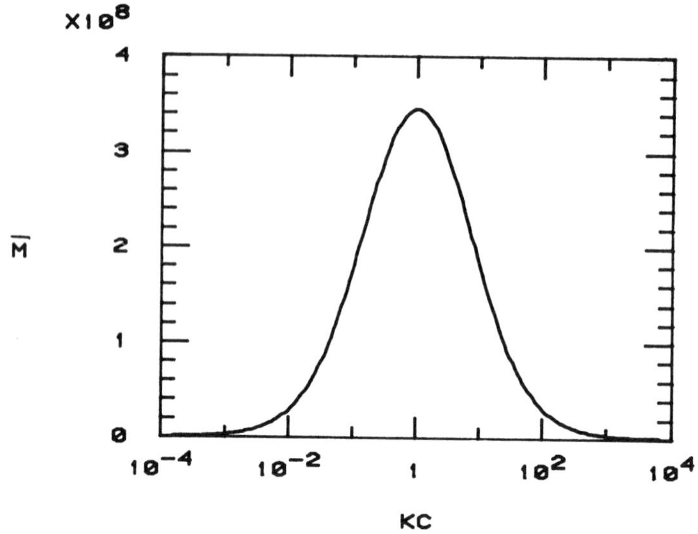

Fig. 3. The equilibrium concentration of doubly bound antigen $\overline{M}$ (cross-links) versus KC, with S_0 and K_i as in Fig. 2.

Some additional information about the state of the cell surface can also be derived from this model. Combining equations (1), (4) and (8) one finds

$$\beta = \frac{\overline{m}}{S_0 - 2\overline{M}} \quad . \tag{9}$$

Therefore, in the state of maximal cross-linking where $\beta = 1/2$,

$$\overline{A} \equiv \overline{m} + \overline{M} = S_0/2 \ . \tag{10}$$

The sum $\overline{m} + \overline{M}$ is the total concentration of the bound antigen, $\overline{A}$. Thus, at equilibrium, with the external antigen concentration C chosen to give maximal cross-linking, the (average) number of antigens bound per cell is equal to the (average) number of bivalent receptor molecules per cell.

For any external antigen concentration, the total amount of bound antigen, A, can be determined by summing (6) and (7). Figure 4 shows that $\bar{A}$ increases monotonically with KC, even though $\bar{M}$ decreases at high antigen levels.

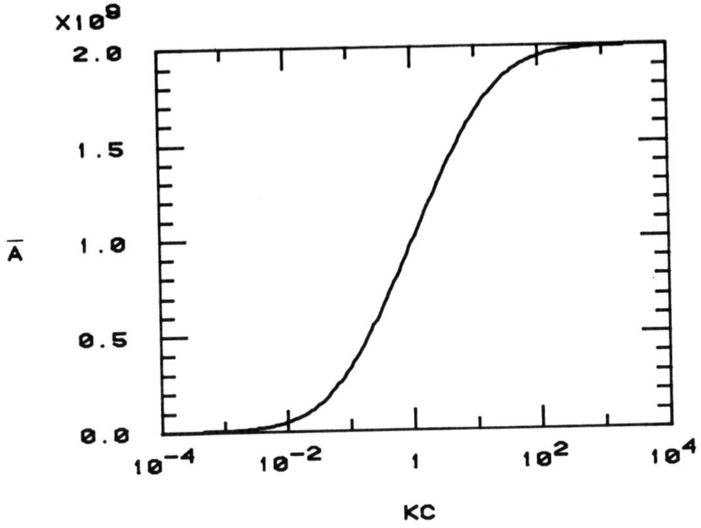

Fig. 4. The equilibrium concentration of bound antigen $\bar{A}$ versus KC, with S_0 and K_i as in Fig. 2.

At very large or very small concentrations of antigen (KC<<1 or KC>>1) one can obtain an approximate expression for $\bar{M}$ as follows. Let $\delta \equiv \beta(1-\beta)K_iS_0$. Then equation (7) can be written as

$$\bar{M} = S_0 \frac{1 + 2\delta - \sqrt{1+4\delta}}{4\delta} . \qquad (11)$$

For $\delta \ll 1$, one can expand $\sqrt{1+4\delta}$ in a Taylor series around 1 and obtain

$$\bar{M} = \frac{S_0\delta}{2}[1 - 2\delta + 5\delta^2 - \ldots].$$

To first order

$$\bar{M} \simeq \frac{S_0\delta}{2} = \frac{\beta(1-\beta)K_iS_0^2}{2} . \qquad (12)$$

Thus, as one might expect, the concentration of cross-links is proportional to the square of the receptor concentration.

The cross-linking of IgE molecules on the surface of mast cells or basophils in the presence of calcium leads to the release of histamine by these cells. Sirganian, Hook and Levine[10] showed that histamine release from basophils can be activated in vitro by simple bivalent antigens. Dembo, et al[5] have argued on the basis of an equilibrium binding model[4] that the amount of histamine released from human, passively sensitized basophils by bivalent antigen is controlled by the equilibrium number of cross-links on the cell surface. Figure 5 summarizes the data of Siraganian et al[10]. Notice the similarity between this curve, Figure 3 of Dembo and Goldstein (this volume), and the one in Fig. 3. All have a single maximum around which the curve is symmetric and all approach zero as C gets very large or very small. Thus the amount of histamine released by basophils appears to be proportional to the number of cross-links on the cell surface. If this is the case, then disrupting the cross-links by adding a cross reacting monovalent hapten should decrease the amount of histamine released. A slight extension of the binding model in Fig. 1 can predict the quantitative effects of added hapten.

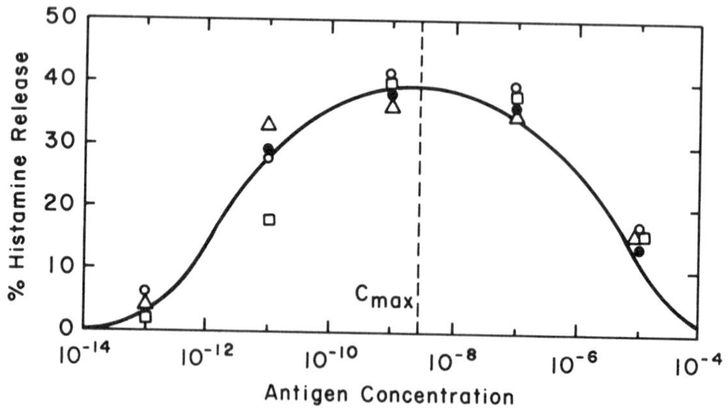

Fig. 5. Data of Siraganian, Hook and Levine[10] on the percent of total intracellular histamine released from rabbit basophils in the presence of bivalent antigens of the form BPO - NH - $(CH_2)_n$ - NH - BPO with n = 3 (•), 6(Δ), 9 (0), or 12 (□). BPO represents the penicillin analog, benzylpenicilloyl.

Consider cells incubated in the presence of bivalent antigen and monovalent hapten. As before, assume that such incubation does not significantly decrease the concentration of the antigen, C or hapten, H. Hapten binds to free receptor sites and thus

$$\frac{dH}{dt} = k_h HS - k_h' H_b, \qquad (13)$$

where H_b is the concentration of bound hapten, and k_h and k_h' are the forward and reverse rate constants, respectively, for the hapten binding reaction. At equilibrium, dH/dt = 0 and

$$H_b = K_H HS \qquad (14)$$

where $K_H = k_h/k_h'$ is a receptor site's affinity for hapten.

Equations (4) and (5) still apply, but now the conservation law (1) becomes

$$S = S_0 - m - 2M - H_b. \qquad (15)$$

Substituting (14) into (15) and solving for S gives

$$S = \frac{S_0 - m - 2M}{1 + K_H H}. \qquad (16)$$

Replacing S in (4) and (5) by (16) and solving the resulting system of equations gives equations identical to (6) and (7) for $\overline{m}$ and $\overline{M}$, with the exception that K and K_i need to be replaced by

$$K' = \frac{K}{1 + K_H H} \text{ and } K_i' = \frac{K_i}{1 + K_H H} \qquad (17)$$

Thus a plot of $\overline{M}$ versus $K'C$ would be identical to Fig. 3 if the numerical value for K_i' were chosen equal to that of K_i in Fig. 3. Notice, however, that if we denote the antigen concentration for maximum histamine release, C_{max}, we have,

$$C_{max} = 1/K' = \frac{1 + K_H H}{K} . \qquad (18)$$

For H = 0, $KC_{max} = 1$ as before. However, as H increases C_{max} increases. Thus, if the histamine release curve were a reflection of the concentration of cross-linked receptors, $\overline{M}$, one would expect the maximum in the histamine release curve to progressively shift to the right as H increased. Further, at high antigen concentrations, the expansion of $\overline{M}$ to first order in (12) is valid and with (8) gives

$$\overline{M} = \frac{K'CK_i'S_0^2}{2(1+K'C)^2} . \qquad (19)$$

For C sufficiently large, $1 + K'C \simeq K'C$ and

$$\overline{M} = \frac{K_i'S_0^2}{2K'C} = \frac{K_iS_0^2}{2KC} , \qquad (20)$$

which is independent of H. Thus, if the amount of histamine released were proportional to $\overline{M}$, one would predict that at high values of C, histamine release curves for all values of H would approach the H = 0 curve. These two strong theoretical predictions of an equilibrium binding model were first noticed by Dembo and Goldstein[4]. Experiments by Dembo et al[5], have verified both of these predictions (see Fig. 6 of Dembo and Goldstein, this volume).

DYNAMICS OF CROSS-LINKING

The most novel use of the binding model embodied in (1) - (3) is in the analysis of the kinetics of receptor cross-linking. Previous models, such as that of Dembo and Goldstein[4,7], have successfully dealt with the equilibrium binding of bivalent antigens. However, such theories become overly complicated when extended to the dynamics of cross-linking. Other workers have considered the kinetic aspects of cross-linking by multivalent antigens which bind irreversibly[1-3,8] or reversibly[11]. Here the effects of reversible binding by bivalent antigen will be treated explicitly. Kinetic studies of receptor cross-linking become biologically important in situations where the response of a cell to the cross-linking of its receptors begins to occur before binding equilibrium has been reached.

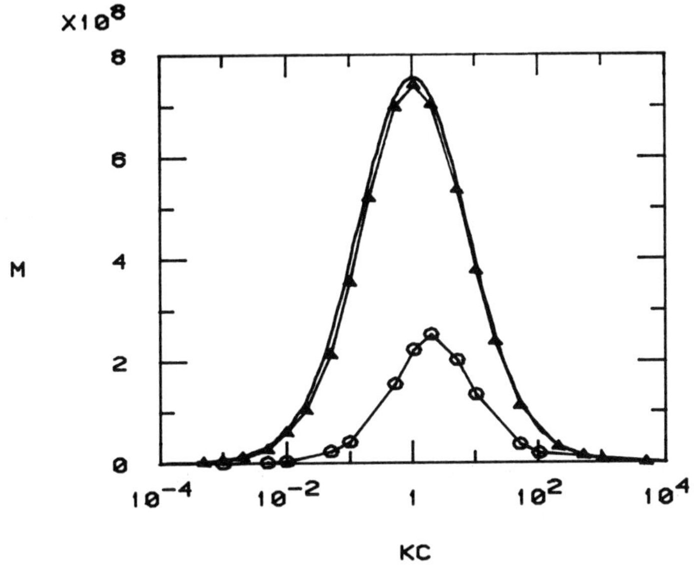

Fig. 6. The dynamics of cross-link formation is illustrated for t = 0.1 sec. (0) and t = 1.0 sec. (Δ). The uppermost solid curve corresponds to equilibrium, i.e.. t = ∞. Here $S_0 = 4.26 \times 10^9$ molecules cm^{-2}, $k_f = 10^8\ M^{-1}\ sec^{-1}$, $k_r = 10\ sec^{-1}$, $k_i^0 = 8 \times 10^{-10}$ molecules^{-1} cm^2 sec^{-1}. and $k_i' = 1\ sec^{-1}$.

Let us begin by noting that (1) - (3) cannot be solved analytically. However, numerical solutions and certain approximate solutions are possible. Figure 6 shows the results of a numerical study of (1) - (3), using an integration routine for solving stiff differential equations based on Gear's method[12]. Notice that for this set of parameters which applies to the binding of $(BPO)_2$ to passively sensitized human basophils[6], equilibrium is approached very rapidly. This would not be the case for antigens which bind the cell more strongly, i.e., have a much smaller value of k_r. Further notice that the symmetry of the cross-linking curve is apparent well before equilibrium is reached. In order to ascertain if symmetry of the non-equilibrium curves in Fig. 6 is a general property, one can derive approximate solutions to (1) - (3). For example, if $k_i S_0 \ll k_r$ and $k_i' \ll k_r$, then the antigen will come on and off the surface many times before cross-linking occurs. Under these conditions, the binding reaction can be treated as being in equilibrium and thus

$$m = KCS \ . \qquad (21)$$

Using this value of m in (1) and (3) gives

$$S = \frac{S_0 - 2M}{1 + KC} \qquad (22)$$

and

$$\frac{dM}{dt} = \frac{4k_i KC\, M^2 - [4k_i KCS_0 + 2k_i(1+KC)^2]M + k_i KCS_0^2}{(1 + KC)^2} \ . \qquad (23)$$

Rewriting (23) in terms of the previously defined variable $\delta = \frac{KCK_i S_0}{(1+KC)^2}$, one finds that the solution to (23), with initial condition $M(0) = 0$, is

$$M(t) = \frac{\lambda_1 \lambda_2 \, [1 - \exp(\xi t)]}{\lambda_2 - \lambda_1 \exp(\xi t)}, \qquad (24)$$

where

$$\xi = 4\delta k_i'(\lambda_1 - \lambda_2)/S_0 \qquad (25)$$

and $\lambda_1(\lambda_2)$ is the larger (smaller) of the two roots of the quadratic equation

$$M^2 - [S_0(1 + 2\delta)/2\delta]M + S_0^2/4 = 0. \qquad (26)$$

Notice that (24) satisfies the initial condition and that as $t \to \infty$, M approaches its equilibrium value, i.e.,

$$M(\infty) = \lambda_2 \equiv S_0 \left[\frac{1 + 2\delta - \sqrt{1+4\delta}}{4\delta}\right] . \quad (27)$$

One very noteworthy feature of the solution (24) is that λ_1 and λ_2 and hence ξ depend on KC only as a function of δ. Therefore, M is also only a function of δ and hence for any fixed time sufficiently large that the equilibrium approximation is valid, a plot of M versus log KC would be symmetrical about log KC = 0. Thus the symmetry found in the computer simulations of (1) and (3) and presumably in experimental histamine release curves is not solely an equilibrium property.

This simple binding model can be used to estimate the efficiency of washing cells in antigen free media in order to remove bound antigen. Again assume $k_i S_0 << k_r$ and $k_i' << k_r$, so that singly bound antigen would be in equilibrium with free antigen if any were present. Under these conditions, (21) implies m = 0, if C = 0. Thus I am assuming that singly bound antigen dissociates rapidly and that the rate at which antigen is removed from the surface is determined by the rate of dissociation of doubly bound antigen. From (3) with m = 0, one obtains

$$M(t) = M(0) \exp(-2k_i' t), \quad (28)$$

where M(0) is the initial amount of multiply bound antigen. This equation predicts that the amount of antigen on the cell surface decreases exponentially with a time constant of $1/2k_i'$. A computer simulation of a system of 21 mass-action equations representing the changes with respect to time of the concentrations of various receptor-antigen complexes on the cell surface by Dembo et al[6], with S_0 = $4 \times 10^9 cm^{-2}$, $k_f = 10^8 M^{-1} sec^{-1}$, $k_r = 10 sec^{-1}$, $k_i = 1.6 \times 10^{-9} cm^2 sec^{-1}$ and

$k_i' = 1\ sec^{-1}$, showed that the amount of cross-linked antibody per cell decreased from 10^4 molecules to 10 molecules in approximately five seconds and that a plot of log cross-linked antibody versus time was linear. With these parameters, equation (28) predicts that this change would occur in 4.6 sec, and also clearly predicts a linear decrease in log M(t) with time. This rapid rate of antigen loss is only applicable when cyclic receptor-antigen complexes are absent from the cell surface. These complexes can stabilize the antigen on the surface, requiring longer washing times for antigen removal[6].

SUMMARY

This model establishes in a simple, straight forward manner some of the general equilibrium and kinetic properties for the cross-linking of receptor molecules by divalent antigen. In particular, it demonstrates that the concentration of receptor cross-links is a symmetric function of antigen concentration, both in chemical equilibrium and during the approach to equilibrium. When such symmetries are found in dose response curves, such as in the example of histamine release, the possibility is presented that the effect being measured is related to receptor cross-linking. The destruction of cross-links by monovalent hapten is also explored as well as the effects of washing cells in antigen free media.

ACKNOWLEDGEMENTS

I wish to thank Byron Goldstein for helpful comments throughout the course of this work and George Bell and Micah Dembo for reading and commenting on this manuscript.

REFERENCES

1. Bell, G.I. (1974) Nature, 248, 430-431.

2. Bell, G.I. (1975) Transplant. Rev. 23, 23-36.

3. DeLisi, C., and Perelson, A. (1976) J. Theoret. Biol. 62, 159-210.

4. Dembo, M. and Goldstein, B. (1978) J. Immunol. 121, 345-353.

5. Dembo, M., Goldstein, B., Sobotka, A.K. and Lichtenstein, L.M. (1978) J. Immunol. 121, 354-362.

6. Dembo, M., Goldstein, B., Sobotka, A.K. and Lichtenstein, L.M. J. Immunol. (in press).

7. Wofsy, C., Goldstein, B. and Dembo, M. (1978) J. Immunol. 121 593-601.

8. DeLisi, C. and Thakur, A. K. (1977) Cell. Immunol. 28, 416-426.

9. Perelson, A. (1979) in Mathematical Models in Molecular and Cellular Biology, Segel, L. A., ed. Cambridge University Press, London, in press.

10. Siraganian, R.P., Hook, W.A. and Levine, B. B. (1975) Immunochem. 12, 149-157.

11. Gandolfi, A., Giovenco, M.A. and Strom, R. (1977) Technical Report R. 77-21, Istituto di Automatica, University of Rome.

12. Gear, C.W. (1973) Numerical Initial Value Problems in Ordinary Differential Equations, Prentice Hall, Englewood Cliffs, New Jersey.

ON THE EFFECT OF CROSSLINKING ON THE ATTACHMENT OF MEMBRANE PROTEINS TO THE CYTOSKELETON

NAHUM D. GERSHON
National Institutes of Health, Division of Computer Research and Technology, Physical Sciences Laboratory, Bethesda, Maryland 20014 (U.S.A.)*
and
Department of Biology, The Johns Hopkins University, Baltimore, Maryland 21218 (U.S.A.)

ABSTRACT

Processes that might lead to attachment of membrane proteins to submembraneous filaments after crosslinking are considered. Because of their size, patches formed by crosslinking proteins might be entangled with the filamentous network to a greater extent than a single protein. In addition, microfilaments might be sensitive to mechanical disturbances and to their rate of change. Areas under patches and caps are likely to be more static and less mechanically disturbed than areas under the rest of the cell surface and thus such delicate filaments can be polymerized more extensively under patches and caps. Also patches might serve as nuceation centers for filament polymerization.

Many kinds of multivalent membrane proteins on cell surfaces can be crosslinked by multivalent molecules to form patches which can be aggregated in one pole of the cell to form a cap[1,2,3]. It has been suggested[4-7] that the cytoskeleton (which includes microfilaments, made from actin and/or myosin, and microtubules) takes an active part in the process of pulling the patches into one location to form a cap. Recently a report has been published[8] that shows that when surface Ig (immunoglobulin) molecules are crosslinked by anti-Ig and form patches and/or a cap they become attached to actin. It has also been observed, in the fluorescence microscope level[9-12], that before crosslinking of certain membrane proteins the actin and/or myosin are distributed evenly under the membrane. After crosslinking the proteins from the outside of the membrane, the actin and/or the myosin appear to be aggregated under the patches and the cap. The most common interpretation is that the attachment of the patches to actin and/or myosin is required in order that actin and myosin can take part in the process of collecting the patches into a cap.

---

*Reprint requests should be sent to the Bethesda address.

We would like to suggest an explanation why more attachment is seen after crosslinking and to point out an alternative possibility which views the attachment of patches to actin and/or myosin as a side effect.

Actin and myosin are presumably present under cell membranes in the form of filaments, and possibly also in a non-polymerized form. It is believed that many types of proteins can be attached to the filament system when they are crosslinked. Therefore the attachment of membrane proteins to the filament network maybe nonspecific. Thus, naturally, physical or other nonspecific forces may play a role in the process of attachment of membrane proteins to the filaments. For example, one can expect a protein molecule which is in contact with a meshwork of filaments to be entangled with them to some extent. A detachment of such a protein molecule from the network should require a certain amount of force and maneuvering. After crosslinking (from the outside of the membrane) many protein molecules into a patch, the portions of the crosslinked molecules which are inside the cell may still interact with the filament network. Thus a detachment of a patch which contains many proteins will require greater forces and more maneuvering than a detachment of a single protein from the filaments.

This is the simplest picture which can be more composite in a number of ways. If a single protein moves laterally in the membrane faster than a patch it might have a chance of being less attached to the filament network.

Another possibility is that the presence of a patch might affect the state of the microfilament network. It is known that some polymerized forms of actin (and/or myosin) (e.g.[13]) are susceptible to mechanical stress such as shear forces, and to the rate rate in which they are changing. If these proteins tend to exist under the membrane in such a polymerized form, the motion of the membrane and/or the proteins embedded in it may cause enough mechanical disturbance to prevent polymerization of actin and/or myosin. Thus under such conditions these delicate filaments will be hardly formed. If the motion of the membrane and/or the membrane proteins is hindered at certain regions in the cell surface, then, in the absence of such mechanical disturbances, actin and/or myosin might be polymerized under these static regions intensely and form such delicate filaments. Examples of such regions are patches and caps, where proteins are essentially fixed together and thus these portions of the membrane should be more static than in areas where the proteins are relatively freely moving. An additional possibility is that patches and caps can serve as nucleation centers for actin and/or myosin polymerization. Any further polymerization in the vicinity of a patch or a cap may increase the entanglement (and so the attachment) of crosslinked proteins to actin and/or myosin, as is explained above.

The attachment of membrane proteins to actin and/or myosin after crosslinking does not necessarily mean that these cytoskeletal elements take an active part in the collection of patches into a cap. This attachment might be a side effect to the actual processes of patching and capping, occurring by different mechanisms. An example to a process which can proceed without using a contractile system is a nucleation process[14]. This situation might occur in a membrane which is overloaded or supersaturated with receptors. The crosslinking of the proteins produces stable patches which can grow spontaneously into a cap by precipitation of the remaining proteins of the same kind.

ACKNOWLEDGEMENTS

I would thank Drs. M. Edidin, G. Ackers and R. Nossal for exciting and fruitful discussions.

REFERENCES

1. Taylor, R. B., Duffus, W. P. H., Raff, M. C. & de Petris, S. (1971) Nature New Biol. 233, 225-229.
2. Schreiner, G. F. & Unanue, E. R. (1976) Adv. Immunol. 24, 38-135.
3. Loor, F. (1977) Prog. Allergy, 23, 1-153.
4. de Petris, S. & Raff, M. C. (1972) Eur. J. Immunol. 2, 523-532.
5. Berlin, R. D., Oliver, J. M., Ukena, T. E. & Yin, H. H. (1974) Nature 247, 45-46.
6. Nicholson, G. L. (1976) Biochim. Biophys. Acta 457, 57-108.
7. Edelman, G. E. (1976) Science 192, 218-226.
8. Flanagan, J. & Koch, G. L. E. (1978) Nature 273, 278-281.
9. Schreiner, G. F., Fujiwara, K., Pollard, T. D. & Unanue, E. R. (1977) J. Exp. Med. 145, 1393-1398.
10. Bourguignon, L. Y. W. & Singer, S. J. (1977) Proc. Natl. Acad. Sci. USA 74, 5031-5035.
11. Toh, B. H. & Hard, G. C. (1977) Nature 279, 795-797.
12. Tabriani, G., Chaponnier, C., Zumbe, A. & Vassali, P. (1977) Nature 269, 697-698.
13. Pollard, T. D. (1976) J. Cell Biol. 68, 579-601.
14. Gershon, N. D . (1978) Proc. Natl. Acad. Sci. USA 75, 1357-1360.

A CORRELATION BETWEEN IMMOBILIZATION AND CAPPING OF SURFACE ANTIGENS IN LYMPHOCYTES

PAUL DRAGSTEN
Laboratory of Theoretical Biology, National Cancer Institute, National Institutes of Health, Bethesda, Maryland 20014

ABSTRACT

We have investigated a possible relationship between membrane antigens' ability to cap and to diffuse in the plane of the membrane after cross-linking with appropriate antibodies. Mouse spleen lymphocytes were labeled with fluorescent antibodies against surface immunoglobulin or Thy-1 antigen in the presence of sodium azide to inhibit capping. The fraction of bound antibody free to diffuse was then measured using the technique of fluorescence photobleaching recovery. In parallel incubations without azide, cells were labeled with the fluorescent antibody and the capping efficiency recorded. We found that the surface antigen is essentially immobile with respect to diffusion (with azide) at concentrations of antibody sufficient to induce capping (in the absence of azide). Two-layer (antibody sandwich) cross-linking showed the same correlation.

INTRODUCTION

Several proposed mechanisms for the capping of membrane antigens suggest direct or indirect attachment of surface-aggregated antigens to cytoskeletal structures, after which contractile elements in the cytoskeleton pull the antigens into a cap[1,2]. Metabolic inhibitors are known to abolish capping, perhaps by eliminating contractile activity in the cytoskeleton. If it were assumed that metabolic inhibition did not interfere with the membrane antigen - cytoskeleton coupling, then cross-linked antigens might be immobilized in the plane of the membrane, being anchored to the presumably static cytoskeletal structures. Thus, in the presence of a metabolic inhibitor and under cross-linking conditions otherwise sufficient to induce efficient capping, one might expect membrane antigens to be immobile with respect to lateral diffusion. We have made measurements that directly test this possibility using the technique of fluorescence photobleaching recovery[3-6].

MATERIALS AND METHODS

Mouse spleen cells were prepared from C57Bl/6J mice (Jackson Laboratories) according to the procedure of Ryan et al.[7].

Antiserum against mouse IgG was raised by hyperimmunizing a goat with mouse IgG. IgG antibodies from this antiserum (henceforth called "anti-Ig") were affinity purified on mouse IgG - sepharose, and eluted with citrate buffer (pH 2.2). Fab fragments were prepared from a DEAE - purified IgG fraction by papain digestion and separation on a sephadex G-100 column.

Rabbit anti-mouse brain antiserum was prepared according to the procedure of Golub[8] and the IgG fraction purified by DEAE chromatography.

Intact antibodies and Fab fragments were labeled with tetramethylrhodamine isothiocyanate according to conventional procedures[9].

The fluorescence photobleaching recovery experiments were done as described previously[3-6]. Briefly, this technique involves fluorescently labeling the cell membrane component to be studied and monitoring fluorescence from a small area of the cell surface. The fluorescence is excited by a laser beam focused to a small spot (0.6 μm radius) on the cell surface using a slightly modified fluorescence microscope. The laser intensity is then increased 500 to 5000-fold for a fraction of a second, irreversibly bleaching a fraction of the fluorophore in the illuminated region. The fluorescence intensity from that region will therefore have decreased immediately following the bleach but, if the fluorescently labeled molecules are free to diffuse in the plane of the membrane, the fluorescence intensity from the illuminated area will recover with time as unbleached label diffuses into the bleached region. The kinetics of recovery of fluorescence permits calculation of the diffusion constant, and the eventual extent of recovery (often less than 100% for labeled proteins) gives the effective fraction of labeled molecules free to diffuse over distances comparable to the dimensions of the illuminated spot.

RESULTS

Mouse spleen lymphocytes were incubated with different concentrations of rhodamine - conjugated anti-Ig for 10 minutes at 0°C in the presence of 3mM sodium azide. Fluorescence photobleaching measurements of the mobility of the fluorescent antibodies were performed (with azide still present to inhibit capping), and the fraction of the bound fluorescent antibodies free to diffuse on the cell surface was measured. The fluorescent antibody labeling was then repeated in the absence of azide, and the percent of labeled cells that proceeded to cap was recorded. The results are shown in figure 1. At low anti-Ig concentrations capping is

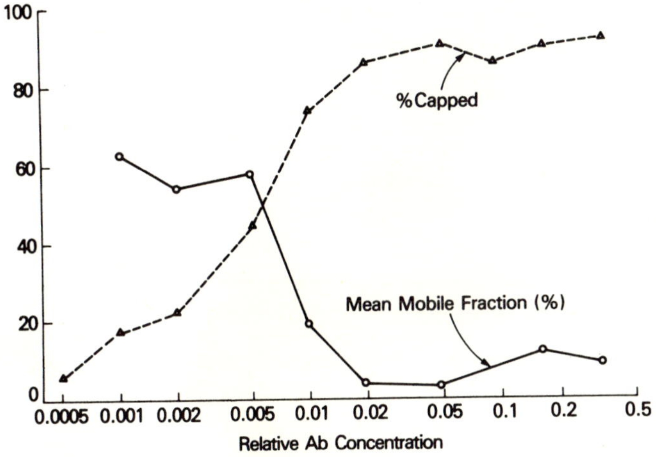

Fig. 1 Capping frequency and degree of mobility of surface immunoglobulin as a function of the concentration of cross-linking anti-Ig (given as fraction of anti-Ig stock concentration). Cells were incubated with rhodamine-conjugated anti-Ig for 10 min. at 0°C with and without the addition of 3mM sodium azide. The azide-free cells were then washed and incubated for 30 minutes at room temperature, after which the capping frequencies of labeled cells were scored (--Δ--). Azide-treated cells were washed with azide-containing medium, and the mean mobile fraction of fluorescent label (——o——) was determined by fluorescence photobleaching recovery measurements.

rare and roughly 50 to 60 percent of the fluorescent label is free to diffuse (as compared with a typical range of 50 to 90 percent mobile fraction in the case of a monovalent Fab label; data not shown). At high antibody concentrations, capping efficiency increases to around 90% as the mean mobile fraction decreases to near zero. It is apparent that, at concentrations of anti-Ig that produce maximal capping, the labeled surface Ig are essentially immobile. The same general relationship holds when the fluorescent Fab anti-Ig, which does not cap, is cross-linked by a second layer of antibody (rabbit anti-goat IgG): the Fab ligand caps and, in the presence of azide, shows no diffusion in fluorescence photobleaching measurements.

We have also measured mobilities of Thy-1 antigen on T cells using a rhodamine-conjugated IgG fraction of rabbit anti-mouse brain, which is known to contain anti-Thy 1 activity. We did not consistently obtain immobilization of this marker at any RAMB concentration, and the cells did not cap this ligand. Addition of a second layer of cross-linking antibody (goat anti-rabbit) did produce both capping and, in the presence of azide, immobilization.

A similar relationship between immobilization and capping has also been observed by Wolf et al.[10] for a non-protein synthetic antigen (a stearoyl-dextran derivatized with TNP groups) inserted into 3T3 cells. This antigen could be capped with anti-TNP antibodies.

In conclusion, our results show that surface immunoglobulin and Thy-1 antigen are essentially immobile with respect to diffusion over distances of a micron or greater under cross-linking conditions sufficient to induce cap formation. This immobilization is consistent with bivalent antibody induced anchorage to cytoskeletal structures although other explanations are also possible, such as the formation of large patches whose diffusion rates are too slow to be detected.

ACKNOWLEDGEMENTS

This work was done in collaboration with Joseph Schlessinger, Pierre Henkart, John Weinstein, and Robert Blumenthal, whose contributions are gratefully acknowledged.

REFERENCES

1. Bourguignon, L.Y.W., and Singer, S.J. (1977) Proc. Nat. Acad. Sci. USA 74, 5031-5035.
2. Schreiner, G.F., and Unanue, E.R. (1976) Adv. Immunol. 24, 37-165.
3. Axelrod, D., Koppel, D.E., Schlessinger, J., Elson, E., and Webb, W.W. (1976) Biophys. J. 16, 1055-1069.
4. Jacobson, K., Wu, E.-S., and Poste, G.(1976) Biochim. Biophys. Acta 433, 215-222.
5. Zagyanski, Y., and Edidin, M. (1976) Biochim. Biophys. Acta 433, 209-214.
6. Schlessinger, J., Koppel, D., Axelrod, D., Jacobson, K., Webb, W.W., and Elson, E. (1976) Proc. Nat. Acad. Sci. USA 73, 2409-2413.
7. Ryan, J.L., Arbeit, R.D., Dickler, H.B., and Henkart, P.A. (1975) J. Exp. Med. 142, 814-826.
8. Golub, E.S. (1971) Cell Immunol. 2, 353-361.
9. Goldman, M. (1968) in Fluorescent Antibody Methods, Academic Press, New York.
10. Wolf, D.E., Webb, W. W., and Henkart, P.A. (1978) Private communication.

DeLisi/Blumenthal, eds. Physical Chemical Aspects of Cell Surface Events in Cellular Regulation

THE CONCEPT OF PASSIVE MODULATION OF MEMBRANE RESPONSES

MEIR SHINITZKY
Department of Membrane Research, The Weizmann Institute of Science, Rehovot (Israel)

There is no doubt today that the protein-lipid interplay is the main determinant of most physiological processes which take place in biological membranes. As determined by the membrane structure, lipids and proteins are at constant thermal motions which, because of structural heterogeneities and stringent boundary conditions, are of formidable complexity. Even though sophisticated experimental tools for monitoring diffusion processes in membranes are now at hand, we are still far from analytical presentations of experimental diffusion data. One approach for overcoming this serious limitation is by simulating the recorded mobility values into an analogous homogeneous macroscopic system. Accordingly, the hydrocarbon core of the membrane lipid layer is treated as being a homogeneous hydrocarbon fluid. Although somewhat remote from reality, this approach enabled us to apply the classical relations of viscous flow to the membrane lipid layer, and thus to obtain empirical comparative scales for fluidity parameters of various biological membranes (for a review see Ref.1).

One of the most intriguing findings, which was promoted by this macroscopic simulation, was that the flow activation energy, ΔE_f, in the lipid domain of most mammalian membranes falls in the narrow range of 7.5±1.5 Kcal/mole and is approximately independent of the lipid microviscosity. Conversely, in artificial liposomes free of proteins ΔE_f ranges approximately from 5 to 15 Kcal/mole and is highly dependent on the lipid microviscosity[2]. This observation was interpreted by assuming that the position of the membrane proteins is displaced towards the aqueous phase when the lipid layer becomes more viscous, a process which will buffer the changes in ΔE_f[2]. This important concept was promptly verified by both spectral[3-5] and chemical methods[6], all indicating that the exposure of membrane proteins can be reversibly modulated by alterations of lipid microviscosity. Increase in the lipid microviscosity will displace the membrane proteins towards the aqueous layers on both sides of the membrane, and *vice versa*. This process is termed as "Vertical Displacement" of membrane proteins and the associated modulation of any membranal activity is termed as "Passive Modulation", since it does not require metabolic energy. Obviously, in the case of proteins which are anchored to the underlying cytoskeleton of the membrane the vertical displacement by passive modulation will be hindered or even completely abolished.

The vertical displacement process

The position of a protein in a membrane can be regarded as a state of minimum free energy of all the intermolecular interactions of the protein with its surroundings. The most prominent are the external protein-water interactions and the internal protein-lipid interactions. Towards proteins, the lipid layer has features analogous to a classical weak solvent, due to the considerable difference in dielectric properties (even uncharged hydrophobic peptides are only weakly soluble in hydrocarbon solvents). This weak solvent-solute like interaction implies that it is mostly entropy driven. In other words, it is highly dependent on the molecular free volume of the lipid chains or on its related parameters - the density and microviscosity. This dependence can be illustrated in simple solubilization experiments of a weak solute (chemically unrelated) in a series of aliphatic hydrocarbon solvents ranging, let us say, from *n*-pentane (d_{20° = 0.626 gr·ml^{-1}) to *n*-hexadecane (d_{20° = 0.773 gr·ml^{-1}). The chemical nature in such a series can be presented by units of $-CH_2-$ and therefore can be regarded as chemically identical. However, the physical nature in this series changes progressively as the average spacing for each $-CH_2-$ adopts a smaller free volume. Indeed, the solubilization capacity of weak solutes gradually decreases in such solvent series.

Another example, which is closely related to biological membranes, is presented in Figure 1. The fluorophore 1,8-anilino-naphthalene-sulfonate (ANS), is of a typical amphipathic structure and is therefore expected to partition freely between lipid and aqueous phases. This partitioning can be readily monitored by the fluorescence yield of ANS, which in water is negligible compared to that of ANS bound to lipids. In addition, the apparent dielectric constant of the binding region can be evaluated from the wavelength of maximum emission. The partitioning characteristics of ANS into a series of single walled liposomes made of a mixture containing egg-lecithin and L-dimyristoyl lecithin (DML) above its phase transition (31°) is shown in Figure 1. In this series of liposomes the only chemical variable is the degree of unsaturation of the acyl chains, which induces an overall progressive change in lipid microviscosity. As clearly demonstrated in the figure the increase in lipid microviscosity decreases the partitioning of ANS into the liposomes.

Supportive evidence for the above observation comes from analogous partitioning experiments of ANS, and the hydrophobic fluorophore N-phenyl naphthylamine (NPN), and phospholipid vesicles above and below the phase transition. The incorporation of these molecules above the phase transition increases by about 2-fold for both ANS[7] and NPN[8]. It should be stressed that below the phase transition these fluorophores are not excluded completely from the lipid as was observed for the incorporation of the spin-label probe TEMPO[9], which is probably an extreme case of fluidity dependent partitioning.

The above described dependence of partitioning on lipid fluidity probably applies as well for other similar small solutes. However, when the molecular size of the membrane "solute" is similar to that of the lipid bilayer itself, changes in microviscosity are expected to result mostly in changes in the degree of exposure of the "solute" because of the numerous sites of solvent-solute interaction. In this respect, the degree of exposure of membrane proteins is analogous to the partitioning coefficient of a small amphipathic solute such as ANS. Alteration in lipid fluidity will therefore change the balance of the opposing interactions with the water and the lipids and thus will displace the protein to a new equilibrium position. When lipid fluidity is decreased the new equilibrium position will be of an overall weaker protein-lipid interaction but with correspondingly greater protein-water associations. All of these arguments imply that vertical displacement of membrane proteins is expected when changes in lipid fluidity take place. In principle, the same concept can also be applied to special amphipathic lipids which are distinctively different from the bulk membrane lipids. Such lipids could be glycolipids to which class the blood group antigens belong.

Membrane proteins and functional lipids can be classified according to their vertical displacement behaviour, as depicted in Figure 2.

Class A. An integral membrane protein which is partially exposed on the outer side of the membrane. The figure describes a membrane receptor or enzyme with a binding site near the hydrocarbon-water interface of the lipid bilayer. Small changes in lipid microviscosity can modulate the activity of such sites, especially towards hydrophilic substrates. Class Ai is an analogous integral membrane protein which is exposed on the inner side of the membrane.

Class B. A glycoprotein or a glycolipid with a polysaccharide chain which is exposed on the outer side of the membrane. This site could be an antigen which by increasing lipid microviscosity becomes more exposed and thus renders the system greater antigenicity. Class Bi is an analogous site which is exposed on the inner side of the membrane (such sites are exceptionally rare).

Class C. A highly non-symmetrical cross-membrane protein with a greater exposure on the outer side of the membrane. The net effect of increase in microviscosity will be an increase of exposure of the protein on the outer surface of the membrane and decrease of its exposure on the inner surface.

Class Ci. A cross-membrane protein but with a greater exposure on the inner side of the membrane. Increasing microviscosity will increase the exposure of the protein on the inner side of the membrane and will diminish its exposure on the outer side.

Class D. A relatively symmetrical cross-membrane protein where increase in lipid microviscosity will mostly increase protein-protein interaction (e.g., receptor-enzyme combination, subunit association or non-specific aggregation).

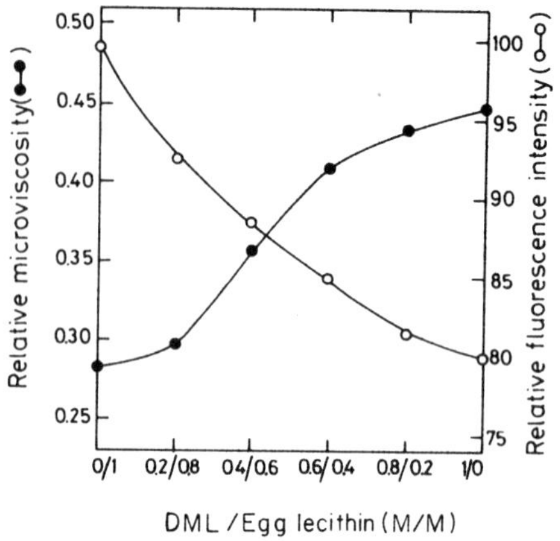

Fig. 1. Partitioning of 1,8 anilino naphthalene sulfonate (ANS) into sonicated liposomes (0.4 mg/ml) of phosphatidylcholine of increasing microviscosity. Liposomes were prepared by 15 minutes sonication at 70 watts at 4° under argon atmosphere of mixtures containing egg lecithin and L-dimyristoyl lecithin (DML) in PBS. Under these conditions over 90% of the liposomes are unilammelar vesicles. Microviscosity of the lipid layer was determined by fluorescence polarisation of DPH as described[1,2]. Incorporation of ANS into the liposome walls was determined by fluorescence intensity (λ_{ex} = 360 nm, λ_{em} = 475±5 nm). All measurements were taken at 31° (above the phase transition of DML) to avoid inhomogeneities due to phase separations.

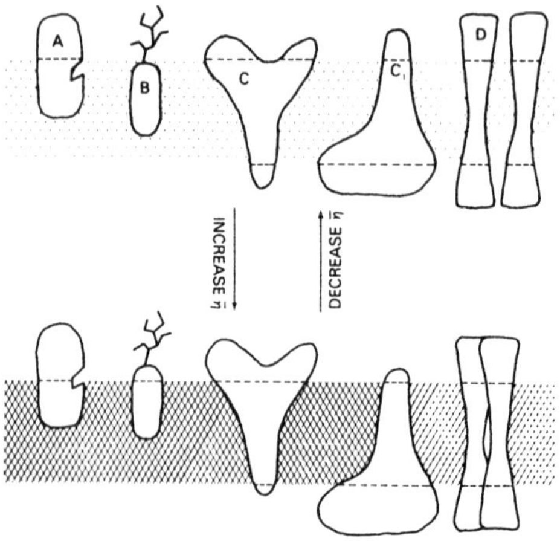

Fig. 2. Vertical displacement of membrane proteins induced by changes in lipid microviscosity. In principle the changes induced by the physical and the chemical effectors should be of a similar magnitude.

Factors affecting lipid fluidity

To a great extent lipid fluidity determines both the lateral and rotational mobilities of membrane sites, as well as their degree of exposure or projection. These loosely dependent parameters are of major importance in determining various membranal activities. A series of physical and chemical factors can affect membrane lipid fluidity. In general terms the physical effects are homogeneous and instantaneous, whereas the chemical effects are regional and vectorial and reach maximum after minutes or hours.

Physical effectors. Membrane lipid domains, similar to macroscopic fluids, decrease their viscosity with temperature according to the simple exponential relation $\bar{\eta} = A \exp \Delta E_f/RT$. Thus, the main parameter which determines the change in microviscosity, $\bar{\eta}$, with temperature is the apparent flow activation energy ΔE_f. Modulation of $\bar{\eta}$ with temperature and the corresponding changes of physiological activities are of great importance in plants and cold-blooded animals.

The lipid microviscosity should correspond to pressure, P, in a similar exponential manner as to temperature, $\bar{\eta} = B \exp P\cdot\Delta V_f/RT$ where ΔV_f is the activation volume of the flow. Changes in $\bar{\eta}$ with P are expected to occur with hydrostatic pressure, as well as with intracellular osmotic pressure.

More relevant to animal membranes are changes in pH in the physiological range. It has been observed that the apparent $\bar{\eta}$ values of the intestinal microvillus membrane[10] and the plasma membrane of neuroblastoma cells[11] increase markedly when the overt pH increases from 6.5 to 7.5. It may be suggested that decrease in intracellular pH, for example by increase of glucose metabolism, can in turn modulate the membrane microviscosity and the cellular activities.

A potent physical effector, which has not yet been explored, is the membrane potential. Changes in lipid microviscosity mediated by transient changes in membrane potential may be one of the physical means which regulate the physiological processes associated with nerve conductance.

Changes in conformation of membrane proteins could cause local changes in lipid fluidity. This process is pertinent to specific binding of ligands, like hormones, and since the associated conformational change derives its energy from the ΔF of binding this process is also classified as a physical effector.

Chemical effectors. Any modification of the chemical composition of the membrane can, in principle, affect the lipid fluidity. For most physiological processes the most prominent chemical effectors are the cholesterol level, generally presented as the mole ratio of unesterified cholesterol to phospholipids (C/PL), and the degree of unsaturation of the phospholipid acyl chains. Two other important effectors are the mole ratios of lecithin to sphingomyelin (L/S) and of proteins to lipids. The levels of all of these factors are controlled by intracellular syntheses and membrane biogenesis. In addition to these processes the C/PL, and to a lesser extent the L/S, are also regulated by passive translocation of lipids from the serum lipoproteins. Such lipid acquisitions are especially pertinent to blood cells.

Cellular processes and membrane fluidity

Virtually all morphological changes which are associated with cell differentiation, maturation or transformation were found to be also associated with changes in membrane fluidity[12-14]. These processes are of relatively long time scale, in the range of several days, and may involve several cell divisions. It is likely that the modulation of lipid fluidity in these processes involves distinct changes in membrane composition. Shorter cellular processes, which occur within one cell division, like blastogenesis[15], the cell cycle[16] and intracellular contacts[17] are also associated with changes in membrane fluidity. The time-scale of these falls in the range of several hours to 1-2 days, and it is reasonable to assume that both physical and chemical effectors take part in the concomitant modulation of lipid fluidity. As suggested by these arguments the contribution of the physical effectors increases as the time-scale of the process decreases. This is especially relevant to receptor modulation which is mediated by the ligand binding, as will be discussed below.

Passive modulation of membrane receptors. It has long been realized that the battery of receptors in the cell surface changes its relative responses along with various cellular processes[18]. In some cases this receptor modulation could be accounted for by the turnover of membrane biogenesis. However, to many other cases this trivial interpretation cannot be applied, especially when the modulation occurs within a relatively short time. These latter processes were generally assumed to occur by active modulation, namely by structural changes in the cytoskeleton network to which the receptors are supposedly anchored[19]. Obviously, this mechanism *a priori* presumes that the modulated receptors are cross-membrane proteins which are chemically bound in the inner side of the membrane to actin-myosin type filaments. This active modulation mechanism also requires metabolic energy and should be blocked at low temperature or in the presence of metabolic inhibitors.

The proposed mechanism of passive modulation of receptor and antigen expressions is simpler and more general than that of the active modulation, and it is likely that this mechanism takes place in a number of physiological processes. However, since the idea of passive modulation is rather new no such cases are yet reported in the literature. The results presented in the following, which clearly demonstrate the actuality of this idea, were recently obtained in our laboratory but are not yet published.

The possibility to passively modulate receptor expression was tested by us in systems containing the receptors for transferrin and insulin. Bone marrow cells, which are rich in transferrin receptors, were found to modulate markedly their capacity to bind transferrin by changing the cholesterol level in the surface membrane[20].

Cholesterol enrichment, which increased by only 10% the average microviscosity of the cell membrane, resulted in 40% reduction of ^{3}H-transferrin binding. Decrease of only 6% in membrane microviscosity by depletion of cholesterol resulted in 40% increase of ^{3}H-transferrin binding. Similar results were obtained with FITC-transferrin by cell sorting, and are shown in Figure 3. The modulations of transferrin binding, which probably relate to changes in the projection of the transferrin receptor, were found to be virtually fully reversible.

Another cell type which was investigated, was the Friend virus erythroleukemic cell. Both the undifferentiated and the DMSO differentiated cells are rich in transferrin receptors. The process of induction of differentiation by DMSO is followed by parallel profiles of changes in membrane fluidity and in transferrin binding. The increase in microviscosity is concomitant with a decrease in transferrin binding, similar to what was observed with the bone marrow cells. Furthermore, a distinct decrease in both membrane fluidity and transferrin binding was observed in the undifferentiated cells when the cell density increased. Upon abrupt dilution these parameters reached the initial values after 6 hours[21]. Following the models of passive modulation which are depicted in Figure 2 the transferrin receptor seems to belong to class Ci. Namely, the results suggest that it is a cross-membrane protein with a massive portion facing the cytoplasm. This conclusion should of course remain tentative until verification by other means is achieved.

A study on passive modulation of the insulin receptor in turkey erythrocytes revealed analogous characteristics to what was observed with the transferrin receptor in bone marrow cells. Specific binding of insulin was increased upon decrease in membrane microviscosity by cholesterol depletion, and *vice versa*. Accordingly, the insulin receptor seems also to be a cross-membrane protein with a significant portion facing the cytoplasm, which is in line with the current views on this receptor. In this respect it is pertinent to mention the observation that when physiological amounts of insulin were added to isolated plasma membranes of rat liver cells a small ($\sim$10%) but consistent increase in the overall lipid microviscosity was observed[22]. A detailed analysis of this observation indicated that upon insulin binding the receptor probably undergoes a conformational change which exerts an increase in the microviscosity of the surrounding lipids. If indeed the insulin receptor is of class Ci, then the unoccupied receptors in the vicinity of an occupied one should be displaced inward due to the local increase in lipid microviscosity. In fact, this simple mechanism could account for the well-established negative cooperativity of the insulin binding to its receptor. It is tempting to speculate even further that the inward displacement may promote the observed process of receptor aggregation after insulin binding[23] which could be an important step in the hormone action.

<u>Passive modulation of membrane antigens</u>. The major blood group antigens are probably the best characterized and detectable antigenic markers known. Manipulation

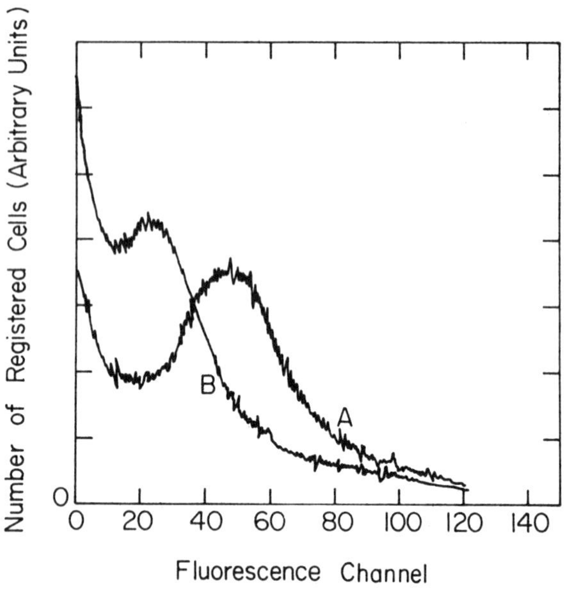

Fig.3. Distribution of fluorescence intensity obtained with a cell sorter of fluorescein isothiocyanate labelled transferrin-Fe_2 attached to bone marrow cells. A - cholesterol depleted cells, B - cholesterol enriched cells. The profiles indicate a substantial increase of transferrin binding upon fluidization of the cell membrane by cholesterol depletion.

of these antigens in human erythrocyte should, in principle, be simple since changes in cholesterol level in these cells can be readily accomplished. We have, therefore, selected the blood group antigens A and Rh for testing the feasibility of passive antigenic modulation. The antigenic expression was qualitatively monitored by agglutinability with the respective antibody, and quantitatively by binding of radioactively or fluorescently labelled antibodies. Indeed, all of these tests have clearly demonstrated that the antigenic expression is augmented upon cholesterol enrichment (increase in microviscosity) and that it is suppressed upon cholesterol depletion. Similar results were also obtained with Θ-antigen on T-cells as assessed by complement lysis after addition of anti-Θ. Thus, it is likely that membrane antigens fall in the category of class B and that their expression can be modulated by the membrane microviscosity.

The concept of passive modulation of membrane antigens opens new avenues to studies of tumor antigenicity and autoimmune reactions. We are now studying the possibility of augmenting the exposure of tumor antigens and its possible resultant effect in increasing the immunological response against tumor cells. We are also studying the converse possibility that autoimmune reaction can be prompted by overexposure of membrane components.

REFERENCES

1. Shinitzky, M. and Barenholz, Y. (1978) Biochim. Biophys. Acta Reviews on Biomembranes, in press.

2. Shinitzky, M. and Inbar, M. (1976) Biochim. Biophys. Acta, 433, 133-149.

3. Borochov, H. and Shinitzky, M. (1976) Proc. Natl. Acad. Sci., 73, 4526-4530.

4. Shinitzky, M. and Rivnay, B. (1977) Biochemistry, 16, 982-986.

5. Humphries, G.M.K. and McConnell, H.M. (1977) Proc. Natl. Acad. Sci. USA, 74, 3537-3541.

6. Borochov, H., Abbott, R.E., Schachter, D. and Shinitzky, M. (1978) Submitted for publication.

7. Sackmann, E. and Täuble, H. (1972) J. Am. Chem. Soc., 94, 4482-4491.

8. Overath, P. and Täuble, H. (1973) Biochemistry, 12, 2625-2634.

9. Shimshick, E.J. and McConnell, H.M. (1973) Biochemistry, 12, 2351-2360.

10. Schachter, D. and Shinitzky, M. (1977) J. Clin. Invest., 59, 536-548.

11. deLaat, S.W., personal communication.

12. Nicolson, G.L. (1976) Biochim. Biophys. Acta, 457, 57-108.

13. Nicolson, G.L. (1976) Biochim. Biophys. Acta, 458, 1-72.

14. Shinitzky, M. and Henkart, P. (1978) Int. Rev. Cytol., in press.

15. Inbar, M. and Shinitzky, M. (1975) Eur. J. Immunol., 5, 166-170.

16. deLaat, S.W., van der Saag, P.T. and Shinitzky, M. (1977) Proc. Natl. Acad. Sci. USA, 74, 4458-4461.

17. Inbar, M., Yuli, I. and Raz, M. (1977) Exp. Cell Res., 105, 325-335.

18. Raff, M. (1976) Nature, 259, 265-266.

19. Edelman, G.M. (1976) Science, 192, 218-226.

20. Muller, C. and Shinitzky, M. (1979) Brit. J. Haem., in press.

21. Muller, C., Volloch, Z. and Shinitzky, M. (1978) Submitted for publication.

22. Luly, P. and Shinitzky, M. (1979) Biochemistry, in press.

23. Schlessinger, J., Shechter, Y., Willingham, M.C. and Pastan, I. (1978) Proc. Natl. Acad. Sci. USA, 75, 2659-2663.

III

TRANSDUCTIVE COUPLING BY CELL SURFACE RECEPTORS I.

TRANSDUCTIVE COUPLING BY CELL SURFACE RECEPTOR - I.

JESSE ROTH
Diabetes Branch, NIAMDD, National Institutes of Health, Bethesda, MD, USA

Chairpersons

I am the chairman of the morning session. Traditionally everyone was a chairman whether male or female. Then everyone became a chairperson, which seemed more equitable. But now we have regressed. The male is called a chairman but the female is called a chairperson. For those of you who think that that is a new situation, I brought a picture of a very ancient chairperson (Figure 1), one from 1100 B.C. Notice that this really is a chairperson.

Fig. 1.

Another thing to notice about the chairperson is that she is distinctly female. Thus having chairpersons as females is a very ancient tradition going back at least 3,000 years.

Hormones and Receptors

For those of you who have not been in the receptor field but have worked in other areas, you may be wondering why we have made so little progress in the receptor area. Hormones and receptors were both conceptualized around 1900. With the hormones, enormous progress has occurred during the intervening 75 or 80 years, whereas with receptors the progress has been slow and intermittent until the last decade. The reasons why the receptors, especially the membrane receptors, have been so slow in developing are outlined in Table 1.

TABLE 1

NATURAL FACTORS THAT FAVORED PROGRESS WITH HORMONES

	Peptide Hormones	Cell Surface Receptors
Highly concentrated in a localized region	yes	no
Soluble in aqueous solvents	yes	no
Biological effect when introduced *in vivo* and *in vitro*	yes	no
Present in blood	yes	no

The first question: is the material highly concentrated in a localized region within the body? A hormone typically is, and it is those hormones with which great progress has occurred; the gland naturally provided a very rich concentrated source of starting material. By contrast there is no place where receptors are especially concentrated, except for the electric organs of a few marine animals; for the acetylcholine receptor, the availability of a highly enriched preparation as the electric organ has clearly accelerated progress but only for that receptor. In fact the exceptions here give strong support to the rule; not only has rapid progress occurred with acetylcholine receptors, where a highly concentrated source is available, but conversely, hormones that lack a concentrated site, either because secretory sites are diffusely distributed or have little stored hormone, have also progressed quite slowly.

Solubility is another important determinant. All of the hormones are soluble in simple solvents. The peptide hormones (which comprise about 80% of all hormones) and the catecholamines are all soluble in aqueous solutions. By contrast, their receptors are integral membrane proteins and are not soluble in aqueous media. For solubilization receptors require detergents, and the technology necessary to work with these types of proteins is a recent innovation. The lack of methodology for this class of proteins has clearly hindered progress.

Another advantage of the hormone is the simplicity of the assays, at least conceptually. You simply inject hormones _in vivo_ or _in vitro_ to get an effect; when you do the same with receptors nothing happens. Thus, assays for receptors, which require much more sophisticated technology, were greatly delayed in development. Similarly hormones, because they are universally present in blood are easily accessible for measurements, whereas receptors are less readily accessible.

Surface Receptors and Intracellular Hormones

The next topic I wish to review appears to be perplexing to many people in the audience and was raised at this meeting yesterday by Martin Raff. Is it a new discovery that peptide hormones can get into cells? I propose to review the development of ideas about mechanisms of hormone action to show how the sequence of experiments molded (or distorted) the collective thought processes. Figure 2 summarizes thoughts about hormone action up until the late fifties and early sixties. Twenty years ago only a few people were considering hormone

Fig. 2.

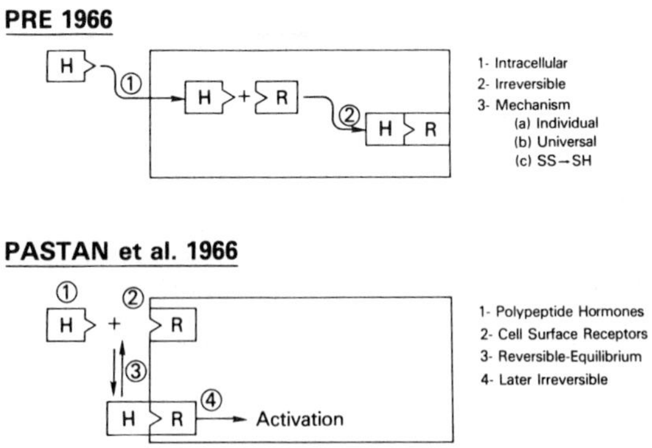

action in the broad general terms that we do today. Among these elite were the late Gordon Tomkins[1] and Oscar Hechter[2] who is with us at this meeting.

As shown in the top half of Figure 2, the early ideas were that all hormones had to get into cells in order to work and that hormones bound irreversibly to some intracellular component. However, in terms of mechanism, there tended to be in most people's minds an individualization or atomization of everything, i.e., every one of the 40 different hormones was thought to have a unique mechanism of action.* Among the few people who thought universally were Hechter and Tomkins. However, they tended not to make clear distinctions between steroid, catecholamine, thyronine or peptide hormones.

The lower half of Figure 2 shows the change in our thoughts about hormone action introduced by studies of Pastan, et al., on TSH, insulin and ACTH in the middle sixties.[4] The first was the notion that all of the polypeptide hormones were one class of hormone that had great similarities in mechanism and were different from the steroids. Secondly these hormones bind to receptors on the outer cell surface. The interaction of hormone with receptor was a reversible equilibrium and the initial reversibility was followed later by irreversibility.[4]

Having finally accepted the evidence that the initial interaction is at the cell surface, the field went further. It went beyond the data and decided that peptide hormones don't get into cells. Accelerating this process were experiments to show that hormone linked covalently to a solid matrix was biologically active.[5,6] Neither of these approaches ruled out entry of hormone into cells independent of receptor binding. Later, evidence for widespread solubilization of hormones from the solid matrices discredited the evidence from the matrix studies[7] so that now there is no evidence at all against receptor-mediated or receptor-independent entry of hormone into the cell.

---

*The only exceptions were insulin, oxytocin, and vasopressin. Each has at least one disulfide bond. It was thought that a disulfide bridge of the hormone opened to form one or more disulfide bridges with SH groups on the target cell and that this was the mechanism for attachment of these hormones to the cell. The synthesis of bioactive analogues of the hormone with a methylene bridge to replace the disulfide bridge terminated this approach. For insulin (and by analogy to the two other hormones) the cell surface was implicated as the site of action because the hormone affected cell surface activities (e.g., permeability and transport).[3] Since the hormones activated processes at the cell membrane, ergo, they must bind to the cell surface. The latter conclusion is probably correct, though today we may find the logic that led to the conclusion somewhat uncertain.

Now we believe that the initial binding of the hormone to the receptor is at the cell surface. In addition, there are specific receptors on other cell membranes including Golgi,[8] the rough endoplasmic reticulum, and possibly even the nuclear membrane.[9,10] Although the initial binding of hormones (and the one that is crucial for hormone action) is at the cell surface, it is also clear that the peptide hormones do enter cells. Let me emphasize that there was never any evidence that hormones didn't get in, and now there is very strong evidence that they clearly all do get into the cells, though we don't understand why.

The other point I should stress is that even the most strenuous devotees of the importance of intra-cellular binding sites agree that the initial interaction is at the cell surface and that if you prevent the binding of the hormone to the cell surface receptors, little else of interest will happen.

Regulatory Pathways

Figure 3 illustrates one other feature of the systems that should be emphasized. First there is hormone binding to receptor to form hormone receptor complexes. A second messenger is generated intracellularly leading to a multiplicity of events (Figure 3, upper). I would call this formulation a straight arrow pathway or a trident pathway where each step follows sequentially the

Fig. 3.

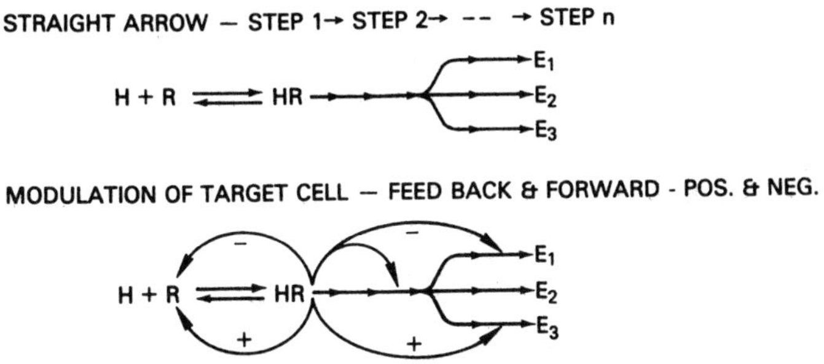

next. What is clear now is that the target cell is highly modulated. De Meyts emphasized that point earlier in the meeting and I shall amplify it, namely, the hormone receptor interaction not only leads to specific biological effects but leads to important modulatory events. Thus the hormone receptor complex feeds back to reduce the affinity of the receptor and by other mechanisms to regulate the concentration of the receptor.[11] There are also positive modulations for some systems. More important, there are feed forward regulatory processes. For example, high concentrations of hormone not only desensitize back at the receptor, but also desensitize later steps. As has been shown in several systems, most extensively by Catt and his group with the gonadotropins,[12] it is clear, that negative and positive feedback modulatory events are major determinants of target cell responsiveness to hormone.[1,13]

Functions of the Receptor

An extension of this approach is shown in Table 2 where I have listed the known functions of the receptor. Clearly there are many more than we had thought. The fundamental concept of a receptor has only two essential functions, recognition and activation. The receptor molecule must recognize the active material and bind it. Then the combination of the two must activate the cell. Clearly the job description of the receptor has been expanded beyond the fundamentals. It has been assigned a large number of other functions in addition to recognition and activation. For example the receptor acts as a reservoir for plasma hormone.[14] In general the peptide hormones are free in plasma, lacking any binding proteins (there are a few exceptions). When the concentrations of these hormones are rapidly rising, the surface receptors actually act as a reservoir to bind plasma hormone. When the hormone concentrations in plasma fall, some of that hormone comes back off the receptor to resupply the plasma pool.[14,15]

TABLE 2

FUNCTIONS OF THE RECEPTOR-HORMONE INTERACTION

A. Fundamental
   1. Recognition
   2. Activation

B. Additional
   1. Reservoir for plasma hormone
   2. Regulate degradation of hormone
   3. Regulate degradation of receptor
   4. Regulate receptor concentration and receptor affinity
   5. Cross-link and translocate hormone-receptor complexes
   6. Regulate post receptor events

The receptor also acts to regulate the degradation of hormone. *In vivo*, there are two major mechanisms for hormone degradation, both at the level of the cell. (For most peptide hormones, no degradation occurs in plasma *in vivo*). One mode of degradation does not require binding to receptor.[14,16] However another major mechanism of hormone degradation does require binding of hormone to the receptors.[17,18] The receptor doesn't perform the degradation but delivers the hormone to a site where the hormone gets degraded. Thus, the hormone-receptor interaction regulates the degradation of both hormone and receptor, as well as regulates receptor concentration and receptor affinity, cross-linking and translocation of hormone-receptor complexes, and regulates post-receptor events. While the hormone-receptor interaction affects all of these events, cause and effect relationships among these functions are as yet uncertain; conclusions on this point should be postponed until more direct data are available.

Information Transfer

Table 3 introduces another issue, which is the question of information transfer. For three classes of receptors -- for the toxins (e.g., cholera toxin, diphtheria toxin or ricin), for the asialo-glycoproteins in plasma, and for the low density lipoproteins, the biological information is all in the ligand. The receptor acts only to concentrate and/or process and/or translocate the ligands, i.e., to expedite entry of the active moiety of the ligand. If these materials enter the cell without the receptor, the biological effect takes place just as well. This has tempted people to extrapolate to hormones and to other ligands and to conclude that some piece of the hormone must get into the cell to bring the information for cell activation.

If you have a hormone-receptor complex activating the cell, where is the information? Is it in the hormone, is it in the receptor, or is it in the combination? These kinds of questions gain answers slowly. It wasn't until the turn of the century that we knew that the egg and the sperm contributed about equally to the information transfer in fertilization. In many of the systems that have been worked out (Table 3), the ligand has the information. For the insulin system,[19] for the IgE system, and probably also for TSH and acetylcholine, the information is in the receptor; the ligand only acts to cause the receptor to express that information.

TABLE 3

INFORMATION TRANSFER

A. Hormone + Receptor ⇄ Hormone-Receptor complex → activation

B. Where is the "information" for the activation -- H ?? R ?? HR ??

C. Examples of ligand-receptor systems

1. Both moieties contribute about equally -- egg and sperm
2. Information is in the ligand; receptor acts to concentrate, process, and/or translocate the ligand to an intracellular site
   a. toxins: cholera, diphtheria, ricin
   b. low density lipoproteins
   c. asialo-glycoproteins
3. Information is in the receptor; ligand acts to bring out the program from the receptor
   a. insulin
   b. IgE
   c. (thyrotropin)
   d. (acetylcholine)

Diseases and Receptors

In closing, I want to emphasize the application of receptors to diseases. Most of the members of this meeting are not physicians nor is their work disease-related. Diseases have exceptional value as experiments, particularly because their realness and relevance are unquestioned. With receptors, diseases have an important place in our understanding the regulation and function of these systems.[20,21] Table 4 shows four major areas where surface receptors relate closely to human disease.

TABLE 4

MAJOR AREAS OF RECEPTOR DISEASES

		Overall Prevalence	Receptor Involvement Among Patients With This Class of Disorder
1. Hyperglycemia, diabetes and obesity	1972	5%	4/5
2. Myasthenia gravis	1973	Common	Most
3. Hyperthyroidism	1974	1%	2/3
4. Atherosclerosis	1974		
Familial Hypercholesterolemia			
(a) heterozygotes		0.2%	all
(b) homozygotes		very rare	all

TABLE 5 - Classification of Receptor-Related Disease States

A. Homologous hormone - receptor concentration inversely related to chronic level of hormone

   1. Insulin receptor

      a. Obesity (± diabetes, ± chronic diet or chronic starvation); thin diabetics; anorexia nervosa
      b. Insulinoma
      c. Uremia
      d. Type A extreme insulin resistance

   2. Other

      a. Glucagon receptor in states of hyperglucagonemia
      b. Probably in states of hCG and thyrocalcitonin excess
      c. Estrogen increase oxytocin binding

B. Heterologous hormone - one hormone acts to alter the receptor of another hormone

   1. Insulin receptor

      a. Growth hormone excess and deficiency
      b. Glucocorticoid excess and deficiency

   2. Other

      a. Thyroid hormone increases concentration of β-adrenergic receptors
      b. Estrogens increase level of progesterone receptors

C. Receptor Autoantibodies - autoimmune process with antibodies that bind to the receptor molecule and either mimic hormone action, block hormone action, or both

   1. Insulin receptor

      a. Type B form of extreme insulin resistance
      b. Ataxia telangiectasia with insulin resistance

   2. Other

      a. Graves' disease (diffuse hyperthyroidism)
      b. Myasthenia gravis

D. Genetic

1. Insulin - none reported

2. Other

   a. Familial hypercholesterolemia
   b. Androgen receptors - multiple
   c. Other steroid receptors

E. Disorders of Receptor Design - One hormone present in excess acts to produce excess biological effects through the receptor of a closely related hormone ("Specificity Spillover")

1. Insulin

   a. Non-islet tumor hypoglycemia - an insulin-like growth factor (IGF-II) in excess acts through the insulin receptor to produce an excess insulin effect
   b. Infants of diabetic mothers - insulin in excess acts through receptors for one or more of the insulin-like growth factors to produce increased skeletal growth (and possibly also the macrosomia, biochemical immaturity, and teratogenicity) characteristic of these patients.

2. Other

   a. The skin darkening associated with ACTH excess is probably mediated through reactivity with MSH receptors
   b. The hyperthyrodism associated with hCG excess is probably mediated through reactivity with TSH receptors
   c. Growth hormone excess without prolactin excess can be associated with the syndrome of amenorrhea, galactorrhea, and suppressed gonadotropic function (typical of prolactin excess) by reactivity of growth hormone with the prolactin receptors
   d. The hypertension associated with glucocorticoid excess may in some cases be due to reactivity of glucocorticoids with mineralcorticoid receptor

In the area of hyperglycemia, there are about a dozen different conditions that go under the heading of diabetes and hyperglycemic states. The overall prevalence in the population is about 5 percent, and about four out of five are primarily receptor problems rather than hormone problems. Thus hormone deficiency accounts for hyperglycemia in only about 20 percent of the cases; the majority of hyperglycemic states are actually receptor problems.

Hyperthyroidism, which is another major metabolic disorder, occurs in about one percent of the population. Two out of three of those patients have antibodies directed against the TSH receptor as the probable cause of their hyperthyroidism. Likewise in myasthenia gravis, a common neurological disorder, the vast majority of the patients appear to have antibodies directed against the acetylcholine receptor.

Atherosclerosis is a very common medical problem that is typically <u>not</u> related to receptors. However, as Goldstein will discuss in detail later, all patients who have familial hypercholesterolemia, a rare cause of atherosclerosis, appear to have a receptor defect. These patients are relatively uncommon but they present a very exciting model of a genetic basis for receptor disorders. In addition to the major form of familial hypercholesterolemia, Brown and Goldstein have recently described another variety of the disease, which again sheds major light on how the receptors work.

In summary, my point was to stress to you that diseases are experiments that have extraordinary value. In this field diseases have been a major way in which much of the physiology and understanding of receptors has been achieved (Tables 4 and 5).

REFERENCES

1. Roth, J , Lesniak, M.A., Megyesi, K., and Kahn, C.R. (1979) in Hormone Receptors, Human Diseases, and Disorders in Receptor Design, Cold Spring Harbor Symposium - Quantitative Biology, in press.
2. Hechter, O., and Halkerston, I.D.K. (1964) in The Hormones, Pincus, G., Thimann, K.V., and Astwood, E.B. eds., Academic Press, New York, pp. 697-825.
3. Levine, R., and Goldstein, M. (1955) Rec. Prog. Horm. Res., 11, 343.
4. Pastan, I., Roth, J., and Macchia, V. (1966) Proc. Natl. Acad. Sci., USA, 56, 1802.
5. Schimmer, B.P., Yeda, K., and Sato, G.H. (1968) Biochem. Biophys. Res. Commun., 32, 806.
6. Cuatrecasas, P. (1969) Proc. Natl. Acad. Sci., USA, 63, 450.
7. Kolb, H.J., Renner, R., Hepp, K.D., Weiss, L., and Wieland, O.H. (1975) Proc. Natl. Acad. Sci., USA, 72, 248.
8. Bergeron, J.J.M., Posner, B.I., Josefsberg, Z., and Sikstrom, R. (1978) J. Biol. Chem. 253, 4058.

9. Kahn, C.R. (1976) J. Cell Biol. 70, 261.
10. Goldfine, I.D., and Smith, G.J. (1976) Proc. Natl. Acad. Sci., USA, 73, 1427.
11. Gavin, J.R.,III, Roth, J., Neville, D.M.,Jr., De Meyts, P., and Buell, D.N. (1974) Proc. Natl. Acad. Sci., USA, 71, 84.
12. Conti, M., Harwood, J.P., Dufau, M.L., and Catt, K.J. (1977) Mol. Pharmacol. 13, 1024.
13. Roth, J. (1979) in Endocrinology, De Groot, L. ed., Grune and Stratton, New York, pp. 2037-2054.
14. Zeleznik, A.J., and Roth, J. (1978) J. Clin. Invest. 61, 1363.
15. Berman, M., McGuire, E.A., Roth, J., and Zeleznik, A.J., submitted.
16. Freychet, P., Kahn, C.R., Roth, J., and Neville, D.M.,Jr. (1972) J. Biol. Chem. 247, 3953.
17. Terris, S., and Steiner, D.F. (1975) J. Biol. Chem. 250, 8389.
18. Kahn, C.R., and Baird, K. (1978) J. Biol. Chem. 253, 4900.
19. Lawrence, J.C., Larner, J., Kahn, C.R., and Roth, J. (1979) in Molecular and Cellular Biochemistry (in press).
20. Kahn, C.R., Neville, D.M.,Jr., Gorden, P., Freychet, P., and Roth, J. (1972) Biochem. Biophys. Res. Commun. 48, 135.
21. Bar, R.S., Harrison, L.C., Muggeo, M., Gorden, P., Kahn, C.R., and Roth, J. (1979) Adv. Intern. Med. 24, 23.

DeLisi/Blumenthal, eds. Physical Chemical Aspects of Cell Surface Events in Cellular Regulation

THE RANDOM HIT MATRIX MODEL OF COUPLING IN MEMBRANES: FUNDAMENTAL FEATURES, UNIQUENESS, AND MOLECULAR IMPLICATIONS

RICHARD N. BERGMAN[1], MARK GREENBERG[1], AND OSCAR HECHTER[2]
Departments of Engineering Sciences[1] and Physiology[2], Northwestern University, Evanston and Chicago Illinois 60201 and 60611

INTRODUCTION

How receptors for hormones (and neurotransmitters) are coupled to effector units in the plasma membrane is a central problem in contemporary biology. Plasma membrane fractions from a wide variety of tissues and cells with hormone sensitive adenylate cyclase activity have been studied in this regard, and a multiplicity of specific mechanisms have been proposed[1,2,3]. The nature of the coupling process remains obscure, in part because little is known about the molecular nature of receptor and cyclase units, possible coupling factors[4], or the structural organization of these elements in the membrane.

Even if the molecular nature of all of elements involved were known in detail, such knowledge, in itself, would not represent a complete understanding of the coupling process. Such an understanding requires that the explicit spatial and temporal relations between the elements be defined so that it is possible to account for the transient and steady state relationship which obtains between hormone binding to receptors, and concomitant adenylate cyclase activation.

In order to evaluate mechanistic proposals of coupling in the neurohypophyseal hormone (NHH) sensitive-adenylate cyclase system in bovine renal medullary membranes we developed a novel modeling approach to account for the data obtained relating the binding of tritiated lysine and arginine vasopressin ([^{3}H] - LVP, [^{3}H] - AVP) and receptor mediated adenylate cyclase activation in bovine membrane preparations studied under similar experimental conditions. Similar studies of the temporal relationships between receptor occupation and enzyme activation had been carried out by Jard and his colleagues[5-9] for porcine renal medullary membranes; the results obtained in bovine and porcine membranes have been compared in a previous publication[10].

The present discussion is limited to the bovine renal membrane case. An apparently homogeneous set of receptor sites in bovine membranes bound [^{3}H] - LVP (or [^{3}H] - AVP) reversibly, and without cooperativity. The Hill plot between hormone concentration and enzyme activation indicated that receptor-mediated activation likewise was non-cooperative. The half-maximal concentrations for adenylate cyclase activation (K_a) for both [^{3}H] - LVP and [^{3}H] - AVP were uniformly lower than the respective hormone concentrations required for half-maximal binding (K_b). The K_b/K_a ratios with both ligands averaged 4.4 (range: 3.7 to 6.2). For both ligands, also, a consistent relationship was observed between normalized (percent maximal) steady-state activation (A) and equilibrium occupation (O) ($A:O_{ss}$) in different membrane preparations. Indeed, the $A:O_{ss}$ relationship was found to be remarkably similar when membranes were studied under a wide variety of experimental conditions. At low occupancy of receptors, the normalized steady-state A:O ratio approached 5:1; with increased occupancy, the steady-state A:O ratio progressively decreased, and maximal cyclase activation was achieved only at full occupancy ($A:O_{ss}$ = 1:1).

It appeared to us that the nonlinear $A:O_{ss}$ relationship of bovine membranes could be explained by a variety of possible coupling mechanisms[10] provided that appropriated _ad hoc_ assumptions are introduced. In contrast, the A:O dynamics observed in membranes appeared to greatly constrain the possible mechanisms for coupling. The dynamics of hormone bindings and enzyme activation in renal membranes was found to be variable and to have certain unexpected features. In certain membranes (e.g. low R membrane, Fig. 1) following hormone addition, activation reached steady-state while receptor occupation of hormone was still increasing. This temporal dissociation, previously designated as the _silent receptor phenomenon_, was dependent upon hormone concentration and particularly marked at high concentrations of hormone. In other bovine renal membrane preparations (e.g. high R membrane, Fig. 1), steady-state activation and equilibrium binding were achieved at approximately the same time (although some temporal dissociation of occupation and activation was evident) and the slope of the trajectories of approach to the final $A:O_{ss}$ value were inversely related to hormone concentration.

The silent receptor phenomenon raised the possibility that receptor and cyclase units might be spatially arranged in certain regions of the membrane, such that receptors respond to hormone by activating a multiple number of catalytic units, so that subsequent occupation of adjacent receptor units would have a diminished overall effect on enzyme activation. To critically test this idea, we considered structural models of molecular organization in membranes which were implemented using a digital computer. Receptor and cyclase units were represented as matrices of R and C units. We systematically introduced specific rules ("specifications") relating to binding, coupling,

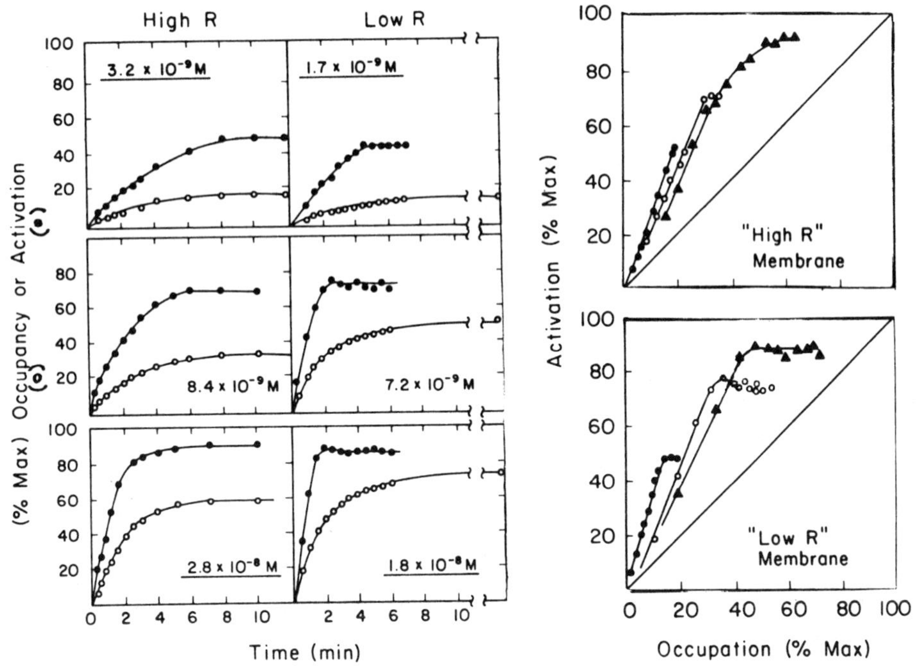

Figure 1. Dynamics of binding and activation of bovine renal medullary membranes. Left panels: ^{3}H-LVP added at 0 time at concentrations shown; "High R" and "Low R" are different membrane preparations. Right panels: Same data plotted as occupation vs. activation, where data is normalized to full occupation and maximal activation, and (•) represents lowest concentration, (o) is intermediate concentration, and (▲) is highest concentration for each preparation. From ref. 10.

and activation, thus generating a series of related models. A specific model with multiple recruitment and overlap features evolved (Model VI), which accounts for the variable dynamics, as well as the steady-state A:O relationships observed in bovine renal membranes[11]. Model VI was also able to account for the form of the nonlinear relationship between intrinsic adenylate cyclase activity and affinity of a large series of NHH analogs[11,12].

In the present report we wish to review the modeling approach which led to Model VI and to consider the following questions: (i) Which features of this model are fundamental to its dynamic behavior? (ii) Is this model an unique explanation for the dynamics of the bovine membrane data? (iii) What are the molecular implications of this model for hormone sensitive-adenylate cyclase systems in membranes?

MATRIX MODELING APPROACH

Receptor units are represented as "R" elements in a square matrix which is associated with a second matrix of "C" elements considered to represent cyclase units. R units are either unoccupied, or occupied by "hormone" (HR). C units are either inactive (C_i) or active (C_a). The interaction of hormone molecules with the receptor field is represented as random bombardment (collision) of the R matrix by "hormone" (H). Collision of H with R instantaneously gives HR ("effective collision"). Collision of H with HR is without consequence. We have demonstrated that simulated binding by random interaction with the R matrix is formally equivalent to the classical bimolecular association of $H + R \underset{kr}{\overset{kf}{\rightleftharpoons}} HR$[11]. Collision rate in the model is equivalent to hormone concentration in membranes; the constant of proportionality is related to kf. The mean lifetime of HR (occupation time, $\overline{T_o}$) is equal to 1/kr of membranes. For the case of LVP in bovine renal membranes, kr was about 0.2 min^{-1}; accordingly $\overline{T_o}$ for LVP is about 5 min (10 min for AVP).

To investigate specific mechanistic proposals for membrane coupling, it is necessary to establish a specific stoichiometric relationship between R and C units (R/C ratio) in the associated matrices and to implement these proposals in terms of a set of rules concerning the spatial and temporal relationships between binding and activation of cyclase units. For example, HR, once formed, may activate a single or multiple C units. Coupling may be instantaneous, or delayed; C_a units may be active for a time interval greater than, less than, or equal to the lifetime of the HR complex. In addition, R and/or C units may either be stationary, or allowed to migrate laterally from one locus to another within their respective matrices. Accordingly, by choosing appropriate specifications, models can be created which correspond to specific mechanistic coupling proposals. The characteristic dynamic behavior of each model may then be compared to data from NHH-sensitive bovine renal membranes.

DEVELOPMENT OF MODEL VI

To account for the silent receptor phenomenon and the nonlinear $A:O_{ss}$ relationship of bovine membranes, we assumed a structural arrangement of equal numbers of R and C units such that each C_i unit could be activated by more than a single HR complex ("overlap"). Thus, each binding event has the potential for activating multiple C_i units in the neighborhood ("interaction field") of the HR complex. Since the maximum slope of the ($A:O_{ss}$) curve of bovine membranes (as determined at lowest occupation) approached 5:1, we specified that each HR unit in the model was able to activate a maximum of 5 "nearest neighbor" C_i units (Figure 2). It was further specified that

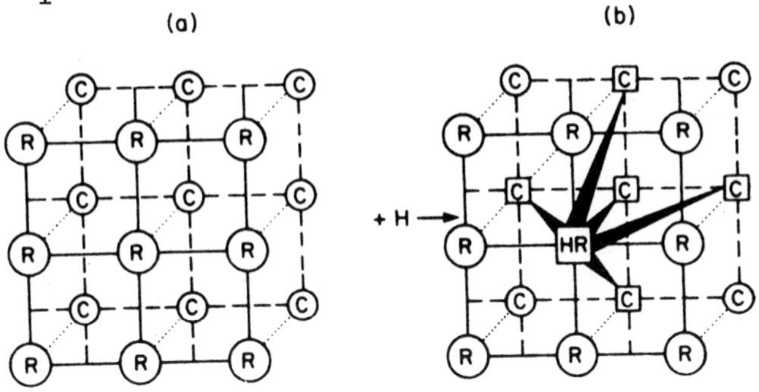

Figure 2a., Arrangement of units in Model VI. The R units represent receptors and C elements represent cyclase units in the membrane model. For real simulations a 20 X 20 matrix was used. b., Stoichiometry of activation. "Collision" of R by hormone (H) forms HR, which sequentially activates up to 5 nearest neighbor C_i units located in the "interaction field" of the HR complex. C_a units remain activated for a short period following dissociation of HR. From reference 11.

dissociation of HR caused deactivation of the C_a units originally activated by the HR complex. It was also necessary to specify the influence of a binding event at a specific R in the matrix ($R_{x,y}$) upon those C_a units in its interaction field. Such a previously activated C_a may revert to an inactive state during the lifetime of $HR_{(x,y)}$. One must therefore specify whether or not such previously activated C_a units are permanently "refractory" to a subsequent binding event in its "neighborhood". The alternative is that the period of activation of such a C_a unit is extended when it reverts to inactive state for the remaining lifetime of $HR_{x,y}$. It was only possible to obtain the correct form of the $A:O_{ss}$ curve in bovine membranes by including the specification that C_a units are refractory to subsequent binding events (Figure 3). Without this specification, the form of the $A:O_{ss}$ curve of the model was invariably different from the bovine membrane case, in that 100% activation was achieved with less than full occupation of R units. The specification that C_a units are refractory limits the time period during which HR can activate C_i units to a small fraction (<15%) of its total lifetime. The temporal dissociation of activation and binding characteristic of the silent receptor phenomenon was observed independent of whether or not one specified refractory C_a units.

To obtain the variable A:O dynamics observed in bovine membranes, we were compelled to introduce into the model a time delay between

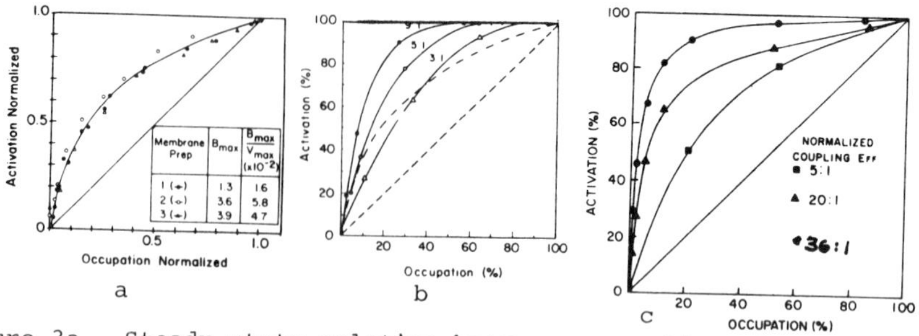

Figure 3a. Steady state relation between occupation of receptors and activation of cyclase ($A:O_{ss}$) in 3 different bovine renal membrane preparations. Bmax: maximal specific binding; Vmax: maximal cyclase activity. b. $A:O_{ss}$ curves predicted by matrix model (⟵) with C_a units not refractory (see text). Dashed curve is data from bovine membranes. Upper and lower solid curves represent increases or decreases in the maximal number of C_i units activated per R unit ("coupling efficiency"). c. Predicted curves from Model VI with refractory C units and variable coupling efficiency. Note that in 3c 100% activation is only achieved at full occupancy. From references 10 and 11.

binding and activation of individual C_i units. Coupling delays for the various units (T_d) varied according to an exponential distribution. By varying mean coupling delay ($\overline{T_d}$) from 3 to 12% of occupation time, it was possible to simulate the dynamics of bovine membranes (Figure 4). Without the coupling delay specification, it was not possible to simulate the variable hormone concentration-dependent trajectories of approach to the $A:O_{ss}$ relation observed in bovine membranes.

With these specifications formation of HR leads to activation of multiple C_i units ("multiple recruitment"), one at a time, where the number of C_i units activated ("coupling efficiency") is determined by the activation state of C units in the interaction field of HR. It is apparent that the quantitative relationship between the time that R is occupied by a given agonist, and the coupling delay, has important consequences vis a vis the ability of a given binding event with that agonist to recruit C_i units. As long as occupation time exceeds coupling delay by a sufficient amount, HR exists long enough to achieve the maximal possible recruitment of C_i units. However, as $\overline{T_o}$ becomes progressively shorter as with analogs of progressively decreasing affinity, and $\overline{T_d}$ remains constant, full occupation by such analogs may produce only partial, or possibly even zero activation. In the Model, a decrease in $\overline{T_o}$ of several orders of magnitude, (simulating a decreased affinity of ligand) was associated with a decreased magnitude of activation at full occupancy (analagous to intrinsic activity in membranes). Similarly, in bovine membranes, there was a tendency for

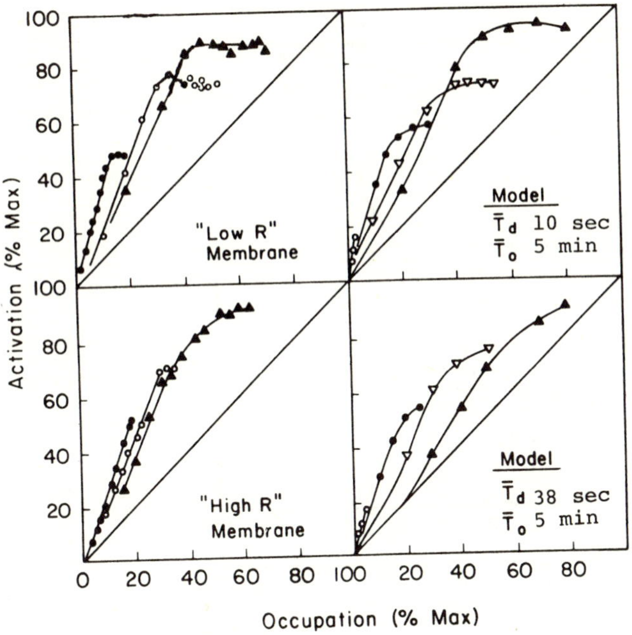

Figure 4. Ability of Model VI to simulate the A:O dynamics of bovine membranes. Left panels: real membrane data (from Figure 1); right panels: predicted trajectories of matrix model. "High R" dynamics and "low R" dynamics (with silent receptor characteristics) were both simulated by introducing a variable mean coupling delay ($\overline{T_d}$); other model parameters were identical. Modeling results obtained are shown for 4 collision rates: 22(○), 133(●), 400(▽), and 1200(▲) (arbitrary units). From reference 11.

intrinsic activity to fall progressively for analogs of steady decreasing affinity (Figure 5)[12]. In order to achieve the precise shape of the inverse relationship between peptide affinity and intrinsic activity observed in bovine membranes, it was necessary to introduce an additional specification into the model: namely, that C_a units remain active for a short period ($\overline{T_p}$; exponentially distributed) beyond the moment of dissociation of the H R complex.

Model VI was able to explain the following characteristics of coupling of LVP and analogs in bovine membranes: (i) the nonlinear steady-state relation between occupation and activation with LVP (and AVP), where at low occupancy the A:O ratios are 5:1 and become 1:1 as occupancy approaches 100%; (ii) variable hormone concentration-dependent trajectories of approach to the $A{:}O_{ss}$ relation, which may be either monophasic or biphasic; and (iii) the decrease in magnitude of

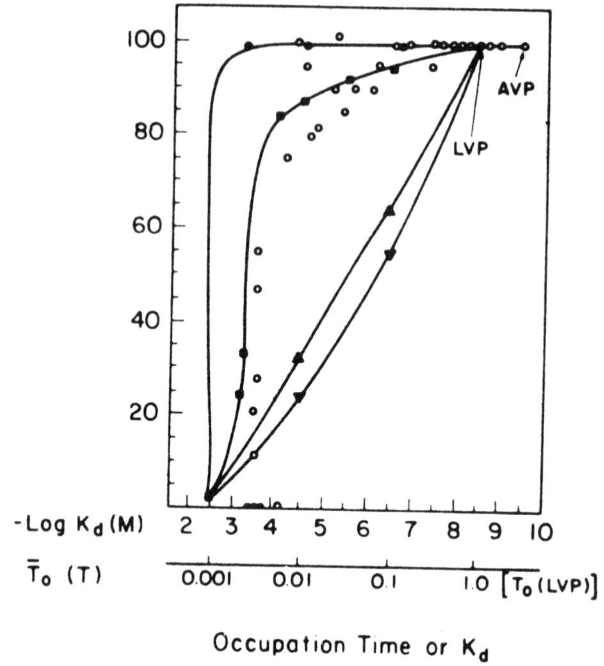

Figure 5. Influence of variable affinity on intrinsic cyclase activity in bovine membranes (○) and as predicted by Model VI (closed symbols). Open circles represent activity data from a large series of NHH analogs[12]. Decreasing affinity in membranes was assumed to be due to a proportional decrease in k_f and increase in k_r; thus, affinity was decreased in the Model by increasing occupation time ($\overline{T_o}$). To get the relationship observed in membranes, activation in the model was prolonged beyond HR dissociation by an interval ($\overline{T_p}$) equal to 8% of $\overline{T_o}$ (■). Other curves are for shorter (▼,▲) or longer (●) prolongation. From references 11 and 12.

adenylate cyclase stimulation (intrinsic activity) associated with decrease in affinity in NHH analogs.

UNIQUENESS OF MODEL VI

Although Model VI is able to stimulate data of bovine membranes, and therefore may represent a series of temporal events which must have their counterparts in any specific hypothesis of membrane coupling. The significance of this model is determined by the degree to which it is unique. To test the uniqueness of Model VI, we have compared its dynamic predictions to two alternative models (one involving heterogeneous R and C units, the other designated as the two step floating receptor model). These alternative models were implemented using the matrix technique. The versatility of the computer simulation method is illustrated by its ability to examine the explicit dynamic predictions of various alternative hypotheses of coupling.

Mobile Receptor Hypothesis. It is generally assumed that if the mobile receptor hypothesis applies to a given membrane system the Scatchard plot for hormone binding will be nonlinear. Since LVP (and

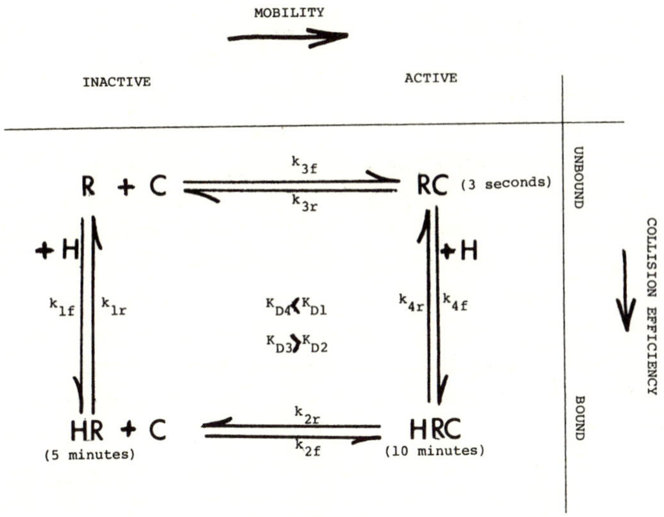

MOBILE RECEPTOR MODEL

Figure 6. Mobile Receptor Model. R and C units are located in a 26 X 26 matrix; R units migrate randomly among the loci. Collision of H with R or RC can yield HR or HRC respectively. Formation rates of RC and HRC are determined by the "mobility" of R and HR. All bound complexes are assigned average "lifetimes", the real time equivalents of which are shown in the Figure (to model $A{:}O_{ss}$ for bovine membranes).

AVP) binding in bovine membranes yield linear Scatchard plots [10], this might be taken as sufficient to exclude the floating receptor hypothesis. However, there are two possible ways of generating a nonlinear $A{:}O_{ss}$ relationship in the Mobile Receptor Model (Figure 6): (i) the binding of H to RC, or to R, may be assigned different affinities; or (ii) HR units can have the ability to "search" for C_i by random migration. Thus, when the R_T/C_T ratio is greater than 1, this second mechanism provides for a high efficiency of coupling on a normalized basis which yields nonlinearity without introducing differences in affinity for H between RC and R. Accordingly, we implemented two versions of the mobile receptor model: one where the affinities of R and RC for H are similar in magnitude (2.5 fold difference); second, where the RC has 100 times greater affinity for H than does R.

The model was simulated with 400 R units and 80 C units distributed randomly in a single matrix of 756 loci. R units migrate randomly within the matrix with a mobility determined by the average residency time ($\overline{T_r}$) at a given locus in the matrix; the C units remain stationary. The appearance of an R unit at a single locus with a C unit causes a short-lived formation of RC, which is responsible for the existence of "basal activity" in this model. The magnitude of "basal activity" is determined by the balance between the mobility of R units, and the lifetime of the RC unit.

Hormone stimulated activity was simulated by random bombardment of the matrix, which causes formation of either HR or HRC. The subsequent conversion of HR to HRC is governed by the mobility of HR, which, in the model, has the same mobility as R. HRC can also be formed by the appearance of HR and C units at a single locus. The HRC complex has a lifetime much greater than that of the RC complex.

The results of implementing the Mobile Receptor Model are shown in Figure 7. In both cases, with "near-homogeneous" or heterogeneous hormone binding, it was possible by selection of appropriate conditions to obtain an $A:O_{ss}$ relationship which was similar to that observed in bovine membranes. However, the trajectories to the $A:O_{ss}$ value in both types of models are very different from the trajectories observed in the membranes. For all doses, in either representation of the model, the trajectories are convex downward, whereas they are convex upwards in membranes. Accordingly, the floating receptor model, with or without heterogeneities in hormone binding, may be rejected on the basis that it cannot reproduce the A:O dynamics obtained from Model VI and observed in bovine membranes.

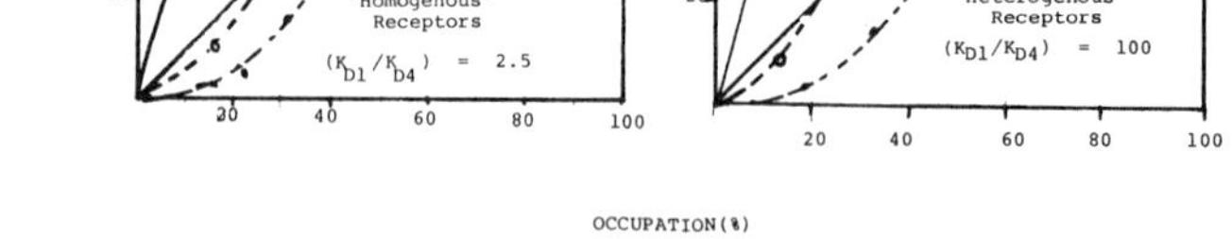

Figure 7. Dynamic and steady-state predictions of the Mobile Receptor Model. Mobility of units was adjusted to give $A:O_{ss}$ curves similar to those of bovine renal membranes. Left panel is near-homogenous case where affinity of R for H (K_{D1}) was only 2.5-fold greater than affinity of RC for H (K_{D4}, Figure 6). "Heterogeneous" receptor case is shown in right panel. $\overline{T_r}$ = average residency time that any mobile unit (R or HR) remains at a single locus, and is chosen to simulate the $A:O_{ss}$ curve of bovine membranes.

Heterogeneous R and C Units. The possibility exists that receptor units in bovine membranes are actually heterogeneous, but that such heterogeneity was not detected by our analytic procedures. In addition, cyclase units may be heterogeneous with respect to catalytic capability, there being two classes, one with high and the other with low catalytic capability. If (i) a small number of "ultra-high affinity" receptor sites were coupled to the cyclase elements of high catalytic capability, and (ii) a large number of receptor sites (detected by us) with "normal" affinity were coupled to cyclase with low catalytic activity, this could result in a nonlinear $A{:}O_{ss}$ relationship, similar to that of bovine membranes. The shape of the $A{:}O_{ss}$ curve would depend on the number of units, and difference of affinity between the normal and ultra-high affinity sites, and the ratio of catalytic activities of the two classes of cyclase elements. By choosing appropriate parameters, the $A{:}O_{ss}$ curve of bovine membranes can be simulated. This occurs, for example if one assigns 90% of the catalytic activity to 20% of the cyclase units, which are coupled directly to ultra-high affinity sites, which account for 20% of the total receptor population; the difference in affinity between the two classes of receptor sites was one order of magnitude (Figure 8).

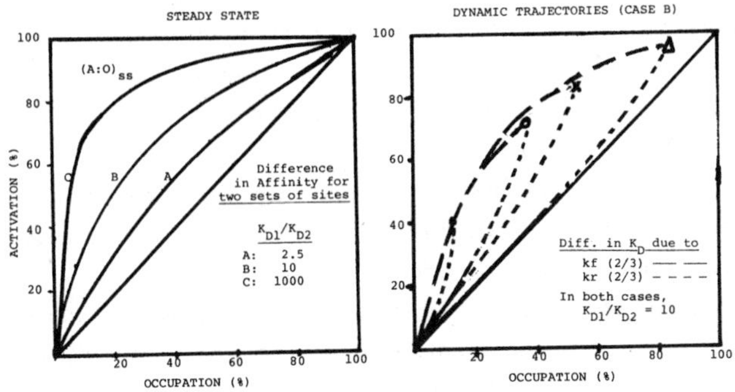

Figure 8. Heterogeneous R and C units. Steady-state and dynamic A:O relationships for a small number of "ultra" high affinity sites, coupled to cyclase elements of high catalytic capability. Left panel: $A{:}O_{ss}$ curves shown for two sets of receptors with 2.5 to 1000-fold differences in affinity. The $A{:}O_{ss}$ curve of bovine membranes is simulated with a 10-fold difference in affinity (Case B). Right panel: Trajectories of approach to $A{:}O_{ss}$ for Model with heterogeneous R and C units. The trajectories predicted differ depending on whether the affinity difference between high and ultra-high affinity receptors is due primarily to a difference in k_r or k_f. K_{D1} = high affinity, K_{D4} = "ultra" high affinity.

However, as was true for the mobile receptor model, it was not possible to simulate the biphasic silent receptor phenomenon observed in bovine membranes and in Model VI. This was true whether the affinity differences for the two classes of sites was induced by changing k_f and k_r equally or by causing unequal changes in the two parameters.

DISCUSSION

The present studies establish that the nonlinear $A:O_{ss}$ curve of bovine renal membranes can be explained in multiple ways, but that the dynamic A:O relationships are not easily accounted for. Thus, the nonlinear $A:O_{ss}$ relationship observed in membranes is predicted by the mobile receptor model, as well as Model VI, but the mobile receptor model fails to simulate the A:O dynamics of membranes. We have additionally shown that it is likewise not possible to simulate the membrane A:O dynamics by postulating heterogeneities of receptor and cyclase units. Thus, of the models tested, only Model VI can account for the variable dynamic patterns of different membrane preparations and particularly the silent receptor phenomenon, and to this extent is unique.

Model VI represents a theory of coupling receptor and cyclase units in the renal medullary membrane, in terms of a series of precisely defined temporal events. The model leaves open the precise molecular interpretation of events involved, but serves to provide a series of constraints, so that certain mechanistic proposals for coupling become unlikely, and others possible. In relating Model VI to the renal membrane, it is therefore necessary to clearly differentiate between those features of Model VI which are fundamental and unique and those characteristics which are arbitrary.

The fundamental features of Model VI requisite for simulation of all dynamic patterns of the bovine membrane data, which apply in principle to the molecular organization of elements in bovine renal membranes are the following: (i) Multiple recruitment with overlap, within a restricted interaction field for R units, which provides a normalized coupling efficiency near 5:1. The particular size, shape and stoichiometry of the matrices we used are arbitrary (not fundamental), and could be replaced by alternative arrangements and stoichiometries. (ii) Time-limited Coupling Effectiveness of HR (refractory C units) so that coupling can only occur for a small fraction (<15%) of the total lifetime of the HR complex. (iii) Coupling delay, where individual units are sequentially activated. (iv) Hysteresis of activated C units so that their lifetimes exceed the duration of occupancy of the "parent" HR complex.

Model VI has the following implications concerning the molecular organization of receptor and cyclase units and the process coupling these units in the bovine renal membrane: (i) Receptor and cyclase units are closely packed, but separate molecular entities, in regions of the membrane. (ii) Enzyme activation is achieved via sequential, independent interaction of the HR complex with nearest neighbor cyclase units, one at a time. (iii) The coupling effectiveness of the occupied receptor is restricted for a small fraction of the total time that the receptor is occupied.

The features of Model VI are consistent with two specific molecular mechanisms of coupling, the chemical nature of which are listed in Table 1. By one mechanism individual enzyme units interact in sequence with a relatively stationary occupied receptor, possessing multiple sites (about 5) for complexing with enzyme (Multivalent receptor formulation). By the alternative mechanism, a univalent hormone-receptor complex would interact briefly, in sequence, with enzyme units, averaging 4-6 per occupation, to "trigger" their individual activation. In the latter Univalent receptor formulation the lifetime of an activated cyclase unit would be essentially independent of the lifetime of the hormone-receptor complex. It is apparent that the latter interpretation of Model VI is a restricted version of the "collision-coupling" mechanism, recently described by Tolkovsky and Levitzki[13]. We are presently engaged in simulation studies to determine whether it is possible to differentiate between the multivalent and univalent receptor interpretations of multiple recruitment by further study of A:O dynamics of membranes.

ACKNOWLEDGEMENTS

This work was supported by the National Institutes of Health (Grant AM20443). We are grateful to Ms. Victoria Nelson for her expert help in preparing the manuscript for publication. M. Greenberg is a Trainee of the National Institutes of Health (Grant GM-   ).

TABLE 1

POSSIBLE MOLECULAR FORMULATIONS FROM MODEL VI

	MULTIVALENT FORMULATION	UNIVALENT FORMULATION
BINDING	$H + R \rightleftharpoons HR$	$H + R \rightleftharpoons HR$
ACTIVATION	$HR + C_i \rightarrow HR(C_a)$ $HR(C_a) + C_i \rightarrow HR(C_a)_2$ ⋮ $HR(C_a)_{n-1} + C_i \rightarrow HR(C_a)_n$	$HR + C_i^{1*} \rightarrow HRC_i^1 \rightarrow HR + C_a^1$ $HR + C_i^2 \rightarrow HRC_i^2 \rightarrow HR + C_a^2$ ⋮ $HR + C_i^n \rightarrow HRC_i^n \rightarrow HR + C_a^n$
DEACTIVATION	$HR(C_a)_n \rightarrow H + R + n(C_a)$ $n(C_a) \rightarrow n(C_i)$	$C_a^i \rightarrow C_i^1$ ⋮ $C_a^n \rightarrow C_i^n$

* Superscript refers to the particular C_i units bound by HR in sequence. Other symbols defined as in the text.

REFERENCES

1. Jard, S. and Bockaert, J. (1975) Physiological Revs. 55, 489-539.
2. Boeynaems, J.M. and Dumont, J.E. (1975) J. Cyclic Nucleotide Res. 3, 123-142.
3. Levitzki, A. (1974) J. Theor. Biol. 44, 367-373.
4. Rodbell, M. and Londos, C. (1976) Metabolism 25, 1347-1349.
5. Bockaert, J., et al. (1973) J. Biol. Chem. 248, 5922-5931.
6. Jard, S. et al. (1975) Adv. Cyclic Nucleotide Res. 5, 31-52.
7. Roy, C. et al. (1975) J. Biol. Chem. 250, 3149-3156.
8. Roy, C. et al. (1975) J. Biol. Chem. 250, 3157-3168.
9. Roy, C. et al. (1975) J. Biol. Chem. 250, 7885-7893.
10. Hechter, O. et al. (1978) J. Biol. Chem. 253, 3219-3229.
11. Bergman, R.N. and O. Hechter. (1978) J. Biol. Chem. 253, 3238-3250.
12. Hechter, O. et al. (1978) J. Biol. Chem. 253, 3230-3237.
13. Tolkovsky, A. and A. Levitzki. (1978) Biochemistry 17, 3811-3817.

DISCUSSION

DR. RODBARD: Your approach to modeling of hormone-receptor binding and activation of adenylate cyclase appears to be very promising and powerful. The "probabilistic" approach avoids the need to write a large family of differential equations and to solve these analytically or by numerical methods. Thus, it provides a practical and economical method for modelling these complex systems. Your model provides one plausible mechanism which may determine the efficacy of structural analogues of the native hormone. However, there are several other mechanisms which may account for the presence of full-agonists, partial-agonists, and mixed agonists-antagonists. One of these alternative models is shown in Figure 1. This is the "multi-subsite" model of a receptor, developed in my laboratory in collaboration with Dr. Andre DeLean and Peter Munson (1). We hypothesize that the "activity" is generated only by pool P_3, i.e. when two (or more) functional moieties of the ligand have bound to separate "subsites" on the receptor area. This can also account for the observed

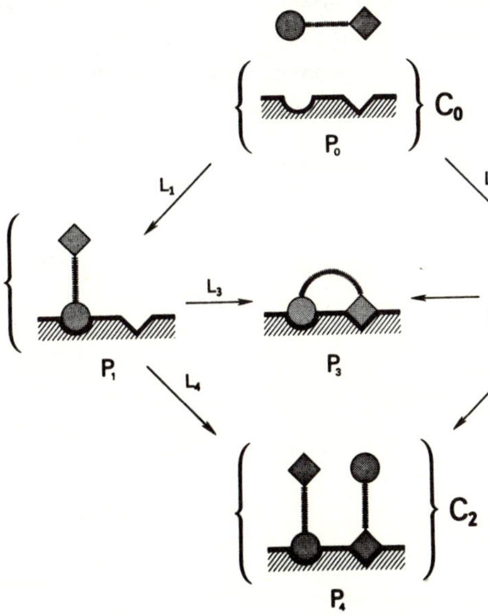

1. DeLean A., Munson, P.J., and Rodbard, D., Molecular Pharmacology, in press, 1979.

discrepancies between binding (which includes pools P_1, P_2, P_3, and two times P_4) and activity. The relationship between binding and activation is controlled by all of the microscopic equilibrium constants (L_1 - L_4). Thus, despite the excellent correlation which you observe in Fig. 5, other models must still be considered.

DR. BERGMAN: I agree that alternative models will have to be considered with respect to their ability to account for the spare receptor, silent receptor, and partial agonist phenomena. It appears reasonable that the Model you have proposed will, in principle, account for partial agonists, and may be consistent with a dynamic dissociation between binding and activation. However, it is not possible at this time to say whether the particular set of variable dynamics of membranes can be explained by the "multi-subsite" model without a careful simulation study. I will remind you that other models (e.g., the mobile-receptor model) were rejected because they were unable to account for the precise A:O dynamics. Such models appeared correct in principle (i.e., intuitively), but failed when subjected to careful, dynamic analysis. Certainly such an analysis is required to analyze the predictions of your Model and compare them with the Matrix-multiple recruitment Model we have proposed.

ELUCIDATION OF THE MODE OF COUPLING BETWEEN HORMONE RECEPTORS AND ADENYLATE CYCLASE BY MODULATION OF MEMBRANE FLUIDITY

EMANUEL HANSKI, GILAD RIMON, SERGEI BRAUN AND ALEXANDER LEVITZKI
Department of Biological Chemistry, The Hebrew University of Jerusalem

The hormone and neurotransmitter dependent adenylate cyclases represent transmembrane regulatory systems in which the receptor unit faces the outside of the cell and the cyclase catalytic unit faces the inside of the cell. Upon binding of the hormone or the neurotransmitter to the receptor, the catalytic unit of the cyclase becomes activated and produces cAMP at the inner surface of the membrane. This sequence of events strongly suggests that the neurotransmitter or the hormone promotes a structural transition in the receptor moiety which propagates to the catalytic moiety through the interaction domain between these two units. This interaction between the receptor regulatory unit and the catalytic cyclase unit represents either of two situations: (a) The receptor and the enzyme are mobile and float in the membrane[1-3]. In this case the activation of the catalytic moiety takes place during the encounter between the receptor and the enzyme. (b) The receptor and the enzyme are permanently coupled to each other, as a typical soluble regulatory enzyme composed of regulatory and catalytic subunits such as aspartate transcarbanylase. In this case the subunit interactions are similar in nature to those found in classic regulatory proteins in solutions. Mechanism (a) is described in the following scheme:

$$\begin{array}{l} HR + E \xrightarrow{k_1} (HRE) \xrightarrow{k_2} HR + E' \\ \quad \Big\updownarrow K_H \\ H + R \end{array} \qquad (1)$$

where H is the hormone, E the enzyme, R the receptor k_1 the bimolecular rate constant, E' the activated enzyme, K_H the receptor-hormone dissociation constant, and k_2 the first order activation constant. (HRE) indicates the ternary complex, hormone-receptor-enzyme. If the ternary complex is short lived and does not accumulate ($k_2 >> k_1$), the bimolecular encounter between the hormone bound receptor and the catalytic unit becomes rate limiting. Such a mechanism of coupling between the receptor and the enzyme has been termed as "collision coupling"[4,5].

The precoupled mode of activation (mechanism b) can be described by the following scheme:

$$\begin{array}{l} HRE \xrightarrow{k} HRE' \\ \quad \Big\uparrow\!\!\Big\downarrow K_H \\ H + RE \end{array} \qquad (2)$$

where RE is the receptor-enzyme complex, H the hormone, k, the first order constant of activation, and K_H the dissociation constant of the hormone to the receptor-enzyme complex.

In the two schemes described above we have eliminated the deactivation process of the active enzyme species back to its inactive state. This is achieved in hormone dependent adenylate cyclases when non-hydrolizable GTP analogs such as guanylylimidodiphosphate (GppNHp) replace GTP in the adenylate cyclase reaction (6,7). Therefore, in the presence of hormone and GppNHp, the enzyme becomes permanently active and accumulates according to[4,5]:

$$B = A\ 1 - \exp(-k_{obs}t) \qquad (3)$$

In the "collision coupling" model (eq. 1), A equals the total concentration of the catalytic units and B the concentration of the units achieved at time t. k_{obs} is the pseudo first order rate constant which is given by:

$$k_{obs} = k_1[R_T]\frac{[H]}{K_H + [H]} \qquad (4)$$

where $[R_T]$ is the total receptor concentration, k_1 is the biomolecular rate constant governing the rate of enzyme activation, and the term $\frac{[H]}{K_H + [H]}$ expresses the fraction of hormone occupied receptors.

The maximal number of the activated catalytic units in that case is independent of the receptor concentration, since every receptor can activate an unlimited number of catalytic units through the collision process. It is apparent from equation 4 that k_{obs} is directly proportional to the receptor concentration.

For the precoupled model (equation 2), A, in equation 3 stands for the total concentration of receptor-enzyme and B represents the concentration of the activated receptor-enzyme unit at time t. k_{obs} in this case is given only by:

$$k_{obs} = \frac{k[H]}{K_H + [H]} \qquad (5)$$

where k is the first order rate constant of activation. The term $\frac{[H]}{K_H + [H]}$ is the fractional occupancy of the receptor of the hormone.

Recently we have shown[4,5] that the coupling of turkey erythrocyte adenylate cyclase to the β-adrenoreceptor is of the type of "collision coupling", whereas the adenosine receptor is permanently coupled to the enzyme in the same cell[8,9]. These findings were based on the observation that the progressive destruction of the β-adrenergic receptor by a specific affinity label results in a proportional decrease in the rate constant of cyclase activation by the receptor, without reduction in the total number of catalytic units which can be activated. In contrast, the progressive destruction of the adenosine receptor by a specific affinity label results in a proportional decrease in the number of catalytic units which can be activated without a change in the rate constant of activation[9]. Since the floating mechanism, in contrast to the precoupled mechanism, necessitates a relative movement of the receptor with respect to the catalytic units in the membrane matrix, a different dependence of the enzyme activation by the hormone on membrane fluidity is expected for epinephrine as compared to adenosine.

To test this hypothesis we have modified the fluidity of turkey erythrocyte membranes and measured the kinetics of enzyme activation by epinephrine and by adenosine. Turkey erythrocyte membranes were fluidized progressively by the insertion of cis-vaccenic acid which has been shown to be one of the most efficient fluidizing agents[10]. The apparent membrane microviscosity was measured by the fluorescence polarization method of Shinitzky et al.[11], using 1.6 diphenyl (1,3,5)hexatriene (DPH) as the fluorescence probe (Figure 1). In parallel, we have measured the time course of adenylate cyclase activation to its permanently active state, by either l-epinephrine (Figure 2) or adenosine (Figure 3), in the presence of saturating concentrations of guanylyl-imido-diphosphate (GppNHp). The first order rate constants for the first order process of adenylate cyclase activation by the two hormones were derived from equation 3, according to the procedure described elsewhere[4,5].

In Figures 2 and 3 a plot of k_{obs} versus membrane fluidity (viscosity^{-1}, 1/n) is shown for both the epinephrine induced activation process (Figure 2) and the adenosine induced activation process (Figure 3). It is apparent that the rate constant of the epinephrine dependent activation increases linearly with membrane fluidity, whereas the rate constant of the adenosine dependent activation is independent of membrane fluidity.

These findings are in complete agreement with the kinetic experiments described above and which directly revealed the two different modes of enzyme activation by the two receptors. Furthermore, the linear dependence of k_{obs} on

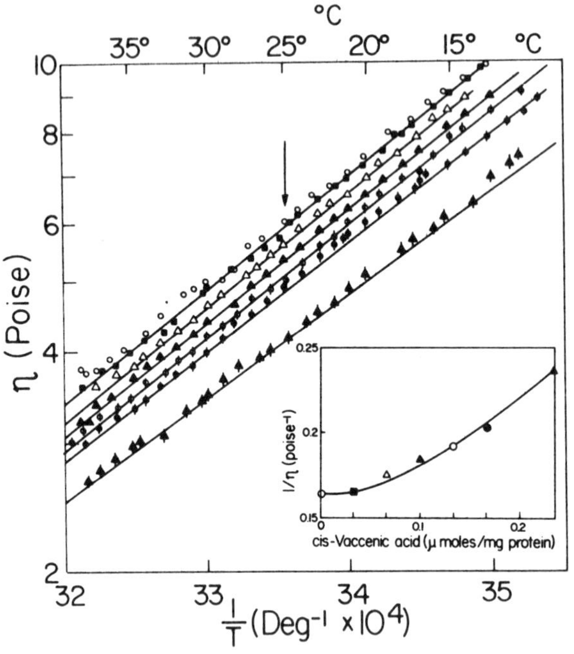

Fig. 1. The fluidization of turkey erythrocyte membranes with cis-vaccenic acid. Turkey erythrocyte membranes were fluidized using cis-vaccenic acid, as described earlier (10,22). The fluidity of the membranes was measured by the fluorescence polarization technique, as described in the text, as a function of temperature. The arrow indicates the temperature at which cyclase measurements were performed.

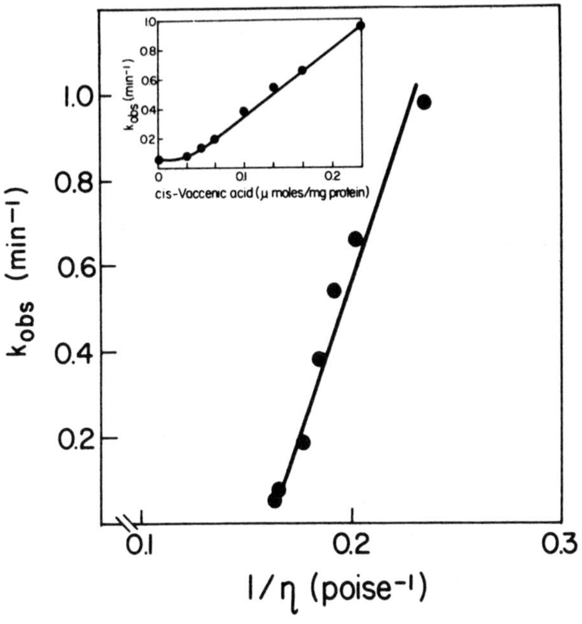

Fig. 2. The dependence of l-epinephrine stimulated adenylate cyclase activity on membrane fluidity. The time course of accumulation of the activated state of adenylate cyclase was measured, as described elsewhere (22,23) according to the integrated form of equation (3). Increasing the membrane fluidity by cis-vaccenic acid results in an increase in the apparent first-order rate constant, k_{obs}. Inset: the increase in k_{obs} as a function of the amount of cis-vaccenic acid incorporated.

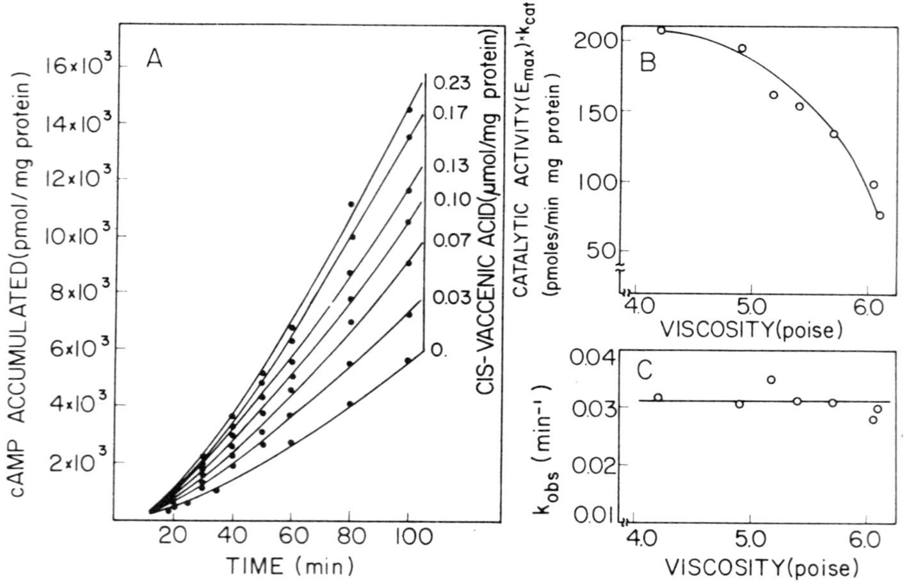

Fig. 3. The dependence of adenosine stimulated adenylate cyclase activity on membrane fluidity.

1/n in the case of the epinephrine dependent activation, strongly suggests that the interaction between the receptor and the enzyme is in fact diffusion controlled. Using the relationship between the bimolecular rate constant k_1 and the viscosity of the medium in a diffusion controlled reaction which occurs in two dimensional space[12], one can calculate that the diffusion coefficient describing the relative motion of the receptor and the enzyme is in the range of 4.0×10^{-11} to 9×10^{-10} cm^2/sec. The value of k_1 can be easily obtained from k_{obs} (equation 4) since $[R_T]$ is known for turkey erythrocytes[13,14]. This range of values is typical for mobile membrane receptors[15,16]. The independence of the adenosine induced rate of enzyme activation on membrane fluidity is in complete agreement with the biochemical studies[8,9] which revealed that this receptor is permanently coupled to the cyclase.

The membrane fluidity also affects the catalytic unit directly, as the maximal activity of the enzyme increases as a function of membrane fluidity. From Figure 2 it is apparent that the adenosine dependent activity, the epinephrine dependent activity, and the NaF stimulated activity increase as a function of membrane fluidity in an identical manner. Similar effects of membrane viscosity on the adenylate cyclase activity were recently reported[17]. Furthermore, the specific activity of other membrane bound systems, such as Ca^{++}-ATPase[18], (Na^+,K^+)ATPase[19] and the β-galactoside transport system[20], were also found to depend on the fluidity of the membrane.

The dependence of the rate of adenylate cyclase activation by a hormone receptor, in the presence of GppNHp, on membrane fluidity can be adopted as a general diagnostic test for probing the mode of coupling between a hormone receptor and adenylate cyclase. Such a procedure can be especially useful when the biochemical approach using an affinity label is cumbersome, as in the base of polypeptide hormones which activate adenylate cyclase. This approach can be extended to other membrane processes which involve the interaction between a receptor and a catalytic moiety, other than cyclase, such as an ionophore.

REFERENCES

1. Boeynames, J.M. and Dumont, J.E. (1975) J. Cycl. Nucl. Res. 1, 123-142.
2. Jacobs, S. and Cuatrecasas, P. (1976) Biochim. Biophys. Acta. 433, 482-495.
3. Levitzki, A. (1978) Biochem. Pharmacol., in press.
4. Tolkovsky, A.M. and Levitzki, A. (197) in Hormones and Cell Regulation, Dumont, J. and Nunez, J. eds., Elsevier/North Holland, pp. 89-105, Vol. 2.
5. Tolkovsky, A.M. and Levitzki, A. (1978) Biochemistry 17, 3795-3810.
6. Sevilla, N., Steer, M.L. and Levitzki, A. (1976) Biochemistry 15, 3493-3499.
7. Levitzki, A. (1977) Biochim. Biophys. Res. Commun. 74, 1154-1159.
8, Tolkovsky, A.M. and Levitzki, A. (1978) Biochemistry 17, 3811-3817.
9. Braun, S. and Levitzki, A. (1978) Biochemistry, in press.
10. Orly, J. and Schramm, M. (1975) Proc. Natl. Acad. Sci. USA 72, 3433-3437.
11. Shinitzky, M. and Inbar, M. (1976) Biochim. Biophys. Acta 433, 133-149.
12. Hardt, S. and Cone, R.A. (1978) Biophys. J., in press.
13. Levitzki, A., Atlas, D. and Steer, M.L. (1974) Proc. Natl. Acad. Sci. USA 71, 2773-2776.
14. Hanski, E. and Levitzki, A. (1978) Life Sci. 22, 53-60.
15. Poo, M.M. and Cone, R.A. (1974) Nature 247, 438-440.
16. Axelrod, D., Ravdin, P., Kopel, D.E. Schlessinger, J., Webb, W.W., Elson, E.L. and Podelski, T.R. (1976) Proc. Natl. Acad. Sci. USA 73, 4594-4598.
17. Klein, I., Moore, L. and Pastan, I. (1978) Biochim. Biophys. Acta 506, 42-53.
18. Warren, G.B., Toon, P.A., Birdsall, N.J.M., Lee, A.G. and Metcalf, J.C. (1974) Biochemistry 13, 5501-5507.
19. Kimelberg, H.K. (1975) Biochim. Biophys. Acta 413, 143-156.
20. Thilo, L., Trauble, M. and Overath, P. (1977) Biochemistry 16, 1283-1289.
21. Steer, M.L. and Levitzki, A. (1975) J. Biol. Chem. 250, 2080-2084.
22. Rimon, G., Hanski, E., Braun, S. and Levitzki, A. (1978) Nature 276, 394-396.
23. Hanski, E., Rimon, G. and Levitzki, A., Biochemistry, in press.

DISCUSSION

DR. DEMBO: I would like to know: If the adenosine receptor is precoupled, how can it, according to reference (8) be using the same pool of cyclase as the β-adrenoreceptor.

DR. LEVITZKY: This is not very easy to answer. What is quite clear is that the data are really showing that the two receptors are using the same pool. I did not imply that the adenosine receptor is physically coupled in a way that it is associated with the enzyme all the time. You must remember that there might also be some verticle movement in the membrane and that these associations are dynamic.

The point is that when the cyclase system is loaded with adenosine, for example, and it happens to collide with a receptor occupied by a beta agonist, it will not be recognized by this receptor. The effect of the two hormones is not additive.

What I have in mind is an analogy between the receptor cyclase system and interactions between antigens and immunoglobulins. Cyclase must have a common domain for many receptors because we know from the fusion experiments of Orly and Schramm and other experiments that the same cyclase pool can be recognized by different receptors. So you must assume that different receptors can recognize the same cyclase. I think most of the people agree with that.

That means that the different receptors will have a common denominator, a constant region that should be able to interact, for example, with a guaninucleotide binding protein, and a variable region which may be the recognition site for the hormone or the neurotransmitter.

This constant region has to recognize a certain region in the guaninucleotide regulatory protein, and that region can accommodate only one receptor at a time. The exact physical mechanism or the molecular mechanism by which it happens, is not known but I don't think it is going to be very difficult to sort out once we know more about the structure and the exact stoichiometry between the three types of components in the system.

DR. SINGER: What I wanted to ask is connected with that. If you do the Orly and Schramm experiment (reference 10) with the adenosine receptor, do you get reactivation by fusion?

DR. LEVITZKI: I suggested this experiment to Dr. Schramm about a year ago and he said that he might do it. I am not involved myself in fusion experiments in Jerusalem, but I hope it will be done.

DR. SINGER: One comment that may be of interest. It is very interesting, I think, that the human erythrocytes have very little apparent mobility of their surface components. For instance con-A induces very little redistribution of surface receptors. But in avian erythrocytes, there is a lot of mobility in the membrane, and with ferritin labeled con-A clustering and endocytosis is quite induced.

DR. LEVITSKI: I just want to say that from the dependence of the rate constant on fluidity, we can calculate the tentative diffusion coefficient which describes the relative mobility of the receptor. It is about $4x10^{-11}$ cm^2/sec. The viscosity of the turkey erythrocyte membrane is about seven-fold higher than most of the membranes of other cell types. Indeed the diffusion coefficients of membrane proteins in the membranes of many cell types is in the range of $4x10^{-10}$ (cm^2/sec) measured by the photobleaching technique (see Schlessinger, this volume). Therefore, I think there is a correlation between these two types of studies.

DR. SINGER: Of course, (human erythrocytes) generally don't show these beta-adrenergic responses.

DR. LEVITZKI: According to the literature human reticulocytes do. Apparently, all erythrocytes which are nucleated, have beta-adrenergic receptors, and human erythrocytes are nucleated in the early stage of development. Also, rat and rabbit erythrocytes possess β-adrenergic receptors.

DR. SINGER: Well, that wouldn't be a reticulocyte. A reticulocyte is already nucleated.

DR. LEVITZKY: In the rats, for example, it is quite well-known that rat reticulocytes have very good beta-adrenergic dependent adenylate cyclase. This has been shown.

DR. SINGER: It has been in the literature but we haven't been able to reproduce that ourselves.

DR. LEVITZKY: This has been shown by Dieter Palm in Frankfurt and has been repeated by Mickey Schramm and his colleagues in Jerusalem, so I think the data are good.

DR. ROSENBURY: If the adenosine receptors are in stoichiometric excess of the cyclase sites to which they bind and activate, and if the adenosine receptors compete with the beta adrenergic receptors for that site, then is it possible that the beta adrenergic receptors still do form a precoupled complex but simply do not show that in the kinetics because they are competitive with the adenosine receptors?

DR. LEVITZKI: There are a number of experiments which suggest that what you are trying to say is not the case. There is no assay for the adenosine receptors yet, but we have by an indirect argument calculated that the number of adenosine receptors in the turkey system is about 70 percent of number the beta adrenergic receptors.

When there is an excess of one kind of receptor, our considerations of the model will be changed. Other observations favor the situation the adenosine receptor is precoupled and the beta receptors is not. Since the adenosine is a much less efficient agonist than the epinephrine, we predict that if an adenylate cyclase assay is done in the presence of GTP (no permanently activation) and in the presence of a given concentration of epinephrine, the activity will go down when adenosine is added it will decrease in a typical fashion because the beta receptor, once it collides with the cyclase that has adenosine, cannot recognize the cyclase-adenosine receptor system. The cyclase activity will then be governed by the adenosine mode of activation, which has a low efficacy. This is indeed the case. (Braun and Levitzki unpublished observations). Other experiments also reveal the competitive relationship between the two receptors and these are now published in Biochemistry Volume 17. The essence of their experiments involves the measurement of the time course of adenylate cyclase activation to the permanently active state (in the presence of GppNHp) as a function of epinephrine and adenosine separately and combined. The data can be fitted only by a model which assumes that the β-receptor is uncoupled to the enzyme, and activates it by "collision coupling", and that the adenosine receptor on the other hand is permanently coupled to the pool of enzyme molecules. All other models fail to account for the observed kinetic patterns. I have no time-now to go into details.

DR. RODBELL: Alex, I certainly enjoyed your talk. Obviously we are in some disagreement because we believe that at steady state, at least, the glucagon receptor is linked to the active form of the enzyme, that is, by our sizing determination.

But aside from that point, you model number 1, collision coupling, really suggests that you don't need the hormone at all. That is, the receptor can actually do everything. It just has to collide.

DR. LEVITZKI: That is not all.

DR. RODBELL: Why? Because the hormone seems not to have to do anything. It just occupies.

DR. LEVITZKI: The receptor will not do anything to the enzyme unless it is in the right conformation and that only happens when it is bound with an agonist.

DR. RODBELL: But the agonist doesn't do anything. It just binds.

DR. LEVITZKI: Of course it does, because the receptor is not in the same conformation, I don't like to use the word "conformation." It is overused and obused. But we must assume that the structure of the receptor is different when it has a agonist or an antagonist or when it is unoccupied.

DR. RODBELL: You have no data to support that, that there is any change. In fact, your data suggests there is no change.

DR. LEVITZKI: No, that is not true, because, we can demonstrate that the steady state level of active enzyme is accounted for by k_{off} over k_{on}. We can also show by performing a k_{on} analysis in the presence of GppNMp as a function of different partial agonists, (we have done it with 11 partial agonists, unpublished) that k_{on} is higher for the better agonist. This, therefore, means that the efficiency of collision depends on the "goodness" of the agonist. So k_{on} can be measured independently, koff can be measured independently, and the steady state level can be measured independently and it is fully accounted for by these rate constants.

What this means, then, is that the efficiency of the encounter between the receptor and the enzyme depends on what agonist is bound to the receptor. This is not surprising because, after all, subunit interactions are different depending on the ligand which is bound to the subunit.

DR. RODBELL: But why do you need occupation to get it to go? If all they have to do is collide.

DR. LEVITZKI: The collision is not efficient as long as the receptor is not in the right conformation. This is elementary physical chemistry.

DR. BLUMENTHAL: You have to bring in conformation,

DR. RODBELL: Yes, you are going to have to imply that there is some structural change in the receptor.

DR. LEVITZKI: Of course there is a structural change. There is an encounter time in which the receptor and the enzyme are spending some time together.

DR. RODBELL: If you say that, then your experiments with cis-vaccenic or any other modulator of the membrane structure could equally be affecting the structure of the receptor as well as the membrane. So you need not invoke necessarily that microviscosity. In fact you may be modulating the conformation of the receptor that is necessary, when occupied by the hormone to give you the collision that is required.

DR. LEVITZKI: The conformation of the receptor doesn't change with cis-vaccenic acid, at least as far as the recognition of the ligands are involved.

DR. RODBELL: But that doesn't say anything because you are obviously not measuring any kind of change in the receptor. You are only measuring binding. I would agree with you if you could measure the change in the conformation and get any change. Otherwise, as far as I am concerned, the collision model says that there is no necessity for the hormone. If you just wait enough for the turkey to live for one year till Thanksgiving, interact as the receptor will interact with the cyclase without the hormone.

DR. ROTH: And there is no place more appropriate than Maryland for turkey erythrocytes.

DR. BLUMENTHAL: I think the Dr. Rodbell has taken up already part of my question namely, I want to react to your experiments on changing microviscosity.

Schlessinger and his co-workers* have shown that the movement of membrane proteins is not very much determined by the microviscosity of the lipid but rather, in most cases, under cytoskeletal control.

DR. LEVITZKY: I must say it is the first time that I heard that.

DR. BLUMENTHAL: You should be talking more with each other.

Secondly, I think you have already answered that question, but I want to bring up what Dr. Shinitzky has said, that by changing microviscosity you can change the affinity to the ligand or conformation of the receptor itself.

DR. LEVITZKI: The affinity of the receptor towards agonists and antagonists is unchanged under these conditions. I showed it in two slides. We also have data for other partial agonists. This is clear.

What we can also measure, the effect of fluidity on the properties of the enzyme itself. For example, the properties of the enzyme such as the affinity to ATP, the affinity to GTP, the affinity to magnesium, as a function of fluidity. They are all unchanged.

If you are talking about something that we cannot measure, of course I can't argue with you. But whatever we can measure is unchanged.

DR. HECHTER: This is an elegant treatment of a single system in great detail, using data to differentiate between models. The question I would put to you is: Don't you feel ultimately that we need a model which explains not only the detailed mechanism of your system, but with minimal modifications applies to the field in a general way? And what would be the nature of the modification that you would have to make to do that kind of thing?

DR. LEVITZKI: I think that there are three types of experiments that are

---

*J. Schlessinger et. al. 1976. Proc. Natl. Acad. Sci 73, 2409.

easy to do in every system, that is: the kinetic analysis of the activation of the system in the presence of non-hydrolyzable analogs of GTP. In every case that we have looked at, it reveals first order kinetics. So you are already confined with a very restricted number of models.

The second type of experiment is to design irreversible affinity labels in order to dissect the dependence of the kinetic parameters on the concentration of receptors. If you can design an affinity label for an enzyme, for instance cyclase -(is has not yet been done -) you can also manipulate the other side of the system, the same it also true for the GTP binding proteins. These tools can be very effective.

The third tool which could be very helpful is to change "micro viscosity." I still maintain that the change in the "micro viscosity" of the membrane can be very instructive and teach us a lot. Now we are even extending this approach to studies, analyzog the temperature dependence of the system, because we know much more about the system. I believe that these approaches are equally valid to every receptor system involving cyclase.

ROLE OF MULTIVALENT BINDING IN THE ACTION OF CHOLERA TOXIN

PETER H. FISHMAN
Developmental and Metabolic Neurology Branch, National Institute of Neurological and Communicative Disorders and Stroke, The National Institutes of Health, Bethesda, Maryland 20014

INTRODUCTION

Choleragen (cholera toxin) is composed of two components, A and B (Fig. 1). The B component consists of 5 identical polypeptide chains and binds to specific receptors on the cell surface; the A component contains two dissimilar peptides linked by a disulfide bond and activates adenylate cyclase[1]. The

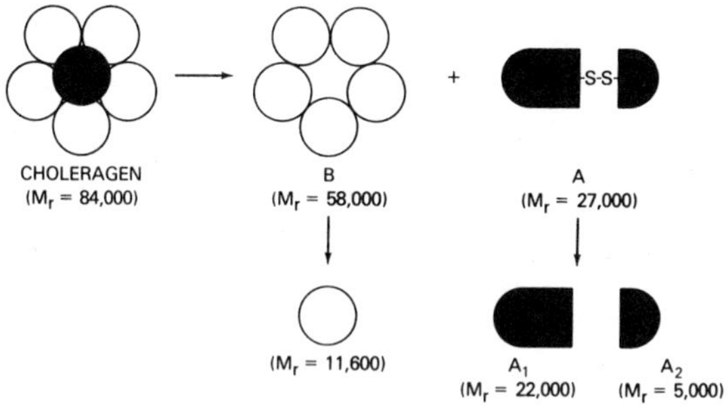

Fig. 1. Subunit structure of choleragen.

receptor for choleragen has been identified as the monosialoganglioside GM1 (Fig. 2)[2]. The oligosaccharide portion of GM1 binds to the toxin and, using a variety of experimental techniques, it was found that each toxin molecule bound 5 oligosaccharides[3]. Thus, choleragen is multivalent and has 5 binding sites, presumably one on each of the peptides of the B component. In addition,

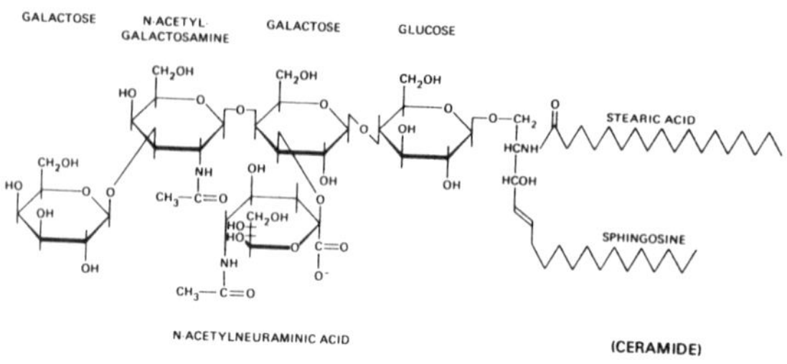

Fig. 2. Structure of the monosialoganglioside GM1

choleragen undergoes patching and capping on lymphocytes[4-6], indicating that each toxin molecule can bind to more than one receptor on the cell surface. In order to assess the role of multivalent binding in the mechanism of action of choleragen, I undertook the following experiments in collaboration with Dr. E. Esat Atikkan, Laboratory of Molecular Biology, NINCDS*. The basic approach was to incubate cells with choleragen at low temperatures where lateral mobility of membrane components would be impaired. Then the oligosaccharide portion of GM1 was added to occupy any empty binding sites on the toxin. Finally, the cells were warmed up to 37°C and the amount of adenylate cyclase activation was determined.

MATERIALS AND METHODS

GM1 and GM1-oligosaccharide were prepared as described previously[3]. ^{125}I-choleragen was prepared by the chloromine-T procedure and binding was determined by filtration on Millipore filters[7]. HeLa, rat C6 glial, and mouse N18 neuroblastoma cells were grown in monolayer culture and harvested by scraping. Friend erythroleukemic cells were grown in suspension and lymphocytes were iso-

*Peter H. Fishman and E. Esat Atikkan, manuscript submitted.

lated from human peripheral blood. All cells were washed several times with phosphate-buffered saline and suspended in minimal essential medium buffered with HEPES, pH 7.4, and containing 0.01% bovine serum albumin. The basic protocol for activation of adenylate cyclase was as follows: 1 ml portions of the cell suspensions were incubated with and without choleragen for 5 min at 4°C; then GM1-oligosaccharide was added to some samples and all were incubated at 37° C for at least 1 h. The samples were then diluted with ice-cold saline and centrifuged. The cell pellet was disrupted and assayed for adenylate cyclase[8].

RESULTS AND DISCUSSION

When HeLa cells were incubated at 4°C with choleragen for 5 min and then, after adding GM1-oligosaccharide (20 μM), at 37°C for 1 h, activation of adenylate cyclase was completely blocked. Inhibition by the oligosaccharide was still effecttive even after 3 h at 37°C or if the cells were incubated for 1 h at 4°C with choleragen prior to adding the oligosaccharide. Inhibition was also observed with rat C6 glial cells and human lymphocytes but not with Friend erythroleukemic or N18 neuroblastoma cells (Table 1). Binding studies indicated that the apparent dissociation constants for ^{125}I-choleragen were similar for

TABLE 1

EFFECT OF GM1-OLIGOSACCHARIDE ON ACTIVATION OF ADENYLATE CYCLASE BY CHOLERAGEN[a]

Cell Line	CT Binding Sites/μ^2	Apparent[b] K_d for CT ($\times 10^{-10}$ M)	Activation of Adenylate Cyclase by CT in presence of GM1-oligo (% of control)[c]
HeLa	21	6.5	0
C6 Glial	21	5.2	10
Human Lymphocytes	184	6.8	0
Friend Erythroleukemic	1310	2.4	100
N18 Neuroblastoma	6680	2.8	100

[a]Details are described under MATERIALS AND METHODS.
[b]Values represent the concentration of choleragen (CT) required for half-saturation of the receptors.
[c]Values were calculated after adenylate cyclase activities were corrected for basal activities which were unaffected by GM1-oligosaccharide (GM1-oligo).

the different cells but the number of binding sites varied (Table 1). To test the possibility that the receptor density determined whether the oligosaccharide was inhibitory, HeLa and C6 cells were incubated with media containing GM1 in order to increase their receptor number[9]. Whereas GM1-oligosaccharide blocked the activation of adenylate cyclase in control cells, it was ineffective with GM1-treated cells (Table 2). The inhibitory effects of the oligosaccharide were both time and temperature dependent. HeLa cells were incubated with choleragen at 37°C and oligosaccharide was added at different times. By 5 min, 50% activation had occurred and within 25 min, no inhibition was observed. Cells were then incubated with the toxin for 15 min at different temperatures, exposed to oligosaccharide and incubated at 37°C for 1 h. There was a sharp transition at 18°C; below this temperature, activation was inhibited by the oligosaccharide and above it, activation increased with increasing temperature.

When the various cell lines were incubated with ^{125}I-choleragen for 10 min at 4°C and then with and without GM1-oligosaccharide for 45 min at 37°C, bound toxin was displaced by the oligosaccharide from some of the cells but not from

TABLE 2

EFFECT OF INCREASED GM1 CONTENT ON INHIBITION OF CHOLERAGEN ACTIVATION OF ADENYLATE CYCLASE BY GM1-OLIGOSACCHARIDE

Cells	Choleragen-Activated Adenylate Cyclase[a] Activity (pmol/mg protein/min) -GM1-oligo	+GM1-oligo	Relative Increase in Choleragen Binding Sites[b]
Control HeLa	227	28	1
GM1-treated HeLa	298	290	458
Control C6 Glial	96	14	1
GM1-treated C6	132	130	78

Details are the same as those described in Table 1 except some of the cells were incubated in media containing GM1 in order to increase the number of toxin binding sites per cell.
[a]Values were calculated after the adenylate cyclase activities were corrected for basal activities.
[b]Determined from the amount of ^{125}I-choleragen specifically bound to the cells.

others (Table 3). The degree of displacement correlated with the density of receptors on the cell and the ability of the oligosaccharide to block the activation of adenylate cyclase. GM1-oligosaccharide did not displace bound toxin from GM1-treated HeLa cells that had an increased number of choleragen binding sites (Table 3). The ability of the oligosaccharide to displace bound choleragen from control HeLa cells also was time and temperature dependent in a manner analogous to its ability to block activation of adenylate cyclase. Finally, although much of the inhibitory effects of the GM1-oligosaccharide are due to its ability to displace choleragen from cells with few receptors, in no case was all of the toxin removed from the cells. With HeLa and C6 glial cells, 5 to 6% of the toxin receptors remained occupied and with human lyphocytes, 22% (see Table 3). This amount of choleragen should have caused significant activation of adenylate cyclase; and, when the oligosaccharide was removed by washing the cells and the cells were incubated further at 37°C, the expected amount of activation occurred. Thus, in the presence of GM1-oligosaccharide, the choleragen that remains bound to the cells with few receptors is in a form incapable of activating adenylate cyclase.

TABLE 3

DISPLACEMENT OF BOUND CHOLERAGEN BY GM1-OLIGOSACCHARIDE FROM DIFFERENT CELLS

Cell Line	Choleragen Bound in presence of GM1-oligo (% of control)
N18 Neuroblastoma	98
Friend Erythroleukemic	70
Human Lymphocytes	22
C6 Glial	6
HeLa	5
GM1-treated HeLa[a]	100

Cells were incubated with ^{125}I-choleragen for 10 min at 4°C; then GM1-oligosaccharide was added to some of the samples and all were incubated for 45 min at 37°C. Binding was determined in triplicate as described in MATERIALS AND METHODS and corrected for non-specific binding (binding in the presence of an excess, 0.2 μM, unlabeled choleragen).

[a]HeLa cells were incubated overnight in medium containing GM1 and bound 275-fold more choleragen than untreated HeLa cells.

These results are consistant with the following model for the mechanism of action of choleragen. The toxin initially binds to one receptor on the cell surface. If the density of receptors is low but the membrane is fluid, the toxin binds to additional receptors by lateral diffusion of the toxin-receptor complex or receptors in the plane of the membrane (Fig. 3). Eventually all of the binding sites on the B component become occupied with receptors and the toxin is converted to a form capable of activating adenylate cyclase. This process may involve a conformational change in the toxin structure[3] which allows the A component to interact with and possibly penetrate the plasma membrane[2]. In broken cell systems, the A component alone can activate adenylate cyclase[1] by transfer of ADP-ribose from NAD to a GTP-binding protein that is associated with and presumably regulates the catalytic component of adenylate cyclase[10,11].

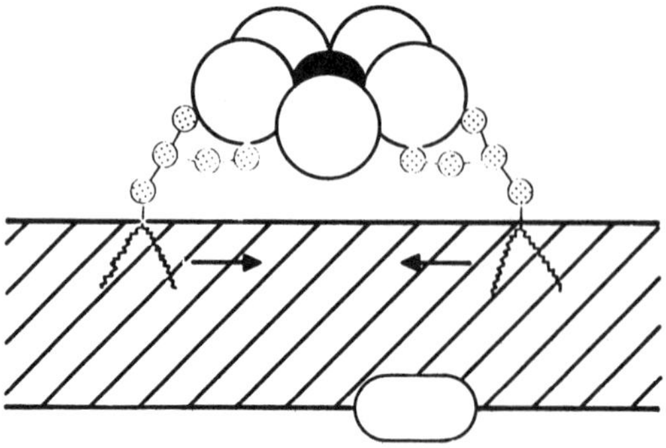

Fig. 3. Multivalent binding of choleragen to several receptors on the cell surface. Arrows indicate lateral movement of lipid portion of GM1 in fluid phase of membrane. This process allows choleragen to bind to several receptors that initially may be widely separated on the membrane. Open oblong represents latent adenylate cyclase on inner surface of plasma membrane.

The initial binding events are both time and temperature dependent when the density of receptors is low. When the cells have a high number of choleragen binding sites, multivalent binding occurs rapidly even under conditions where the cell membrane is not very fluid and mobility of membrane components is hindered. Thus, binding of choleragen to several receptors on the cell surface appears to be an essential step in its subsequent ability to activate adenylate cyclase. In contrast, extensive lateral movement of toxin-receptor complexes may only be required to achieve multivalent binding to cells with few receptors.

REFERENCES

1. Gill, D.M. (1977) *Adv. Cyc. Nuc. Res. 8*, 85-118.
2. Fishman, P.H., and Brady, R.O. (1976) *Science 194*, 906-915.
3. Fishman. P.H., Moss, J., and Osborne, J.C. Jr. (1978) *Biochemistry 17*, 711-716.
4. Craig, S., and Cuatrecasas, P. (1975) *Proc. Nat. Acad. Sci. USA 72*, 3844-3848.
5. Revesz, T., and Greaves, M. (1975) *Nature 257*, 103-106.
6. Sedlacek, H.H., Stark, J., Seiler, F.R., Zeigler, W., and Wiegandt, H. (1976) *FEBS Lett. 61*, 272-276.
7. Cuatrecasas, P. (1973) *Biochemistry 12*, 3547-3558.
8. Tallman, J.F., Smith, C.C., and Henneberry, R.C. (1977) *Proc. Nat. Acad. Sci. USA 74*, 873-877.
9. Moss, J., Fishman, P.H., Manganiello, V.C., Vaughan, M., and Brady, R.O. (1976) *Proc. Nat. Acad. Sci. USA 73*, 3480-3483.
10. Cassel, D., and Pfeuffer, T. (1978) *Proc. Nat. Acad. Sci. USA 75*, 2669-2673.
11. Gill, D.M., and Meren, R. (1978) *Proc. Nat. Acad. Sci. USA 75*, 3050-3054.

IV

TRANSDUCTIVE COUPLING BY CELL SURFACE RECEPTORS II.

ANTIBODIES AS CELL SURFACE RECEPTORS

HENRY METZGER
Chief, Section on Chemical Immunology, Arthritis and Rheumatism Branch, National Institute of Arthritis, Metabolism and Digestive Diseases, National Institutes of Health, Bethesda, Maryland

There are several reasons why antibodies acting as cell-surface receptors are of interest. Firstly, antibodies form the central antigen-specific recognition units in all phases of a highly sophisticated vertebrate protective system. The functioning of this system in addition to being of direct concern to the clinician and epidemiologist, presents a fascinating series of problems to the experimental biologist in the areas of genetics, protein structure and biosynthesis, inter-cellular communication, cell differentiation, growth regulation and still others.

Of more immediate concern to the subject of this conference, antibodies are by far the best characterized glycoproteins which serve as cell surface receptors. This is because these particular glycoproteins are not only incorporated into membranes but are also secreted. The extensive amount of information on sequence, genetic variablity, 3-dimensional structure including detailed knowledge of combining sites, is available for no other cell-surface receptors. It is true that with certain membrane-bound antibodies e.g. those on T-cells, only little is known of their structure and that with others, e.g IgMs on B-cells, there is reason to believe that an as yet poorly defined small carboxy-terminal sequence, not present on secreted antibodies, may exist, which is required for insertion into membranes[1]. Nevertheless it is unlikely that this hydrophobic tail sequence critically affects the antigen recognition and response functions of the antibody[2]; it seems more probable that the extra piece helps to localize where the cell recognition and response shall take place. In other instances e.g. in the case of IgE (below), the antibody bound on the membrane of mast cells is identical to the secreted antibody.

A third reason why antibodies acting as cell surface receptors are of interest is that they appear to act by a different mechanism than many other receptors. This difference allows us to understand more broadly how biological systems solve the problem of trans-membrane stimulation. On the other hand, there are already observations that suggest that the mechanism used by antibodies is utilized in other instances also[3,4].

Success in experimental biology is highly dependent on the choice of system. This is certainly true in studying membrane receptors in general and in studying antibody receptors in particular. Reference to Table I explains why certain groups including our own have concentrated on the IgE-mast cell system compared for example to the B-cell system.

Table I

	IgE-mast cell system	B-cell system
cell-type	homogeneous (more or less!)	heterogeneous (marked)
large number of 'identical' cells available	yes	no (progress being made in this direction)
types of response	limited	multiple[1]
kinetics of response	rapid	slow[2]
measurement of response	simple (chemical)	complicated (biological)
requirements	ligand + Ca^{++}	ligand + variety of macromolecular or even cellular helper substances

[1] B cells appear to have a variable history after stimulation. They may become unreactive (tolerant), proliferate or differentiate.

[2] While there must be rapid initial biochemical events which follow contact with ligand, most of the distinctive responses studied take 24-48 hours or even much longer to demonstrate, and in some cases require prolonged exposure to ligand.

Our orientation has been that the IgE system is sufficiently complex to be challenging but apparently simple enough to yield answers at the molecular level. It seems likely that the experience gained will be helpful in exploring the more complicated problems posed by other systems which are stimulated by membrane-bound antibodies.

The IgE-Mast Cell System

IgE is a 4-chained immunoglobulin synthesized and secreted by plasma cells as are all other secreted immunoglobulins[5]. It has a special affinity for the surface membranes of tissue mast cells (and the related basophils) to which it binds as monomer with high affinity[6]. The binding is via that region of the molecule which is common to all IgE: the Fc region. The binding while tight ($K \geq 10^{10}$) is reversible. The bound IgE behaves much like a peripheral membrane protein[7]. Since the bound IgE has its combining sites for antigen exposed it can react with extra-cellular ligands. Physiologically those ligands which were responsible for the stimulation of IgE production to begin with, are the most important. Under appropriate circumstances interaction of the IgE with ligand leads to a rapid non-cytolytic exocytosis of the mast cell granules. It has been known for decades that the ligand must be multivalent to be effective[8]. This suggested that aggregation of the IgE must be important. It was logical to see if simply reacting cells with artifically aggregated IgE (in the absence of ligand) would suffice. It did[9]. There was increasing evidence that the size of such aggregates had to be small[10,11] and it was recently rigorously shown that dimers were sufficient[12].

The IgE is bound to a glycoprotein, (R_ε), which behaves like an integral membrane protein[13,14,15]. Several groups are in the process of characterizing this component. As yet, not enough has been learned about the structure per se to allow one to draw mechanistic inferences. It has been learned, however, that R_ε does more than bind IgE. In fact it has now been determined that it is the surface R_ε whose aggregation provides the critical initiating signal. The IgE appears to function simply as the link (albeit the critically important one) by which external ligands can dimerize the R_ε protein. The evidence for this is a simple experiment which has been performed on normal mast cells by one group[16] and on a line of cultured tumor cells by ourselves[17]. In both cases it was shown that bivalent antibodies to R_ε itself could stimulate the mast cells. Figure 1 shows the decisive experiment performed by Drs. Taurog and Isersky in our own work[17].

This is the point to which we have arrived. We have learned that the initial critical event in at least one antibody receptor mediated stimulation of a cell, is the dimerization of that component with which the membrane-bound antibody interacts. In the succeeding article Drs. Dembo and Goldstein describe various aspects of the dimerization process. The immediate biochemical sequelae of the dimerization remain unknown and is the subject of our future work. As shown by other presentations in this conference, in other systems stimulated via membrane-bound antibody receptors, aggregation appears to be important. It seems likely that certain common elements will emerge as we learn more.

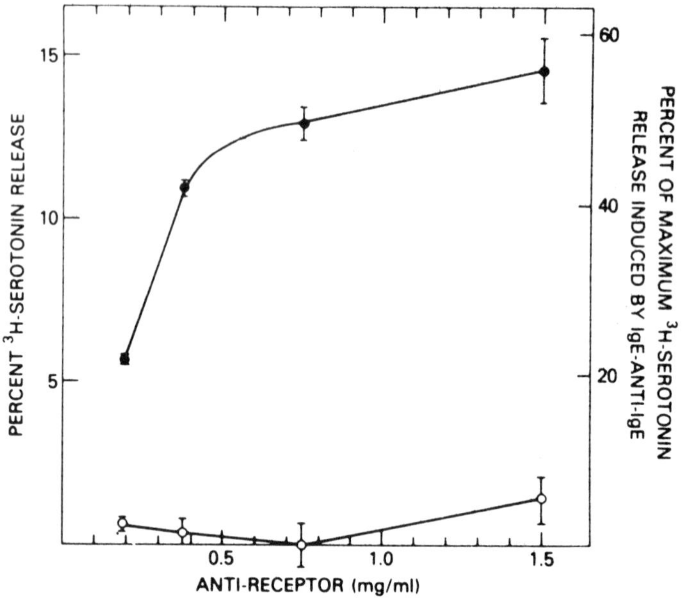

Figure 1. Release of ^{3}H-serotonin from tumor cells reacted with the IgG fraction of an antiserum directed to the cell receptor for IgE. Secretion from cells in the absence of IgE ●-●; secretion from cells which had been pre-treated with IgE and then washed prior to exposure to the anti-receptor antibody O-O. The inhibition by IgE provides strong evidence that it is the antibodies directed to the receptor (and not other possible antibodies) which are causing the reaction.

Reprinted from Reference 17 with permission of the copywrite owner.

REFERENCES

1. Williams, P.B. and Grey H.M. (1978) Fed. Proc. 37, 1838 (Abstract #3115).
2. Metzger, H. (1978) Contemporary Topics in Molec. Immunol. 7, 119-153.
3. Orgiazzi, J., Williams, D.E., Chopra, I.J. and Solomon, D. (1976) J. Clin. Endocrinol. Metab. 42, 341-354.
4. Kahn, C.R., Baird, K., Flier, J.S. and Jarrett, D.B. (1977) J. Clin. Invest. 60, 1094-1106.
5. Bennich, H.H., Johansson, S.G.O. and Von Bahr-Lindstöm, H. (1978) in Immediate Hypersensitivity, Bach, M. ed., Marcel Dekker, Inc., New York, pp. 1-36.
6. Metzger, H. (1977) in Receptors and Recognition, Cuatrecasas, P. and Greaves, M.F. eds., Chapman and Hall, London, Ser A., 4, 75-102.
7. Singer, S.J. (1974) Adv. Immunol. 19, 1-66.

8. Ishizaka, K. and Ishizaka, T. (1971) Ann. NY Acad. Sci. 190, 443-456.

9. Ishizaka, K., Ishizaka, T., and Menzel, A.E.O. (1967) J. Immunol. 99, 610.

10. Ishizaka, K. and Ishizaka, T. (1968) J. Immunol. 101, 68-78.

11. Siraganian, R.P., Hook, W.A. and Levine, B.B. (1975) Immunochemistry 12, 149-157.

12. Segal, D.M., Taurog, J.D. and Metzger, H. (1977) Proc. Nat'l. Acad. Sci. 74, 2993-2997.

13. Conrad, D.H., Berczi, I. and Froese, A. (1976) Immunochemistry 13, 329-332.

14. Kulczycki, Jr., A., McNearney, T.A. and Parker, C.W. (1976) J. Immunol. 117, 661-665.

15. Metzger, H., Elson, E.L., Isersky, C., Mendoza, G., Newman, S., Rossi, G. and Schlessinger, J. (1976) in Proc. IX International Congress Allerg., Buenos Aires, Argentina, Excerpta Medica, Amsterdam, pp. 119-125.

16. Ishizaka, T., Chang, T.H., Taggart, M. and Ishizaka, K. (1977) J. Immunol. 119, 1589-1596.

17. Isersky, C., Taurog, J.D., Poy, G. and Metzger, H. (1978) J. Immunol. 121, 549-558.

DeLisi/Blumenthal, eds. Physical Chemical Aspects of Cell Surface Events in Cellular Regulation

THE MECHANISM OF HISTAMINE RELEASE FROM HUMAN BASOPHILS

MICAH DEMBO and BYRON GOLDSTEIN
Theoretical Division, Los Alamos Scientific Laboratory, University of California, P.O. Box 1663, Los Alamos, New Mexico 87545.

INTRODUCTION

Basophils are a type of white blood cell which comprise approximately one-half of one percent of the total blood leukocytes. Dispersed in the cytoplasm of bashophils are numerous membrane enclosed granules which contain histamine and other mediators of inflammation. The mechanism by which these granules are released is an easily studied example of the general phenomenon of cell activation and secretion. Recently, Dembo, Goldstein, Sobotka and Lichtenstein[1] have proposed a theory of the mechanism of histamine release. In this paper we will summarize the principal elements of this theory and explain the evidence on which these elements are based.

CELL SURFACE IgE

The plasma membrane of basophils has receptors which are capable of binding Immunoglobulin E (IgE) with very high affinity[2-6]. When we speak of "cell surface IgE," the presence of the Fc receptor in complex with the IgE molecule is always implied.

Each cell surface IgE molecule has two identical antigen combining sites located on the Fab fragments of the molecule. The serum of individuals allergic to a particular antigen always contains a certain fraction of IgE (specific IgE) which binds to this antigen. Basophils from such allergic individuals bind the specific IgE in vivo and are said to be actively sensitized. Basophils from many nonallergic individuals can be sensitized in vitro by incubation with serum from an allergic individual; such cells are said to be "passivly sensitized."

Spontaneous disociation of the IgE-Fc receptor complex is very slow, with a half life on the order of days. Consequently, on the time scale of the typical histamine release experiment the binding of IgE to the Fc receptor can be treated as irreversible. Further, under usual experimental conditions, the IgE in the sensitizing serum is negligibly depleted by binding to the Fc receptors. When irreversibility and negligible depletion is assumed, the amount of IgE bound to human basophils after a certain time of passive sensitization is given by[6]

$$X_S(t) = \frac{C_S}{C_T} R_f[1 - \exp(-r_+ C_T t)] \qquad (1)$$

In this equation, X_S is the number of specific IgE molecules bound per cell, C_S and C_T are the concentrations of specific and total IgE respectively during passive sensitization, r_+ is the forward rate constant for binding of IgE to the Fc receptor; R_f is the number of free Fc receptors per cell at the start of sensitization and t is the time of exposure of cells to serum.

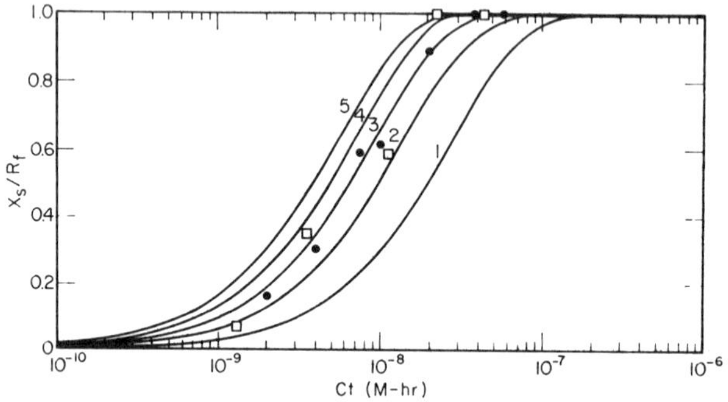

Fig. 1. Binding of IgE to Fc receptors on human basophils. Solid curves are predictions of Eq. (1) for r_+ = 1,2,3,4 and $5x10^4 M^{-1} S^{-1}$. Reproduced from Ref. 6.

Figure 1 shows the agreement of Eq. (1) with some data on the binding of IgE to a purified basophil preparation obtained from a patient with 75% basophilia[4,6]. The value of r_+ deduced from these data is shown in Table I. The values of R_f for various cell donors can be determined by measuring the amount of IgE bound to cells at saturating serum concentrations and subtracting the amount present on the cells prior to the start of sensitization. It has been found that there are large variations between different individuals in the value of R_f[4,5]. Finally, methods are available for the quantitation of C_S and C_T in various allergic sera[8], so that all of the parameters and variables in Eq. (1) can be independently determined.

THE ROLE OF CALCIUM

When sensitized cells are exposed to allergen in the presence of millimolar amounts of calcium, histamine is released with a time constant on the order of 1-10 minutes. The release of histamine is accompanied by the extrusion of the histamine containing granules of the basophils through a classic exocytotic process[9].

Considerable evidence indicates that the requirement for extracellular calcium, [Ca]ex, arises because a change in membrane permeability towards calcium and a consequent increase in the cytoplasmic calcium concentration are both necessary and sufficient for histamine release[1]. Furthermore, the coupling between changes in [Ca]ex and corresponding changes in the rate of histamine release is apparently very rapid. Thus, chelation of free calcium by addition of excess EDTA causes cessation of histamine release wthin a matter of seconds[10,11]. Conversly, readdition of calcium (but not magnesium) causes release to start again, also on a very rapid time scale.

DESENSITIZATION

As discussed above, histamine release does not occur in the absence of extra-cellular calcium. It has been observed however, that if calcium is added to basophils after some period of preincubation with allergen, the amount of histamine ultimately released progressively decreases to zero as the time of preincubation increases[12-14]. This progressive decrease in histamine release due to incubation with allergen in the absence of calcium is known as desensitization. As with release, desensitization has a time constant on the order of 1-10 minutes. Clearly, however, unlike release, desensitization does not require the presence of extra-cellular calcium. It has also been found that desensitization is less sensitive to temperature than is release and that in cells of some individuals desensitization can occur at very low or so called "subthreshold" allergen concentrations, which are too low to produce detectable release[10].

Under certain circumstances, cells which have been desensitized by exposure to a particular antigen lose their ability to respond to other antigenic stimuli as well. This is known as "nonspecific desensitization." In other cases only the ability to respond to the desensitizing antigen is lost, a phenomena known as "specific desensitization"[13,14]. Neither specific nor nonspecific desensitization can be explained as a simple loss of viability of the cells, since desensitized cells retain their ability to release histamine normally as a result of non-IgE mediated stimuli such as calcium ionophores[15].

A factor found to be associated with whether desensitization will turn out to be specific or nonspecific is the number of IgE molecules per basophil. Cells with large amounts of specific IgE on their surface desensitize nonspecifically. Cells with small amounts of specific IgE, such as passivly sensitized cells, or cells from allergic donors who have a low level of specific IgE exhibit antigen specific desensitization. An additional distinction between specific and nonspecific desensitization is that specific desensitization can be reversed by incubation of the desensitized cells with allergic sera whereas nonspecific desensitization cannot be reversed by this method.

In the present discussion we shall restrict ourselves to systems where desensitization is specific. Nonetheless, it is possible to generalize the theory to explain how desensitization can go from specific to nonspecific in a continuous transition as the number of IgE antibodies per basophil increases. A particular example of how such a generalized theory can be constructed is discussed in the conclusion of reference 1.

CROSSLINKING

It has been found that simple binding of allergen to cell surface IgE is not sufficient to produce either release or desensitization[14,16,17]. In addition to binding the allergen must "crosslink" cell surface IgE. Crosslinking in this context means that the allergen must bind to at least two IgE molecules simultaneously and thus link them together. Formation of crosslinks between cell surface IgE molecules is possible even if there is a very low surface density of IgE because cell surface IgE is capable of diffusing laterally in the plane of the plasma membrane[18].

The involvement of crosslinking of cell surface IgE in release and desensitization arises because normal initiaion of both these processes requires that two or more Fc receptors be held in close proximity to each other. The main evidence for this is that nonphysiological agents such as anti-IgE, concanavalin A or anti-Fc receptor, that are capable of bringing Fc receptors into close proximity by essentially artificail means, cause histamine release[19,20,21].

These observations argue strongly that it is the purely steric event of crosslink formation and not an allosteric change in the Fc receptor which is the first signal for the initiation of histamine release and desensitization[22].

BIVALENT HAPTENS

Bis benzylpenicilloyl 1,6 diaminohexane or $(BPO)_2$ is an example of what we shall call a symmetric bivalent hapten.* Such molecules consist of two identical haptenic groups connected by flexible polymer chains. These molecules are of interest since they are clearly the simplest chemical compounds which are capable of crosslinking cell surface IgE.

Siraganian et al[16], have shown that basophils from rabbits immunized with penicillin antigen will release histamine when exposed to $(BPO)_2$ in vitro. The bivalent structure of $(BPO)_2$ is essential for this releasing activity since monovalent BPO haptens, such as ε-BPO formyl-L-lysine $(BPO)_1$, did not cause histamine release. In fact, $(BPO)_1$ was found to specifically inhibit $(BPO)_2$ induced release, but did not affect release due to other antigenic stimuli.

THEORY OF CROSSLINKING OF CELL SURFACE IgE BY $(BPO)_2$

Since $(BPO)_2$ is bivalent, the only type of complexes that can be formed when it interacts with IgE are linear chains or rings as illustrated in Figure (2). This Figure also shows the additional states formed when a monovalent hapten such as $(BPO)_1$ is added to the reaction. Because of the limited variety of complexes shown in Fig. (2), it is possible to make realistic mathematical models of the equilibrium

---

*BPO ≡ Benzylpenicilloyl group; $(BPO)_2$ ≡ BPO-NH-$[CH_2]_6$-NH-BPO

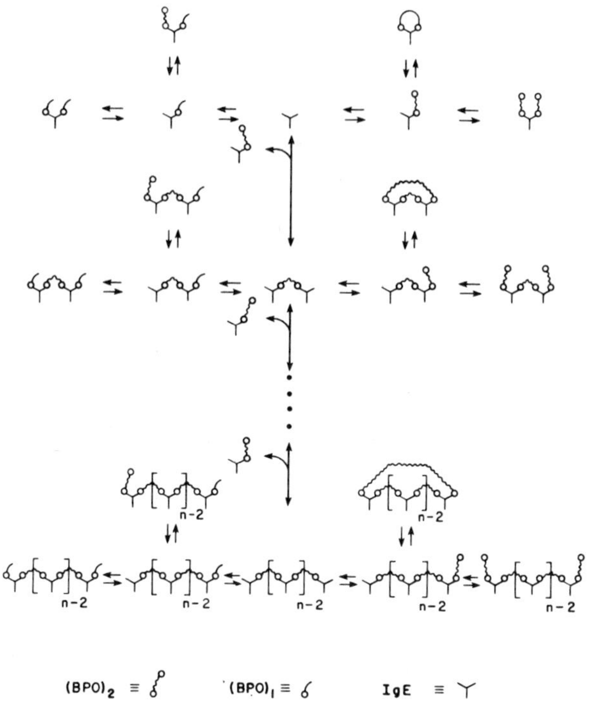

Fig. 2. States formed when a symmetric bivalent hapten, $(BPO)_2$, crosslinks cell surface IgE in the presence of a monovalent form of the hapten, $(BPO)_1$. Reproduced from Ref. 1.

and kinetic properties of the reaction of $(BPO)_1$, $(BPO)_2$ and cell surface IgE[17,23,24]. The principal assumptions of these models are that the anti-BPO IgE on the basophils is homogeneous with regard to its ability to bind BPO and that the binding of $(BPO)_1$, $(BPO)_2$ or additional IgE to linear chains does not depend on the size of the chains.

The results of computer simulations of the time course of the crosslinking reaction indicate that unless the cyclic complexes (rings) of $(BPO)_2$ and IgE are extremely stable, thermal equilibrium is reached in a matter of seconds[17]. This means that in the absence of highly stable cyclic complexes it should be possible to remove $(BPO)_2$ from the cell by simple washing. Indeed, washing in buffer was found to stop histamine release due to $(BPO)_2$[17]. Similar washing had no effect on release due to antigen E. This indicates that binding of antigen E is much tighter and less readily reversible than binding of $(BPO)_2$.

Computer simulations also indicate that addition of a large excess of $(BPO)_1$ to the reaction of $(BPO)_2$ and cell surface IgE causes the breakup of crosslinks within a matter of seconds even if cyclic states of extreme stability are assumed to exist. In accord with this, addition of a large excess of $(BPO)_1$ leads to cessation of histamine release on a time scale comparable to that involved in stoppage of histamine release by addition of excess EDTA[14]. (Also see Figure (5)).

Taken together these results indicate that the binding of $(BPO)_2$ to cell surface IgE is rapidly equilibrating compared to the time scale of release and desensitization of basophils and that cyclic states are negligible. It is thus unlikely that dynamic, as opposed to equilibrium properties of crosslinking will have any effect on histamine release due to $(BPO)_2$.

If ring formation is neglected, then according to the model of crosslink formation[23], the number of crosslinked antibody molecules per cell at thermal equilibrium is given by

$$X_{poly}(A,B,X_S) = X_S \frac{(2\delta^2-2\delta-1+\sqrt{1+4\delta})}{2\delta^2} \qquad (2)$$

where

$$\delta = \frac{2K_xK_a(X_s/S)A}{(1+K_aA+\frac{1}{2}K_bB)^2}$$

In this equation, X_S is the total number of anti-BPO IgE molecules per cell, S is the surface area of the cell and A and B are the concentrations of $(BPO)_2$ and $(BPO)_1$ respectively.

The parameters K_a (and K_b) which appear in Eq. (2) are defined in terms of the equilibrium constants for addition of a $(BPO)_2$ (or $(BPO)_1$) to previously unliganded IgE molecules,+

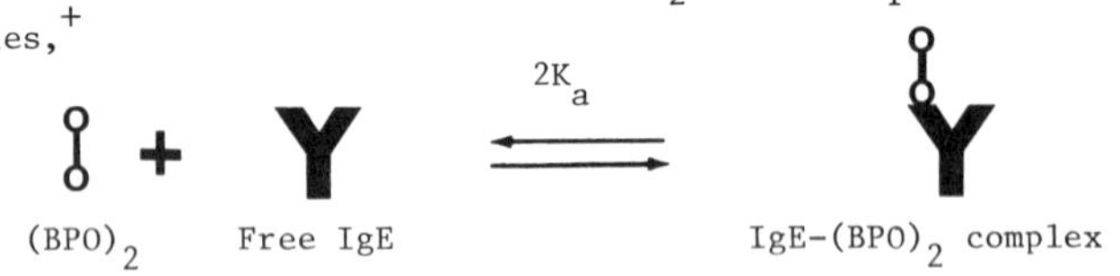

and

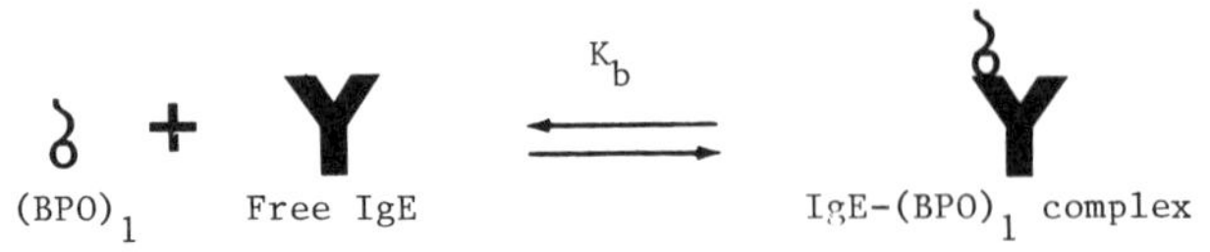

Note that we include a factor of two in the definition of K_a. This is done so that if the two BPO groups on $(BPO)_2$ each bind with the same affinity as the BPO group on $(BPO)_1$, then $K_b = K_a$.

K_x is defined as the equilibrium constant for the crosslinking reaction between free IgE and $IgE\text{-}(BPO)_2$ complex

+In ref. 23 we use a different notation for the equilibrium constants. In terms of our previous notation K_a=H and K_x=K.

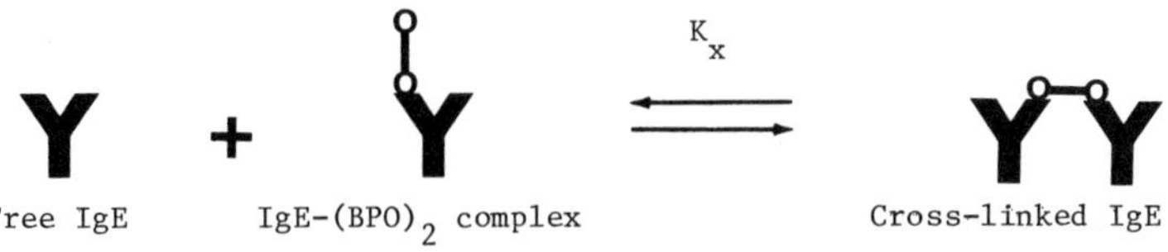

Analysis of Eq. (2) reveals that there are several geometric properties of a plot of X_{poly} versus logA (i.e. the cross-linking curve) which are independent of the choice of parameters. These are 1) X_{poly} approaches zero in the limit of both large and small A, i.e.,

$$\lim_{A\to\infty} X_{poly} = \lim_{A\to 0} X_{poly} = 0 \qquad (3a)$$

2) X_{poly} has a single maximum which occurs at a $(BPO)_2$ concentration, A_{max}, given by

$$A_{max} = \frac{1}{K_a} + \frac{1}{2}\frac{K_b}{K_a} B \qquad (3b)$$

3) Increasing the monovalent hapten concentration always decreases X_{poly}, i.e.

$$X_{poly}(A_1B_1) > X_{poly}(A_1B_2) \text{ if } B_2 > B_1 \qquad (3c)$$

4) If $A >> 1/2(K_b/K_a)B$ then X_{poly} becomes asymtotically independent of B, i.e.

$$X_{poly}(A,B) \approx X_{poly}(A,0)$$
$$\text{for } A >> 1/2(K_b/K_a)B. \qquad (3d)$$

5) The dependence of histamine release on antigen concentration is usually displayed as a plot of the percent histamine released against the $\log_{10}$ of the antigen concentration. We call such plots "histamine release curves." Analogous plots of X_{poly} versus $\log_{10}$ of antigen are called crosslinking curves. It can be shwon that according to Eq. (2) the crosslinking curve is symmetric with respect to a line drawn through A_{max} parallel to the y axis, i.e.

$$X_{poly}(\log(\frac{A}{A_{max}})) = X_{poly}(-\log(\frac{A}{A_{max}})) \qquad (3e)$$

It has been found[25] that histamine release curves for $(BPO)_2$ in the presence or absence of $(BPO)_1$, for cells of various donors and for cells passivly sensitized with various concentrations of serum, display all the properties predicted for crosslinking curves. This observation strongly argues that Eq. (2) is a good approximate designation of crosslinking and that, at least for $(BPO)_2$, histamine release from passivly sensitized human basophils is a monotone increasing function of X_{poly}. In other words, has a maximum when X_{poly} has a maximum, increases when X_{poly} increases, and decreases when X_{poly} decreases.

If this conclusion is accepted, then Eq. (3b) provides a simple means of determining the parameters K_a and K_b. K_a was determined to be $1.2 \pm .2 \times 10^7$ ℓ/mole by simply taking the reciprocal of the concentration of (BPO)2 at which maximum histamine release was observed, A_{max}. The ratio (K_b/K_a) was then found to be 0.38 from the shift in A_{max} produced by adding fixed amounts of $(BPO)_1$[9].

The value of K_x can be estimated on the basis of a statistical mechanical model of crosslink formation presented previously[25,26]. This method of estimating K_x utilizes as input parameters, the collision radius of cell surface antibody, the depth of the antibody combining site, the extended length of $(BPO)_2$ and the value of K_a. We feel that the resulting value of K_x (see Table I) is probably an overestimate although not by a large factor. The implications of the uncertainty in K_x in terms of the compensating errors introduced into the other parameters of the model have been discussed elsewhere[1].

The final unknown parameter in Eq. (2) is the surface area of the basophil, S. We have taken this quantity to be 4.7×10^{-6} cm^2 on the basis of published microscopic studies of basophils[17].

THE EQUATION FOR SPECIFIC DESENSITIZATION

The theory of Dembo, Goldstein, Sobotka and Lichtenstien[1] proposes a simple mechanism to explain the occurance of specific desensitization and the dependence of specific desensitization on crosslinking. This is done by postulating the existence of an enzyme in the basophil which causes the degradation of cell surface IgE molecules if they are crosslinked to other IgE molecules. In this way, only the IgE molecules specific for the allergen are lost during desensitization.

If we assume the enzyme operates according to Michelis-Menten kinetics, then the rate at which specific IgE is lost due to the action of the enzyme is given by

$$\frac{d}{dt} X_S = -V_m X_{poly}/(K_m + X_{poly}) \quad . \tag{4}$$

Since desensitization occurs in the absence of calcium, the theory also assumes that Eq. (4) is unaffected by the calcium concentration. In addition, since equilibration of IgE with $(BPO)_2$ is very rapid compared to desensitization, the crosslinking reaction will be in quasi-equilibrium. Consequently, X_{poly} will be given as a function of A, B and X_S by Eq. (2). Thus Eq. (4) can be solved to predict that the time course of loss of specific IgE from the cell surface during desensitization. As yet, this prediction has not been tested because of the difficulty of directly measuring the loss of specific cell surface IgE.

Examination of Eq. (4) reveals that as with release, desensitization is a monotone increasing function of X_{poly}. Thus Eq. (4) requires that the maximum rate of desensitization occur at the same $(BPO)_2$ concentration as the maximum of the release curve. Furthermore, as with release, the amount of desensitization after a fixed time should decrease in a symmetric manner on either side of its maximum. Both of these predictions of Eq. (4) have been previously verified[17]. (Also see Fig. 4).

THE EQUATION OF HISTAMINE RELEASE

From the existence of histamine containing granules, and from the correlation between release of the granules and histamine release, we conclude that histamine release occurs in quantal bursts. This is also in accord with phase microscopic observations of degranulation in which individual phase dark granules are observed to suddenly disappear to be replaced by phase pale vesicles[9,17,28,29]. We therefore define the notion of a quantum of histamine as being a small unit of histamine, the release of which occurs in an instantaneous burst or event. Morphologically, a quantum of histamine may be a single granule, some part of a granule, or a cluster of granules, as long as its contents are released more or less as a unit.

Since the number of quanta is large, the total rate of quantum release will be the product of the number of "releasable" quanta, Q_R, and the probability per unit time that a releasable quantum will be released, P_R,

$$\frac{d}{dt} Q_R = -Q_R P_R \quad . \tag{5}$$

We speak of "releasable" quanta in order to take account of the fact that a certain number of the quanta in a population of basophils may be contained in dead or damaged basophils or they may be incompetent for some other reason. We shall always assume that the total number of quanta in a population of basophils, Q_T, may be thought of as the sum of a number of releasable quanta, Q_R, and a number of nonreleasable quanta, Q_{NR}.

In light of Equation (5) the question arises as to whether or not P_R depends on Q_R. If there is some dependence of P_R on Q_R then it means that release of different quanta are not "independent" events but that release of a given quantum can influence the probability that subsequent quanta will be released. For example, it is possible that release of a quantum could require the utilization of some chemical which is present in limited supply. In this case, release of a quantum would have a negative influence on release of subsequent quanta. Alternatively, one can conceive of situations in which quanta compete with each other for access to a limited number of release "sites." In this case release of a quantum would relieve the competition and facilitate the release of subsequent quanta.

One way in which to tell whether or not the quanta interact with each other is to examine the histamine release produced by two sequential antigenic stimuli. It is easy to show that if granule release is independent, and the signals produced by the two stimuli do not interfere with each other, then the overall release to two sequential stimuli is given by[1]

$$H_{R12} = H_{R1} + \frac{H_{R2}}{f_R} [f_R - H_{R1}] \quad . \tag{6}$$

In this equation, H_{R1} is the release produced when the concentration of antigen two is zero; H_{R2} is the release produced when stimulus one is omitted and f_R =

$Q_R(0)/Q_T(0)$, is the fraction of the original number of quanta which are releasable. As mentioned above, one should be aware that Eq. (6) is only applicable if the first stimulus does not interfere with the ability of the cell to receive the second stimulus. Thus the first signal should not be of a kind which nonspecifically desensitizes the cell and the second signal should not be an antigen which cross-reacts with the first.

As shown elsewhere[1], Eq. (6) is in good agreement with data. This result supports the hypothesis that the histamine quanta of basophils are independent of each other. We note, that a particularly strong prediction of Eq. (6) is that if all the releasable histamine is released by the first stimulus then the second stimulus will do nothing. The fact that this prediction holds up even for values of f_R considerably less than one, strongly indicates that a certain fraction of the histamine content of basophils is nonreleasable.

From the previous discussion of the role of calcium in histamine release, we see that the evidence indicates that the crosslinking signal must be transduced into a change in the calcium permeability of the cell membrane and subsequently into an increased level of cytoplasmic calcium in order to induce degranulation. This clearly implies that P_R must be related to both the number of crosslinked IgE molecules on the cell surface, X_{poly}, and the extracellular calcium concentration, [Ca]ex.

In order to deduce the exact form of this relationship we first note that there is a negligible time delay between a change in X_{poly} or [Ca]ex and a corresponding change in the rate of histamine release. In other words, the transduction of a change in X_{poly} or [Ca]ex into a corresponding change in P_R must be virtually instantaneous. This means that the various intermediates in the transduction of the signal are all in rapid equilibrium with X_{poly} and [Ca]ex. Consequently, we can neglect the relaxation times associated with these intermediates and assume that P_R can be expressed as an explicit function of X_{poly} and [Ca]ex.

Given that P_R is some explicit function of X_{poly}, the data dictates several properties which this function must possess. These are: 1) P_R must be a monotone increasing function of X_{poly} in order to have any possibility that histamine release will also be a monotone function of X_{poly}; 2) As $X_{poly} \to \infty$ there must be some point at which P_R becomes limited by factors other than crosslink availability. Thus P_R must be a saturable function of X_{poly}; 3) In order that there be a possibility of subthreshold desensitization and in order to insure that the basophil will be stable in the face of very small levels of antigen, there must be some region near the origin within which the rate of desensitization dominates the rate of release. This means that P_R must approach zero as some power of X_{poly} greater than one.

Taken together, these features imply that there is a sigmoid dependence of P_R on X_{poly}. We note, that such a relationship can be easily achieved mechanistically if a positively cooperative enzyme is involved in catalyzing some step in the transduction of a change in X_{poly} into a change in P_R.

Due to our ignorance of further details of the exact functional relationship between P_R and X_{poly} the theory empirically assumes that P_R can be parameterized by a Hill function.

$$P_R(X_{poly}) = F_{Hill} = \frac{\sigma\left(\frac{X_{poly}}{\Gamma}\right)^{\alpha}}{1+\left(\frac{X_{poly}}{\Gamma}\right)^{\alpha}} \tag{7}$$

The choice of this function is motivated by its mathematical simplicity and by the fact that it satisfies the necessary criteria discussed above.

In addition to its dependence on X_{poly}, P_R must also be a direct function of [Ca]ex, the extracellular calcium concentration. Furthermore, since there is no histamine release in the absence of calcium we must require that $P_R \to 0$ as [Ca]ex $\to$ 0. In light of Eq. (7), this means that $\sigma \to 0$ and/or $\Gamma \to \infty$ as [Ca]ex $\to$ 0.

The quantity normally reported in histamine release experiments is the fraction of the total histamine content which is released, H_R. H_R is related to Q_R and Q_{NR} by the equation

$$H_R(t) = 1 - \frac{Q_R(t) + Q_{NR}}{Q_R(0) + Q_{NR}} = f_R - \frac{Q_R(t)}{Q_T(0)} \tag{8}$$

By substituting from Eq. (7) and (8) into equation (5) we obtain the fundamental equation governing H_R according to the theory

$$\frac{d}{dt} H_R = \frac{(f_R - H_R)\sigma\left(\frac{X_{poly}}{\Gamma}\right)^{\alpha}}{1 + \left(\frac{X_{poly}}{\Gamma}\right)^{\alpha}} . \tag{9}$$

QUANTITATIVE COMPARISON BETWEEN THEORY AND EXPERIMENT

Equations (1), (2), (6) and (9) in conjunction with the understanding that $P_R \to$ 0 when [Ca]ex $\to$ 0, are the fundamental equations of the theory[1]. This system of equations can be solved numerically for a given choice of parameter values to predict the outcome of a variety of histamine release experiments in which the antigen is $(BPO)_2$[1]. As discussed earlier the parameters included in Equations (1) and (2) can be independently determined or deduced from very elementary properties of the histamine release curve. The remaining six parameters of the theory, V_m, K_m, σ, Γ, α and f_R have been determined by means of fitting the theory to the data shown in Figs. (5) and (6)[1]. A complete set of parameter values determined in this way are given in Table I. In Figures (3) through (6) the solid curves show the numerical solution of the model equations for the parameter values in Table I. Thus the degree to which the theory simulates experiment can be ascertained from these Figures.

In order that the parameters of the theory be the same for all the data used for data fitting, a standard methodology was employed. Details of this methodology have appeared elsewhere[14,25]. In addition, all data were obtained using a single source of cells and a single source of anti-BPO serum. Consequently, the parameter values of Table I refer to these sources of cells and serum and may not apply to other sources.

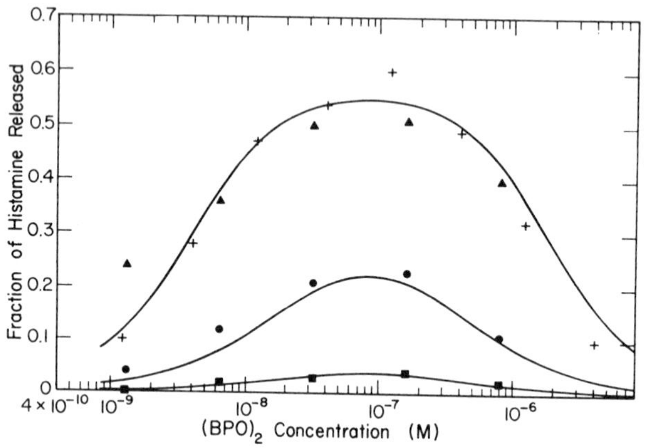

Fig. 3. Histamine release curves for cells sensitized with various dilutions of serum. △ and + ⇒ 1/5 dilution. ● ⇒ 1/10 dilution and ■ ⇒ 1/20 dilution. The concentration of IgE in undiluted serum is given in Table I. Reproduced from Ref. 1.

Figure (3) shows the results of varying the concentration of serum used to sensitize the cells. Cells were incubated for ninety minutes in 1/5, 1/10 or 1/20 dilution of serum. The cells were then washed and exposed to various concentrations of $(BPO)_2$ for thirty minutes in the presence of 0.6 mM calcium. As a control, one set of cells (data indicated with +) were allowed to incubate for 15 minutes in the absence of calcium before addition of $(BPO)_2$ and calcium. The fact that such treatment has no effect on histamine release demonstrates that desensitization requires the presence of antigen and is not an artifact due to incubation without calcium.

As can be seen, the release curves in Fig. (3) are symmetric and all have the same maximum values. These features are clearly simulated by the model which is in good agreement with these data.

Figure (4) shows the effect of $(BPO)_2$ concentration on the extent of desensitization. Cells were sensitized in 1/5 dilution of serum for ninety minutes. Following this, cells were washed and exposed to various amounts of $(BPO)_2$ in the absence of calcium for fifteen or thirty minutes. After additional washing, cells were exposed to 10^{-7}M$(BPO)_2$ for thirty minutes in the presence of 0.6 mM calcium. Also shown in this Figure for purposes of comparison is the histamine release curve for 1/5 dilution of serum from the preceding figure.

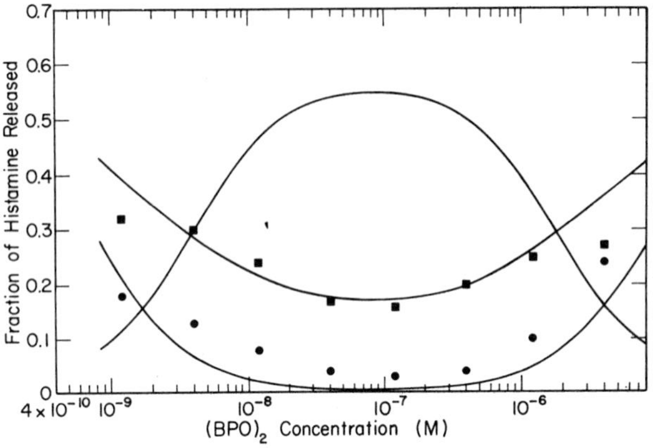

Fig. 4. Histamine release at 10^{-7} M$(BPO)_2$ after desensitization for 15 min., ■, or 30, ●, at various concentrations of $(BPO)_2$. Reporduced from Ref. 1.

Note, that the minima of the desensitization curves occur at the same $(BPO)_2$ concentration as the maximum of the release curve, and that all the curves are symmetric. As discussed earlier, this suggests that cell surface IgE molecules are only degraded by the desensitizing enzyme when they are crosslinked to other IgE molecules. Also note that the minima of the desensitization curves are very broad and flat so that near the center of the curves there is a very weak dependence of desensitization on the $(BPO)_2$ concentration. According to the model this is because X_{poly} is large near the center of the curves and the desensitizing enzyme is operating at maximum capacity.

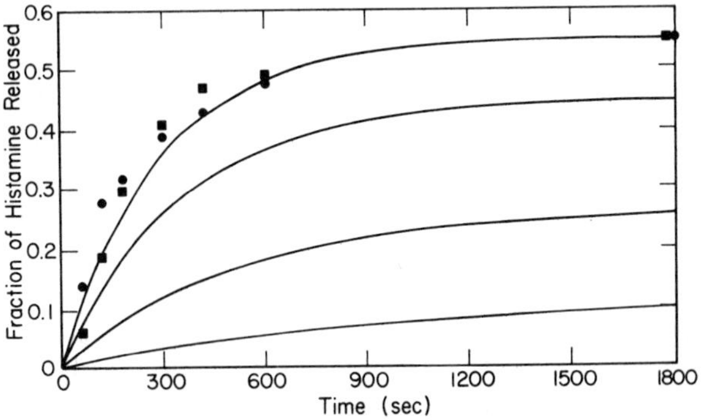

Fig. 5. Time course of histamine release at 10^{-7} Molar $(BPO)_2$ ● ⇒ release stopped by addition at 10^{-4}M$(BPO)_1$. ■ ⇒ release stopped by addition of excess EDTA. Reproduced from Ref. 1.

Figure (5) shows measurements of the time course of histamine release at $10^{-7}M(BPO)_2$. Cells were sensitized exactly as in the preceding experiment although no desensitization was carried out. Following sensitization the cells were exposed to $10^{-7}M(BPO)_2$ in the presence of 0.5mM calcium. Release was stopped at various times by addition of excess EDTA or by addition of a large excess of $(BPO)_1$. As discussed earlier the fact that data obtained by these two methods is in agreement demonstrates that there is negligible time delay between a change in X_{poly} and the consequent change in P_R.

The various other theoretical curves in Figure (5) show the predictions of the theory for the time course of histamine release at 10^{-8}, 3.2 x 10^9 and $10^{-9}M(BPO)_2$.

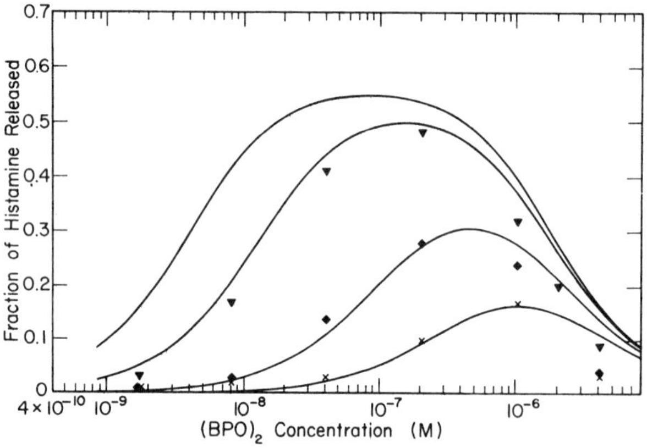

Fig. 6. Histamine release curves in the presence of $(BPO)_1$. x ⇒ $(BPO)_1 = 5x10^{-6}M$, ♦ ⇒ $(BPO)_1 = 2x10^{-6}M$, ∇ ⇒ $(BPO)_1 = 3.5x10^{-7}M$. Also shown is the theoretical release curve for $(BPO)_1 = 0$ from Fig. 3. Reproduced from Ref. 1.

Figure (6) shows the effect of adding a constant amount of $(BPO)_1$ to the release stage of an experiment. Cells were sensitized as in the two preceding experiments. Following this the cells were washed and exposed to various amounts of $(BPO)_2$ in the presence of 0.6mM calcium and either 5 x 10^{-6}, 2 x 10^{-6}, or 3.5 x 10^{-7} molar $(BPO)_1$.

As can be seen, the presence of $(BPO)_1$ causes the histamine release curves to decrease and shift to the right at low $(BPO)_2$ concentrations, but has a negligible effect on release at high $(BPO)_2$ concentrations. Furthermore, the concentration of $(BPO)_2$ for maximum release, A_{max}, is shifted to the right by $(BPO)_1$ as specified by Eq. (3b). Also note that despite being shifted towards the right, the release curves remain fairly symmetric about their maxima. All these features of these data are simulated by the theory due to the fact that addition of $(BPO)_1$ has comparable effects on X_{poly}.

CONCLUSION

The theory of histamine release of Dembo, Goldstein, Sobotka and Lichtenstein[1] is able to explain a variety of kinetic experiments in a self consistent and quantitative way. The principle elements of the theory are:

1) The histamine content of basophils is divided into a large number of discrete and independent "quanta."

2) Release of the contents of a quanta is a stochastic event.

3) A certain fraction of the quanta in a population of basophils are nonreleasable.

4) The probability per unit time, P_R, of the release event is a sigmoid function of the number of crosslinked IgE molecules on the cell surface. In addition, P_R goes to zero when the extracellular calcium concentration goes to zero.

5) Specific desensitization is due to the action of a degradative enzyme on cell surface IgE molecules which are crosslinked to other cell surface IgE molecules. Furthermore, this enzyme is unaffected by the calcium concentration.

TABLE I

PARAMETERS OF HISTAMINE RELEASE

Parameter	Symbol	Value±error	Units
concentration of IgE in undiluted serum	C_T	1.2×10^{-9}	$\frac{\text{liters}}{\text{mole}}$
concentration of anti-BPO IgE in undiluted serum	C_S	1.1×10^{-9}	$\frac{\text{liters}}{\text{mole}}$
number of free fc receptors per cell at the start of an experiment	R_f	$7.3 \pm \times 10^4$	$\frac{\text{molecules}}{\text{cell}}$
forward rate constant for binding of IgE to fc receptors	r_+	$3. \pm 1. \times 10^4$	$\frac{\text{liters}}{\text{mole-sec}}$
binding constant of $(BPO)_2$	K_a	$1.2 \pm 0.2 \times 10^7$	$\frac{\text{liters}}{\text{mole}}$
binding constants of $(BPO)_1$	K_b	$4.6 \pm 1.0 \times 10^6$	$\frac{\text{liters}}{\text{mole}}$
cross-linking constant	K_x	$8. \times 10^{-10}$ ± factor of 3	$\frac{\text{cm}^2}{\text{molecule}}$
surface area of bsophil	S	$4.7 \pm 1.5 \times 10^{-6}$	cm^2
maximum velocity of desensitization	V_m	1.7 ± 0.2	$\frac{\text{molecules}}{\text{cell-sec}}$
Michaelis constant of desensitization	K_m	$45. \pm 5.$	$\frac{\text{molecules}}{\text{cell}}$
maximum velocity of release	σ	$4.4 \pm 0.4 \times 10^{-3}$	$\frac{1}{\text{sec}}$
threshold of release	Γ	$360. \pm 30.$	$\frac{\text{molecules}}{\text{cell}}$

TABLE I Continued

Parameter	Symbol	Value±error	Units
hill coefficient of release	α	1.5 ± 0.05	dimensionless
fraction of original histamine content of the cells which is releasable	f_r	0.6 ± 0.03	dimensionless

REFERENCES

1. Dembo, M., Goldstein, B., Sobotka, A.K. and Lichtenstein, L.M. (1978). A theory of the mechanism of histamine release from human basophils: Systems that desensitize specifically. (Submitted for publication.)
2. Sullivan, A.K., Grimley, P.M. and Metzger, H. (1971). Electron microscopic localization of immunoglobulin E on the surface membrane of human basophils. J. Exp. Med. 134: 1403.
3. Conroy, M.C., Adkinson, N.F. and Lichtenstein, L.M. (1977). Measurement of IgE on human basophils: Relationship of serum IgE and anti IgE induced histamine release. J. Immunol. 118: 1317.
4. Malveaux, F., Conroy, M.C., Adkinson, N.F. and Lichtenstein, L.M. (1978). IgE receptors on human basophils: Relationship to serum IgE concentration. J. Clin. Invest. 62: 176.
5. Kulczyki, Jr., A. and Metzger, H. (1974). The interaction of IgE with rat basophilic leukemia cells II. Quantitative aspects of the binding reaction. J. Exp. Med. 140: 1676.
6. Goldstein, B., Dembo, M. and Malveaux, F. (1978). Some quantitative aspects of IgE binding to human basophils. (Submitted for publication.)
7. Ishizaka, K., Ishizaka, T. and Lee, E.H. (1970). Biologic function of the Fc fragments of E myeloma protein. Immunochemistry 7: 687.
8. Schellenberg, R.R. and Adkinson, N.F. (1975). Measurement of absolute amount of antigen-specific human IgE by a radioallergosorbent test (RAST) elution technique. J. Immunol. 115: 1577.
9. Hastie, R., Chir, B., Levy, D.A. and Weiss, L. (1977). The antigen-induced degranulation of basophil leukocytes from atopic subjects studied by electron microscopy. Lab. Invest. 36: 173.
10. Lichtenstein, L.M. and Osler, A.G. (1964). Studies of the mechanism of hypersensitivity phenomena. IX. Histamine release from human leukocytes by ragweed pollen antigen. J. Exp. Med. 120: 507.
11. Osler, A.G., Lichtenstein, L.M. and Levy, D.A. (1968). In vitro studies of human reaginic allergy. Adv. Immunol. 8: 183.
12. Lichtenstien, L.M. (1971). The immediate allergic response: In vitro separation of antigen activation, decay and histamine release. J. Immunol. 107: 1122.
13. Sobotka, A.K., Dembo, M., Goldstein, B. and Lichtenstein, L.M. (1978). Anti-specific basophil desensitization. Federation Proceedings 37: 1667.
14. Sobotka, A.K., Dembo, M., Goldstein, B. and Lichtenstein, L.M. (1978). Antigen-specific desensitization of human basophils. J. Immunol. (in press).
15. Lichtentstein, L.M. (1975). The mechanism of basophil histamine release induced by antigen and by calcium ionophore A23187. J. Immunol. 114: 1692.
16. Siraganian, R.P., Hook, W.A. and Levine, B.B. (1975). Specific in vitro histamine release from basophils by bivalent haptens: Evidence for activation by simple bridging of membrane bound antibody. Immunochemistry 12: 149.

17. Dembo, M., Goldstein, B., Sobotka, A.K. and Lichtenstein, L.M. (1978). Histamine release due to bivalent penicilloyl haptens: The relation of activation and desensitization of basophils to dynamic aspects of antigen binding to cell surface antibody. J. Immunol. (in press).

18. Schlessinger, J., Webb, W.W., and Elson, E.L. (1976). Lateral motion and valence of Fc receptors on rat peritoneal mast cells. Nature 264: 550.

19. Becker, K.E., Ishizaka, T., Metzger, H. Ishizaka, K. and Grimley, P.M. (1973). Surface IgE on human basophils during histamine release. J. Exp. Med. 183: 394.

20. Ishizaka, T. and Ishizaka, K. (1978). Triggering of histamine release from rat mast cells by divalent antibodies against IgE-receptors. J. Immunol. 120: 800.

21. Magro, A.M. and Bennich, H. (1977). Concanavalin A induced histamine release from human basophils in vitro. Immunology 33: 51.

22. Ishizaka, T., Ishizaka, K., Conrad, D.H. and Forese, A. (1978). A new concept of triggering mechanisms of IgE-mediated histamine release. J. Allergy Clin. Immunol. 61: 320.

23. Dembo, M. and Goldstein, B. (1978). Theory of equilibrium binding of symmetric bivalent haptens to cell surface antibody: Applications to histamine release from basophils. J. Immunol. 121: 354.

24. Wofsy, C., Goldstein, B. and Dembo, M. (1978). Theory of equilibrium binding of asymmetric bivalent haptens to cell surface antibody: Application to histamine release from basophils. J. Immunol. 121: 593.

25. Dembo, M., Goldstein, B., Sobotka, A.K. and Lichtenstein, L.M. (1978). Histamine release due to bivalent penicilloyl haptens: Control by the number of cross-linked IgE antibodies on the basophil plasma membrane. J. Immunol. 121: 354.

26. Dembo, M. and Goldstein, B. (1978). A thermodynamic model of binding of flexible bivalent haptens to antibody. Immunochemistry. 15: 307.

27. Anderson, P. and Uvnäs, B. (1975). Selective localization of histamine to electron dense granules in antigen-challenged sensitized rat mast cells and to similar granules isolated from sonicated mast cells. Acta. Physiol. Scand. 94: 63.

28. Uvnäs, B. (1971). Quantitative correlation between degranulation and histamine release from mast cells. In: Biochemistry of the Acute Allergic Reations. Oxford: Blackwell, p. 175.

29. Hastie, R. (1971). The antigen-induced degranulation of basophil leukocytes from atopic subjects studied by phase contrast microscopy. Clin. Exp. Immunol. 8: 45.

DISCUSSION

DR. BLUMENTHAL: With respect to this model, if in the nonspecifically desensitized state a calcium ionophore is added, does one still get activity?

DR. DEMBO: Yes.

DR. FEWTRELL: Could I ask you what led to the suggestion that it was the IgE enzyme that was degrading the IgE on the cell surface rather than any other possible suggestion, and whether you could suggest any other ways in which you might get the same phenomenon, the specific desensitization?

DR. DEMBO: I first would point out that we do not actually hypothesize that the Fc receptor is inactivated in some way. All we really know for sure is that cross-linking causes some process which leads to the functional loss of the cross-linked IgE. As to other possible mechanisms, a number of people have suggested some kind of pinocytosis of the cross-links. I have no real objection to that so long as they can figure out some way that pinocytosis would be able to discriminate between dimeric cross-linked IgE, which has a half-life on the order of one second and monovalent IgE with antigen bound to its antigen combining sites.

Also, they would have to figure out some reason why pinocytosis should be saturable. As I said, it could be some exotic mechanism but we believe an enzymatic mechanism is the simplest thing consistent with this data.

DR. METZGER: I think it has the virtue of being a relatively easily testable thing, and it will certainly be tested.

DR. BARZILAY: I wanted to ask: Do you have any information about the size of the divalent penicillin?

DR. DEMBO: Yes, we know the dimensions. The total molecule is about 35 angstroms long, extended.

DR. METZGER: Actually, partially in response to that, if one takes some of the data of Vern Schumacher, who looked at various DNP ligands, and looks at the data that he got in terms of the ease of dimerization by various sized ligands and then uses some of the old Levine data on the length of the bivalent ligands which give maximal response, there is almost a 1 to 1 correspondence.

DR. RODBARD: Is the length between the two heads of the double-headed ligand sufficient to prevent binding simultaneously to the two binding sites on the same immunoglobulin -- In other words, I just would comment that if there is intramolecular cross-linking, that would lead to many of the same phenomena.

DR. DEMBO: It is too short to do that. In addition, the fact that we can remove Bivalent Hapten from the cell surface with great ease, indicates that formation of such intramolecular cyclic states, as I pointed out, is negligible in this particular system. Not only is the bivalent hapten incapable of connecting two antigen combining sites on the same IgE molecule, but also the two ends of the dimer or the trimer.

If such cyclic states had appreciable stability the calculations indicate that it would be very difficult to wash the $(BPO)_2$ off the cell surface, if they exist at all.

DR. WEBB: Have you tested that with a larger molecule?

DR. DEMBO: It could be done if we had a larger molecule.

DR. Metzger: It is, in fact, again referring to some of Vern Schumacher's results, very difficult getting these circular complexes. If you use longer molecules, they tend to fold up on themselves, and it has been rather difficult finding antigens suitable for forming circular complexes.

PHYSICAL CHEMICAL AND BIOLOGICAL IMPLICATIONS OF RECEPTOR CLUSTERING

CHARLES DELISI

DCBD, NCI, NIH, Bethesda, Maryland 20014

ABSTRACT

A simple model for the clustering of cell surface receptors by divalent ligand is presented. In the first part of the paper, the physical chemical consequences of the model are pursued, and its thermodynamic features - in particular, the forms of Scatchard plots and displacement curves (label displaced by increasing concentrations of cold ligand) - are shown to be indistinguishable from predictions based on negative cooperativity and heterogeneity. The kinetic features, however, are very different for the three models, and the crosslinking model is shown to quantitatively fit kinetic experiments which the other two, in their simplest forms, cannot.

The second part of the paper develops two examples in which crosslinking serves to connect binding with biological activity. It is proposed on the basis of a variety of experimental evidence, that in certain cellular responses, most notably those involving the release of histamine from basophils and mast cells, ligand binding to cell surface receptors can simultaneously generate conflicting signals which are different non linear functions of the degree of receptor clustering. The parameters which affect the spatial distribution of receptors in the presence of ligand; e.g. affinity, membrane fluidity, and receptor density and geometry, are therefore expected to control the balance between signals, and to influence the observed dose response patterns. The model thus allows quantitative predictions of the effects of changing manipulable experimental parameters on the control of histamine release and desensitization at the cell surface level. Because of the interplay between signals, the predicted release patterns do not always exhibit the simple symmetric rise and fall expected of crosslinking, but sometimes show, in agreement with experiment, bimodality, a simple plateau, or a peak and decline followed by a plateau.

The second application also involves competition, but at a different level. In this example, the degree of biological activity observed when agonists and antagonists are present simultaneously is considered in terms of a requirement for crosslinking, rather than simple binding. Methods used to analyze experiments involving competitive antagonism; in particular the Schild plot, are briefly discussed in view of this requirement.

INTRODUCTION

Plasma membrane fluidity, the mobility of cell surface molecules, and the local and large scale receptor redistributions attendent upon exposing cells to various exogenous and endogenous ligands, have all been documented to various degrees in a variety of systems. In a few instances clear evidence for the sufficiency of receptor crosslinking as a condition for biological activity has been shown[1-3] (see also the review by Schlessinger[4] and references cited therein) and considerable effort is currently being expended, much of it reviewed in this volume, on structural and physical chemical studies of cell surface receptors and their interactions.

If ligand induced receptor aggregation is in fact a general phenomenon, occurring on cells of the endocrine and nervous systems, as well as on those of the immune system, then one can expect that many of the patterns of binding and dissociation involving hormones and neurotransmitters will have been influenced by the induction of receptor clustering. In this paper the physical chemical theory of clustering[5-11] is first reviewed and then applied. The first two applications, to dissociation kinetics, indicate that clustering has an important influence on dissociation patterns, and it must therefore be explicitly considered in the interpretation of such patterns and the kinetic parameters deduced from them. In the following applications I discuss the relation between receptor aggregation and biological activity, using basophil degranulation and desensitization to focus the development, and show how the model can be used to elucidate and quantitate the physical chemical parameters which control release and desensitization.

THE MODEL

The mathematics can be kept relatively uncomplicated by limiting the development to events involving divalent ligands interacting with divalent receptors. There are a number of experimental systems which are known to conform to this requirement, most notably those involving cells with immunoglobulin receptors directed against various synthetic ligands. The physical chemical aspects of the results presented below may therefore be applied quantitatively to such systems.

The initial interaction between ligand in solution and cell surface receptor will, in general, be followed by a sequence of additional reactions within the plane of the membrane, resulting in the formation of linear receptor-ligand aggregates. Receptors will therefore be distributed among a number of physical chemical states; e.g. completely free, dimeric, trimeric and so fourth. This paper will not be concerned with the entire distribution; in part because its explicit introduction is not needed to discuss the thermodynamics and kinetics of the <u>mean</u> number of bound ligands (the observed quantity) and in part because precise connections between biological observables and particular cluster sizes cannot currently be made. Instead, the focus will be on an average property of the distribution ; <u>viz</u>, the concentration of crosslinked receptors.

For bookkeeping purposes it is simpler to develop equations in terms of the states of ligand rather than of receptor. As shown in Fig. 1, a ligand may be free in solution (concentration c), singly bound (concentration m_1) or doubly

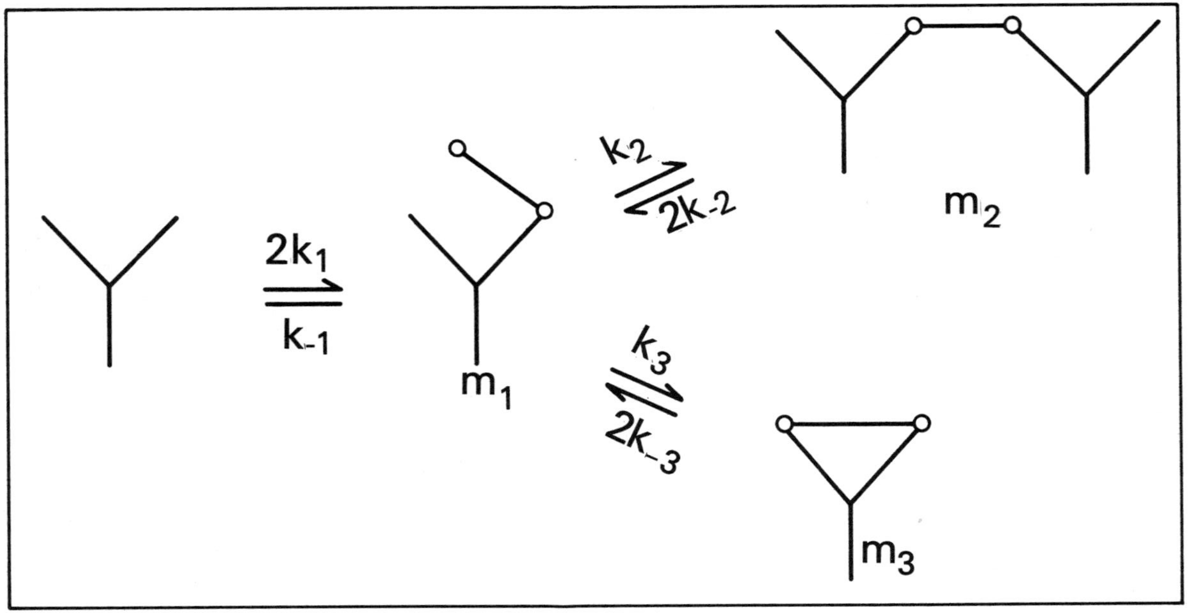

Fig. 1. k_1 is the rate constant for association of solution ligand with a free receptor site to form singly bound ligand, and k_{-1} is the rate constant for dissociation of this complex. The 2 arise because there are two free sites on each molecule of ligand in solution. k_2 is the forward rate connstant for ligand site - receptor site interaction within the plane the membrane and k_{-2} is the site-site rate constant for dissociation a doubly bound ligand which is crosslinking two receptors. If m_2 is in molecules/area, then k_2 will be in units of $(\text{molecule-sec/area})^{-1}$. The measured quantity, however, is k_2m_5 and this has units of sec^{-1}. If dissociation rates are not diffusion limited, (more precisely is they reflect only bonding energies) one expects that $k_{-2} \cong k_{-1}$. k_3 $(\text{molec-sec/area})^{-1}$ and k_{-3} are the rate constants for formation and dissociation of the intramolecular complex. The equilibrium constants are defined as $K_1 \equiv k_1/k_{-1}$; $K_2 \equiv k_2/2k_{-2}$; $K_3 \equiv k_3/2k_{-3}$.

bound. There are two energy states associated with the latter class. Ligands crosslinking two receptors will be denoted by m_2, and ligands bound intramolecularly (each site bound to the same receptor) will be denoted by m_3.

Often two types of ligand, in competition for a single class of receptor, are present. The ligands may differ only in the presence of a label on one which is absent from the other, or they may be physically and biologically distinguishable as in the case of an agonist-antagonist pair. The concentration of this second class will be distinguished by a star, so that c^*, m_1^*, m_2^* and m_3^* denote concentrations analogous to those defined above. According to the definitions of the equilibrium constants given in Fig. 1, the concentration of each state can be written in terms of the free receptor concentration m_4, as follows:

$$m_1 = 2K_1 c m_4 \qquad 1$$

$$m_1^* = 2K_1^* c^* m_4 \qquad 2$$

$$m_2 = 2K_1 K_2 c m_4^2 \qquad 3$$

$$m_2^* = 2K_1^* K_2^* c^* m_4^2 \qquad 4$$

$$m_3 = 2K_1 K_3 c m_4^2 \qquad 5$$

$$m_3^* = 2K_1^* K_3^* c^* m_4^2 \qquad 6$$

If m_5 represents the total receptor site concentration, then an expression for m_4, in terms of equilibrium constants and the concentrations of free ligands can be obtained by utilizing the conservation condition:

$$m_5 = m_1 + m_1^* + 2(m_2+m_2^*+m_3+m_3^*) + m_4 \qquad 7$$

Substituting eqs 1-6 into eq 7, and solving the resulting quadratic eq, one finds that

$$m_4 = -e_1/2e_2 \{1-[1+4e_2 m_5/e_1^2]^{1/2}\} \qquad 8$$

where

$$e_1 \equiv 1+2K_1 c+2K_1^* c^*$$

$$e_2 \equiv 4K_1(K_2+K_3)c+4K_1^*(K_2^*+K_3^*)c^*$$

THERMODYNAMIC CHARACTERIZATION OF BINDING IN THE PRESENCE OF CLUSTERING

When only a single type of ligand is present, the ratio of bound to free ligand B/c is, according to eqs 1 and 3, given by

$$B/c = 2K_1 m_4 + 2K_1(K_2+K_3)m_4^2 \qquad 9$$

If the free receptor site concentration, m_4, could be written as a linear function of B, eq 9 would lead to a simple quadratic generalization of the Scatchard plot from which the best estimates of the parameters, K_1, K_2+K_3, and m_5 could readily be determined by a least square fit to a parabola. Unfortunately the generalization is not possible since the number of bound ligands does not by itself determine the number of bound receptor sites. The best way to proceed in this case is by the use of computer codes which can fit rather general non linear functions to a set of data.[12,13] In addition, rough estimates of the parameters of interest can be obtained from the intercepts and limiting slopes of a standard Scatchard plot, and these may in fact be used as initial guesses in a curve fitting routine. Thus, defining S_h and S_ℓ as the slopes at high and low antigen concentration, respectively, and I as the intercept on the abscissa, one can show that

$$I = m_5 \qquad (10)$$

$$S_h = -2K_1 \qquad (11)$$

and

$$S_\ell \cong -4K_1(K_2+K_3)m_5 = 2S_h I\ (K_2+K_3) \qquad (12)$$

It is evident from eqs 11 and 12, and also clear on physical grounds, that the effective affinity decreases as the concentration of bound ligand increases. This is because the crosslinking step becomes less favorable as the bound ligand concentration increases and the number of free receptors available for crosslinking decreases. The Scatchard plot must thus have a negative slope which

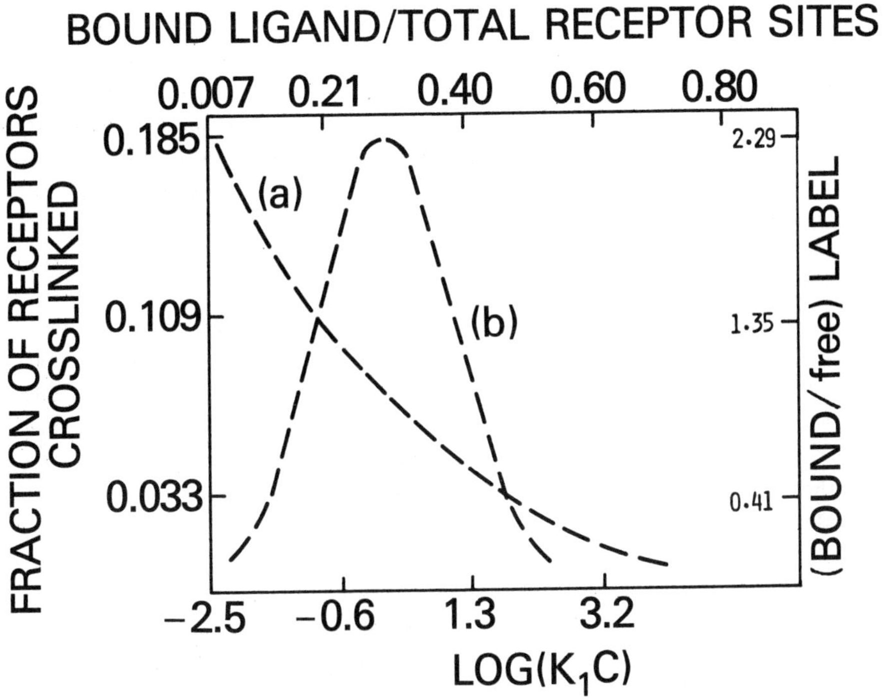

Fig. 2(a). A Scatchard plot simulated using eq 9 with $K_1m_5=K_1^*m_5=0.4$, $K_2m_5=K_2^*m_5=2$, and $K_3=K_3^*=0$. The appearance is the same as would be expected of a heterogeneous or negatively cooperative system. (b) Concentration of crosslinks as a function of free ligand concentration.

decreases (in absolute value) as ligand concentration increases (Fig 2); the same as would be found for a heterogeneous or negatively cooperative system[14]. The natural question therefore is whether thermodynamic and kinetic models based on these three possibilities make distinguishable predictions.

CLUSTERING IN THE PRESENCE OF COMPETITIVE INTERACTIONS

Distinctions between the models will generally be difficult to make on the basis of thermodynamic information alone. In fact Fletcher, et. al.[14] have shown that in the absence of competition, models of heterogeneity and negative cooperativity are thermodynamically equivalent. Crosslinking, however, may not be formally equivalent to the other two. For example, with any model, the weight for a binding event in the absence of other interactions is K_1c. The weight for a subsequent cooperative binding event is K_2c, whereas the weight for crosslinking is K_2m_4, m_4 being a function of c. One might therefore speculate that there is no equivalence, but this needs to be demonstrated rigorously. Even if the models are not fully equivalent, however, simulations indicate that their characteristics are similar enough to make them practically indistinguishable. I will therefore not focus on the thermodynamics of displacement, but only indicate two of the more noteworthy results expected of crosslinking.

First, the displacement curves are monophasic (i.e. bound labeled ligand decreases monotonically) even with K_1^* and $K_2^*m_5$ differing by several orders of magnitude, and with comparable concentrations of singly bound and crosslinked ligands initially present. This means that singly bound label is not, as one might expect intuitively, displaced preferentially over crosslinked label. The reason is that as cold ligand binds, the concentration of free receptor sites is depleted and this must reduce the concentration of crosslinking label (the total concentration of crosslinks, may, however, be increasing over some low concentration range of ligand). The kinetics of this reduction might be slow, and it will be if the reverse rate constant, k_{-2}, is small. However, the labeled ligand will still be displaced since it is kinetically identical to the unlabelled.

Similar results were obtained for displacement from heterogeneous receptors in the absence of crosslinking. The model simulated involved two receptor populations; one which bound ligand with affinity K_1 and the other which bound it with affinity $K_2<K_1$. Intuitively one might expect ligand from the low affinity receptors to be displaced first. However, a careful analysis indicates that this is not the case. Although the low affinity receptors bind c less tightly than do the high affinity receptors, they also bind c* less tightly. Hence displacement is just as difficult as from the high affinity receptors and the resulting thermodynamic curve is monophasic.

In many experiments, labeled ligand is not displaced by the native form, but by some analogue. The crosslinking model predicts a characteristic relationship between the native and analogue displacement curves; specifically, the shift in the curve will be asymmetric, being more pronounced at high degrees of binding then at low (Fig. 3). The reason is that when most bound label has been displaced, few receptors will be crosslinked since most sites will be occupied by c. The difference between two displacement curves in this region will therefore

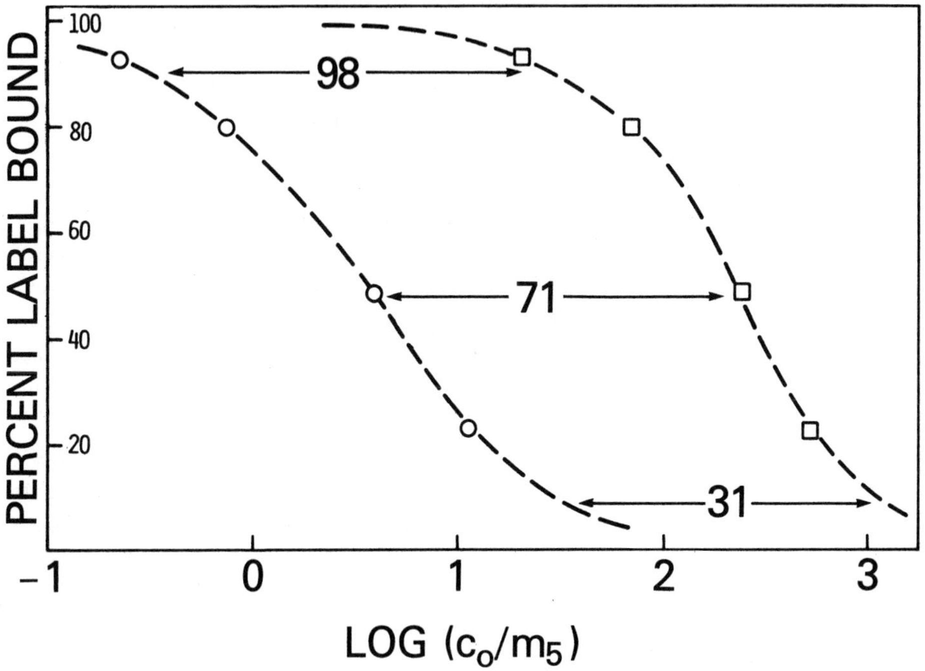

Fig. 3. Displacement of labeled ligand having K_1m_5 = 0.06 and K_2m_5 = 19 by an identical cold ligand (-0-), and by a ligand with K_1m_5 = 0.006 and K_2m_5 = 0.019 -□-. c_o is the total concentration of unlabeled ligand.

reflect primarily the difference between K_1 and K_1^*. On the other hand, at relatively low values of c, most of the label will still be bound, and depending upon total occupancy, a substantial amount of crosslinking is possible. Consequently, the curve in this region will generally reflect both the difference between K_1 and K_1^* and the difference between K_1K_2 and $K_1^*K_2^*$. Therefore as K_2^*/K_2 increases, the asymmetry should become more pronounced, and this has been shown by simulation.

Similar results are expected for heterogeneous and negatively cooperative systems. For example, in the former case, at low values of total ligand concentration (c_o) and assuming the system has been prepared so that only a small percentage of sites is occupied at c_o=0, the high affinity receptors will be preferentially occupied and hence the difference between displacement curves will reflect the difference between K_1^* and K_1. On the other hand, at high c_o almost all receptors will be bound and the labeled ligand will therefore occupy a smaller portion of the high affinity sites. The difference between curves in this region can therefore reflect all four equilibrium constants, and this can give rise to an asymmetry depending upon the values of the equilibrium constants. It thus appears that from the point of view of distinguishing different models, there is little to be learned from equilibrium studies.

Kinetics. The experiments of interest involve the dissociation of labeled ligand from receptors, either as the result of dilution alone, or as the result of dilution and the addition of a high concentration of cold ligand. It is evident that in the simplest situation, with homogeneous receptors, single step binding kinetics (cooperativity and crosslinking are both absent) and dissociation not limited by diffusion, the time course of dissociation should not be influenced by the presence of a high concentration of cold ligand.

If receptors can be crosslinked by ligand, the dissociation curve will generally consist of a sum of exponentials and, moreover, the presence of a competitor will speed the dissociation (Appendix A). The reason for the increased dissociation rate in the presence of a competitor, stems from the fact that two steps are required for doubly bound ligand to leave the cell. The competitor acts to speed dissociation after the first bond breaks, by binding the receptor site which has been freed, and preventing the receptors from recrosslinking.

The crosslinking model (Appendix A) has been applied to two sets of kinetic data: the dissociation of labeled insulin from cultured human lymphocytes[15] and the dissociation of nerve growth factor from sympathetic ganglia[16]. Although the equations are generally non linear, the linear regime is pertinent under conditions outlined in Appendix A and these are likely relevant to the experiments under consideration. Under such conditions there are five unknown parameters: K_{-1}, k_{-2}, k_{-3}, k_2M_5, and k_3M_5, and these also uniquely determine the initial concentrations expressed relative to total bound label.

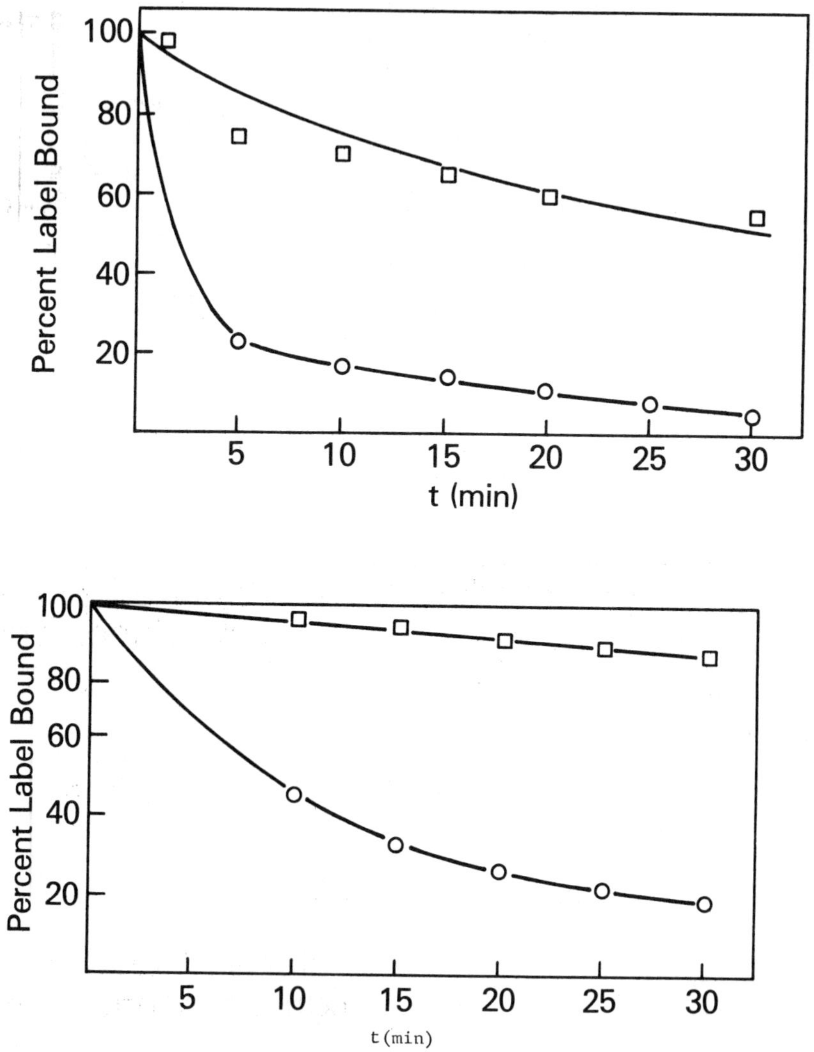

Fig. 4(a). Percentage of the initial labeled insulin bound as a function of time for dilution alone (- -) and dilution plus a large excess of native insulin -0-[15]. The smooth curves are calculated using $K_{-1}=0.18\ min^{-1}$; $k_{-2}=0.57\ min^{-1}$; $k_{-3}=0.016\ min^{-1}$; $k_2m_5=43.2\ min^{-1}$; and $k_3m_5=0.77\ min^{-1}$. (b) Best least square fits of the theory for nerve growth factor dissociating from chick embryo sympathetic neurons[16] $k_1=2.3\ min^{-1}$, $k_{-2}=0.86\ min^{-1}$, $k_{-3}=0.028\ min^{-1}$, $k_2m_5=100\ min^{-1}$ and $k_3m_5=1.4\ min^{-1}$.

linked and intramolecularly bound states is unknown. Hence there are a total of four adjustable parameters: the initial state ratio, k_{-1} k_{-2} and k_{-3}. These were determined by fitting the dissociation kinetics obtained with excess ligand. The fit was constrained by the requirement that the slow (and relatively minor) component be due to intramolecular binding. In this way the adjustable parameters leading to the best fit were determined at 3.3; 0.14 ± 0.01, 0.36 ± 0.10 and 0.0045 ± 0.0006, respectively. With these numbers the forward rates are determined at $k_2m_5=43\ \text{min}^{-1}$ and $k_3m_5=0.16\ \text{min}^{-1}$. Thus all parameters are determined.

There are two important quantitative predictions which can immediately be tested. First, with no further parameter adjustment, the kinetics of dissociation by dilution alone can be predicted (upper curve, Fig. 4a). Second, for forward reaction rates which are diffusion controlled, k_2m_5 can be related to the translational diffusion coefficient[17,7]. An approximate relation for diffusion in a plane is[17]

$$D = \frac{(k_2m_5)d^2}{120\pi}\,\ell n(d/r_o) \qquad 13$$

where d is the mean spacing between receptors when they are randomly distributed and r_o is the distance between their centers when they have reacted. Using a

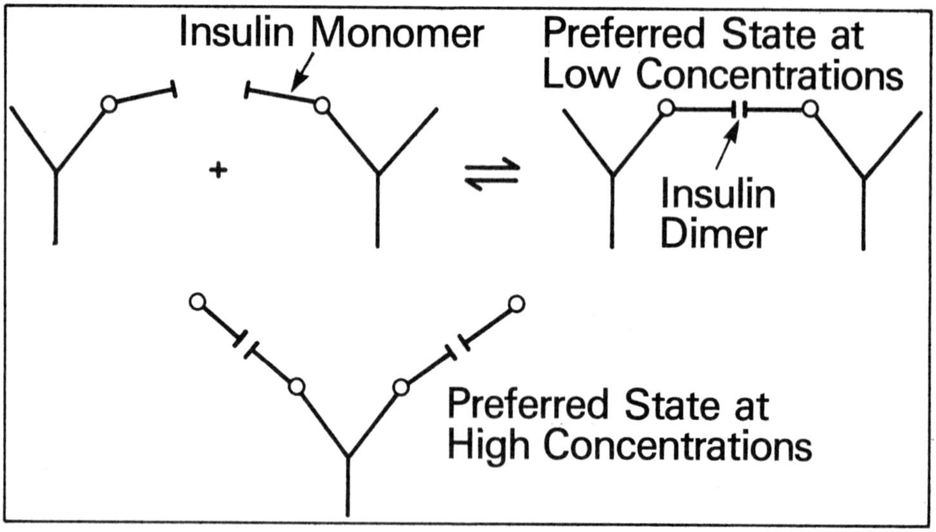

Fig. 5. Because the concentrations of molecules on the cell surface may be enhanced by several orders of magnitude over their concentrations in solution[19], receptor bound insulin may dimerize at concentrations at which solution insulin does not, thus crosslinking receptors. The model is equivalent to that in Fig. 1 when the insulin dimer dissociation time is long compared to the time of the experiment.

cell radius of 10μ, $2x10^4$ receptors per cell and r_o=100Å, eq 13 predicts $D \cong 3x10^{-10}$ cm^2/sec; a number which compares favorably with measured values[4].

These results are not dependent upon the detailed mechanism by which receptors are crosslinked. Specifically insulin need not be divalent, but may be monovalent with two (or more) different combining sites for different receptors sites, or it may be effectively divalent on the membrane because of the increased probability for dimerization[18] there (Fig. 5). The results obtained for NGF are shown in Fig. 4b with the upper curve again being a prediction.

These results differ somewhat from expectations based on negative cooperativity. In this latter case when most bound labeled ligand is in the high energy configuration (i.e. c_o^* is sufficiently small) the addition of cold ligand may destabilize the interaction between labeled ligand and receptor, thereby speeding dissociation. The increased dissociation rate in this case, however, precedes by a different mechanism than it does under a crosslinking model: here dissociation involves only a single step and the kinetics should therefore consist of a single exponential.

A model in which receptors are heterogeneous will generally predict multiple exponential decay, except perhaps a very low values of c_o^* where there may be strong preferential binding of the high affinity group. However, for a purely heterogeneous model with no interaction between receptors the addition of competitor will not effect the dissociation rate.

The predictions of these three "pure" models are summarized in Table I. It would not be surprising if a real system involved some mixture of these e. g. negative cooperativity and crosslinking. Consequently, although the relation between the conclusions reached above and observed dissociation kinetics can be used to eliminate certain pure models, they cannot definitively establish any particular pure model. Thus, an observation that dissociation is enhanced by the addition of a competitor is not compatible with a model of pure heterogeneity; but it cannot rule out heterogeneity in the presence of another effect. Conversely however, if there is no observed enhancement, a Scatchard plot with a positive second derivative would suggest something approaching pure heterogeneity.

CROSSLINKING AS AN INITIAL EVENT IN SIGNAL TRANSDUCTION

The hallmark of a simple model which has activity related to crosslinking is a biphasic dose response pattern: as ligand concentration increases at low doses, the number of crosslinks increases; but as the dose continues to increase the number of bound ligands approaches the total number of receptor sites and the number of crosslinked receptors must therefore drop to zero. As indicated below, however, a dose response pattern of this sort does not necessarily implicate crosslinking - there are other plausible explanations - nor does the absence of such a pattern imply that crosslinking is not playing a major role.

Characteristics of Dissociation Kinetics

MODEL C_o^*	CROSSLINKING	NEGATIVE COOPERATIVITY	HETEROGENEITY	CHARACTERISTIC
HIGH	1 none	1 none	2 or more none	number of exponentials effect of competitor
INTERMEDIATE	2 or more increased dissociation two or more exponentials	2 increased dissociation single exponential	2 or more none	number of exponentials effect of competitor
LOW	2 or more increased dissociation 2 or more exponentials	1 increased dissociation single exponential	1 none	number of exponentials effect of competitor

If receptor crosslinking signals a biological response, then a competitive reaction which reduces the number of crosslinks should reduce the magnitude of the response. Such a reduction may be achieved in a number of ways; e.g. by adding increasing amounts of a monofunctional form of the ligand to the mixture of cells and divalent ligand. Alternatively it is possible that in drug related systems, antagonists and agonists differ in the sorts of complexes they favor, and thereby differ in biological activity. For example, the former might be more flexible than the latter and thus favor intramolecular complexation at the expense of crosslinking.

In this section two examples of activity blockage will be developed within the context of a simple crosslinking model. The first will involve pure competitive antagonism such as one finds in studies of β-adrenergic receptors and cyclase activity.[20] The second, related to basophil degranulation, will involve more than just competition for a given site, but antagonism between conflicting biological signals.

A common method of assessing the potency of an antagonist is to determine the quantities necessary to maintain a biological effect of a given magnitude in the presence of increasing agonist concentrations. To be more specific, suppose $c_2(x)$ is the antagonist concentration needed to maintain a given effect (e.g. 50% maximum cyclase activity) when the agonist concentration is increased from c_1 to xc_1 $(x>1)$. Then a standard procedure is to plot $\log(x-1)$ as a function of $\log[c_2(x)]$. This is known as a Schild plot:[21] a slope of 1 is generally interpreted as true competitive antagonism and the intercept on the abscissa is assumed to be the free energy for antagonist - receptor interaction. The validity of these interpretations requires that the results be independent of c_1. For the particular model on which the analysis is based, one can argue that the converse is also true, viz, if the results are independent of c_1, the interpretations are valid.

The two main features of the model upon which this type of analysis is based are: (1) single site attachment of both agonist and antagonist to univalent receptors and (2) a biological effect which is linearly proportional to the amount of bound agonist. A crosslinking model is of course, incompatible with both these assumptions and the question of interest is how the interpretation of such an analysis would change if the biological effect were assumed to be proportional to the concentration of crosslinked receptors, rather than just the concentration of bound receptors. Since a detailed analysis of this question will be presented elsewhere[22], only a few aspects of the results will be outlined here.

In order to develop a specific example, suppose that the primary difference between agonist and antagonist lies in the flexibility of the latter[23], so that intramolecular bonding is more favorable than crosslinking for the antagonist, whereas the reverse is true for the agonist. The antagonist may then bind tightly, but it does not crosslink receptors and therefore does not activate the cell. If $m_4(c_1,c_2)$ represents the concentration of free receptor sites at concentrations c_1 and c_2 of agonist and antagonist respectively, then

$$c_2(x) = \frac{m_5 - m_4(c_1,0)\sqrt{1/x}\,(1+2K_1c_1x) - 4K_1K_2c_1m_4^{\,2}(c_1,0)}{2K_1^*\sqrt{1/x}\,m_4(c_1,0) + 4K_1^*K_3^*m_4^{\,2}(c_1,0)/x} \qquad 14$$

The * denotes antagonist affinities. The result is evidently considerably more complicated than the simple

$$c_2(x)K_1^* = x-1 \qquad 15$$

which is obtained for the Schild plot. Moreover, it displays the dependence on c_1, which is sometimes observed experimentally and was not explained previously. It is interesting, however, that under conditions which are expected to occur with reasonable frequency, eq 14 reduces to a relatively simple and intuitively direct generalization of equation 15. In particular, when multisite attachement is much more favorable than single site, and nearly maximum activity is obtained at very low degrees of receptor occupancy, eq 14 reduces to

$$c_2(x) = \frac{m_5 x}{4K_1^*K_3^*m_4^{\,2}(c_1,0)} \qquad 16$$

It is evident that over the range of concentrations for which eq 16 is valid, a plot of $\log[c_2(x)]$ against $\log(x)$ will be a straight line with a slope of 1, and that the intercept on the abscissa will be proportional to the affinity for multipoint antagonist receptor interaction. In order to obtain the actual affinity, a family of straight lines must be obtained as a function of c_1, and extrapolated to $c_1=0$.

To summarize, a crosslinking model provides a simple generalization to the single site attachment models used in the past. Under appropriate conditions, the equations reduce to those derived by Schild. The extent to which such a model is applicable to a particular system might be ascertained by methods elaborated in the previous section, or elsewhere[22].

The final example involves the reduction of an observed biological effect both by receptor blockage, and by antagonism between conflicting signals. This is exemplified by studies of histamine release from basophils currently being carried out in Reuben Siraganian's laboratory at NIH. A particularly simple and characteristic example of a dose response pattern expected on the basis of a model in which response is proportional to the number of crosslinks is shown in Fig. 6 in which release is stimulated by a bivalent hapten BPO_2** and inhibited by a monovalent analogue. As pointed out by Dembo and Goldstein[9], the curves not only rise and fall in a fairly symmetric fashion, but more signifi-

**Abbreviations: BPO_2 - Bis Benzypencilloyl 1,6 diaminohexane, BPO_{22} HSA - BPO_{22}Human serum albumin, OV- Ovalbumin cantly, the pattern of inhibition: the convergence at high concentrations and the rightward shifts in the peaks; are precisely what one expects on the basis of crosslinking.

Dose response curves often depart radically from the simple, symmetric biphasic appearance which one expects of a crosslinking function. In animal models, especially, curves rarely drop montonically to zero after reaching a peak. Instead they may plateau, or drop to a non zero plateau after peaking, or rise again after an initial decline (Fig. 7). Even when curves do fall to zero

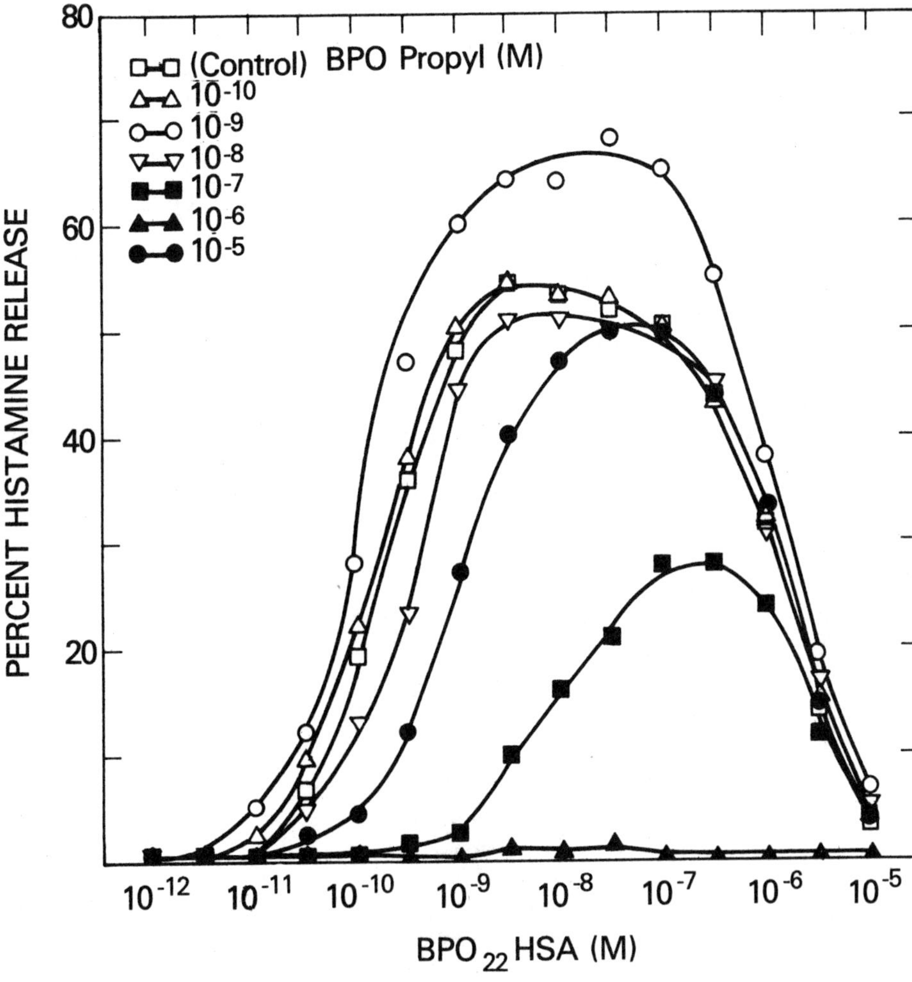

Fig. 6. Histamine release from rabbit basophils in response to $BPO_{22}HSA$. The family of curves was obtained by adding BPO propyl starting at 10^{-5} M, in decrements of ten (from ref. 24).

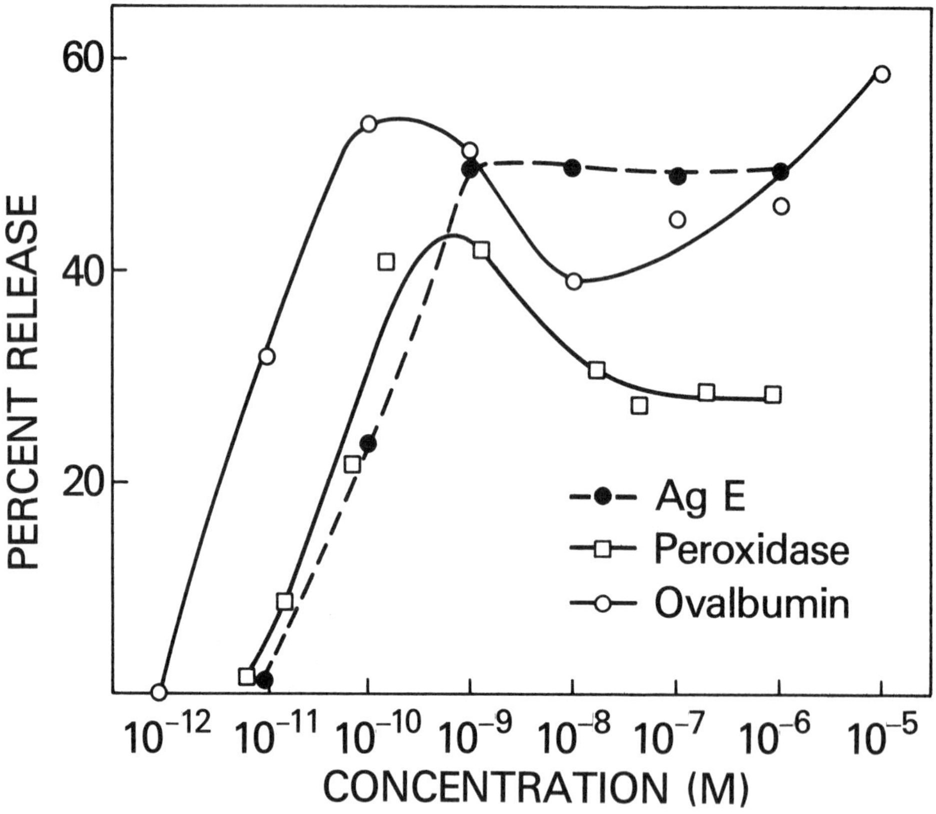

Fig. 7. Illustrates three other types of dose response patterns. Crosslinking presumably provides the signal for release with these antigens as well as for the penicilloyl hapten (from ref. 24).

at high doses, the decline does not always reflect insufficient crosslinking. Becker _et al_ [25] found by electron microscopy studies that as the size of IgE-anti IgE clusters on human basophils increased, the amount of histamine released decreased. Moreover, as shown by Magro[26], the addition of anti IgE Fab at the high dose portion of the descending limb _increased_ release. Since addition of the monomer tends to reduce crosslinking, an increase in release when the monomeric ligand is added suggests that too much crosslinking initiates an inhibitory signal. Siraganian and Siraganian[27] obtained similar results using con A to induce release and its succinilated form as a blocker of crosslinks. The results taken collectively suggest that crosslinking generates both stimulatory and inhibitory signals, the former dominating at low degrees of crosslinking and the latter at high degrees. In connection with this, it is interesting to note that since the concentration of clusters with N receptors is roughly proportional to the Nth power of the forward rate constant for collision on the membrane, small changes in membrane fluidity may be amplified into a substantial biological effect, if such effects require formation of large aggregates and a non equilibrium situation prevails. (Appendix B)

Other indications of an inhibitory signal which increases with the degree of crosslinking come from desensitization studies of rabbit basophils carried out in Siraganian's laboratory. The experiments rely on the fact that Ca^{++} is required for release, and consist of perincubating basophils in the absence of Ca^{++} for thirty minutes, followed by addition of Ca^{++} and further incubation. The result is a reduction in the amount of histamine released, with the most pronounced effect at the portion of the curve which normally shows maximum release. Since we expect that for BPO_2 or BPO_{22}, the peak in the dose response function corresponds to the maximum number of crosslinks, the desensitization results for these haptens suggest that the inhibitory signal rises and falls pretty nearly as the number of crosslinks[29]. Although these results bear on histamine release from basophils, there are numerous indications that the idea of a biological response as an interplay between independent, antagonist and separable signals may be applicable to other systems (see, for example, the review by Bretcher[30] on B-cell activation and reference 31 on muscle desensitization).

The observations that both release and desensitization rise and fall as the number of crosslinks suggest a simple interpretation of the variability in dose response patterns; viz, since an observed pattern is the result of an interplay between antagonist signals, parameters which affect the balance between these will affect the form of the curve. In order to determine whether this idea has quantitative validity, and to help pursue its consequences systematically, we have developed a simple mathematical model for release and desensitization. In keeping with observation, two types of desensitization are considered: specific, in which the cell does not release histamine in response to the desensitizing antigen, but will release it in response to any other antigen for which it is primed; and non specific in which the desensitizing antigen renders the cell unresponsive to all ligands which would normally stimulate it to release histamine.

The nature of the biochemical processes responsible for non specific desensitization are not known. It is interesting, however, that muscle contraction in response to various ligands can also be non specifically blocked, in a dose dependent manner, by preincubation in the absence of Ca^{++} [31]. This suggests the possibility that non specific desensitization of basophils may result from blocking the action of the cytoskeletal musculature. We are still a long way, however, from being able to pursue the details of such speculation experimentally, and fortunately progress on understanding the phenomological aspects of coupling between cell surface events and biological activity is not dependent upon this information.

With respect to specific desensitization, a current idea which we find attractive is that crosslinking forms Ca^{++} channels which disappear with time; e.g. crosslinked receptors may be degraded, endocytosed or shed. Such a mechanism leads to specific desensitization and possibly also to termination of release prior to complete depletion of histamine. It will not, however, lead to variable dose response patterns unless loss of channels is cooperative: i.e. unless clusters of channels are lost more or less simultaneously. Such cooperative loss is in fact what one might expect if receptor clusters are endocytosed or shed. In the model, which will be presented in detail elsewhere[32], we therefore keep track of the total concentration of clusters (Appendix B), as well as the total concentration of crosslinks, and let each <u>cluster</u>, rather than each crosslink, disappear independently. We also require, on biological grounds, that the release and desensitization signals plateau as the number of crosslinks becomes large so that the function controlling these signals may have the form, for example, of a hyperbolic function or a modified Freundlich isotherm.

The model leads to four coupled non linear differential equations: two for chemical reaction on the cell surface, one for the concentration of released histamine, and one for the concentration of non specifically desensitized histamine. If the chemical reactions come to equilibrium rapidly compared to changes in histamine content, the steady state solution to the model can be written very simply. Denoting the total initial histamine content by H_o, the concentration released, H_1, can be written in terms of a hypergeometric function:

$$H_1/H_o = \frac{\alpha}{a\lambda} \; \frac{u}{(1+f/a\lambda)(1+u)^{\alpha/a\lambda}} \; {}_2F_1(1-\alpha/a\lambda,\ 1+f/a\lambda;\ 2+f/a\lambda;-u) \qquad 17$$

where $u \equiv [Am_3(1-m_3/m_5)]^a$ and

m_3 denotes the number of crosslinks per cell immediately after chemical equilibrium is established, but prior to loss of clusters. A and a determine the midpoint and breadth of the modified Freundlich isotherm[33] α is the rate constant for histamine release, λ the rate constant for cluster loss and f the signal controlling non specific desensitization. f is itself expected to be a function of m_3. When non specific desensitization is absent, f=0 and eq 17 reduces to

$$H_1/H_o = 1 - (1+u)^{-\alpha/a\lambda} \qquad 18$$

Detalied interpretations of the qualitative features of variable dose response patterns have been presented elsewhere[24] based on a somewhat simpler model. Preliminary quantitative results obtained using eq 18 to fit release patterns of cells from a single rabbit are shown in Figs 8 and 9. Except for affinity, the same parameter set fits divalent and multivalent penicillinoyl induced release (Fig 8) suggesting that lattice formation, which can occur only for the multivalent form, does not play an important role. The lower affinity of the multivalent molecule (a factor of 35 after correcting for valence) is probably the result of a smaller forward rate constant. Part of the difference can be attributed to the slow diffusion of HSA relative to BPO_2-$(CH_2)_6$, but this cannot account for a 35 fold decrease. The remainder must be attributed to an activation enthalpy; perhaps an HSA related hydration effect, which is present for the protein but not for the hapten. A radius of 36Å for HSA[34] and 7Å for $BPO_2(CH_2)_6$[35], implies that the former effect contributes a factor of five and the latter a factor of seven.

On the membrane, differences in radii between the hapten and HSA will lead to a negligible difference in forward rate constants since both are attached to the more massive IgE molecule. Thus, with no other influences, the multivalent form should react eleven times faster, (probably less, since one can expect some sites to be hidden). Application of the enthalpy correction estimated above leads to the conclusion that the values of K_2m_5 should be roughly equal. For the purpose of completing the discussion, I note that with 5000 monospecific IgE molecules per Basophil and using $(1 \lesssim k_{-1} \lesssim 100)$ sec^{-1} as a typical range[36,37] for reverse rate constants, a value of 1.5 for K_2m_5 implies a diffusion coefficient in the range of $(10^{-11}$-$10^{-9})cm^2/sec$.

Finally, the results presented in Fig 9 indicate that Ovalbumin has an affinity for IgE which is two orders of magnitude smaller than that of BPO, and a crosslinking equilibrium constant (K_2m_5) which is two orders of magnitude <u>larger</u>. Consistency with the above discussion requires that the affinity difference between the protein and hapten be attributed to differences in their forward rate constants. In fact measured forward rate constants for Ova[37] are $(10^5$-$10^6)$ $(M\text{-}sec)^{-1}$; two orders of magnitude below those of haptens. There would, in any case, be no <u>a priori</u> reason for a reverse rate constant difference since there is presumably a representative repertoire of antibodies from which ovalbumin can select. We are thus led to the prediction (details presented elsewhere)[32] that the difference in dose response patterns between ovalbumin and the penicillinoyl hapten stems from a cell bound IgE concentration which is one to two orders of magnitude larger for the former than the latter; <u>i e</u> an increased number of IgE molecules preferentially increases the signal for specific desensitization, thus driving the release curve down at the point where crosslinking is peaking.

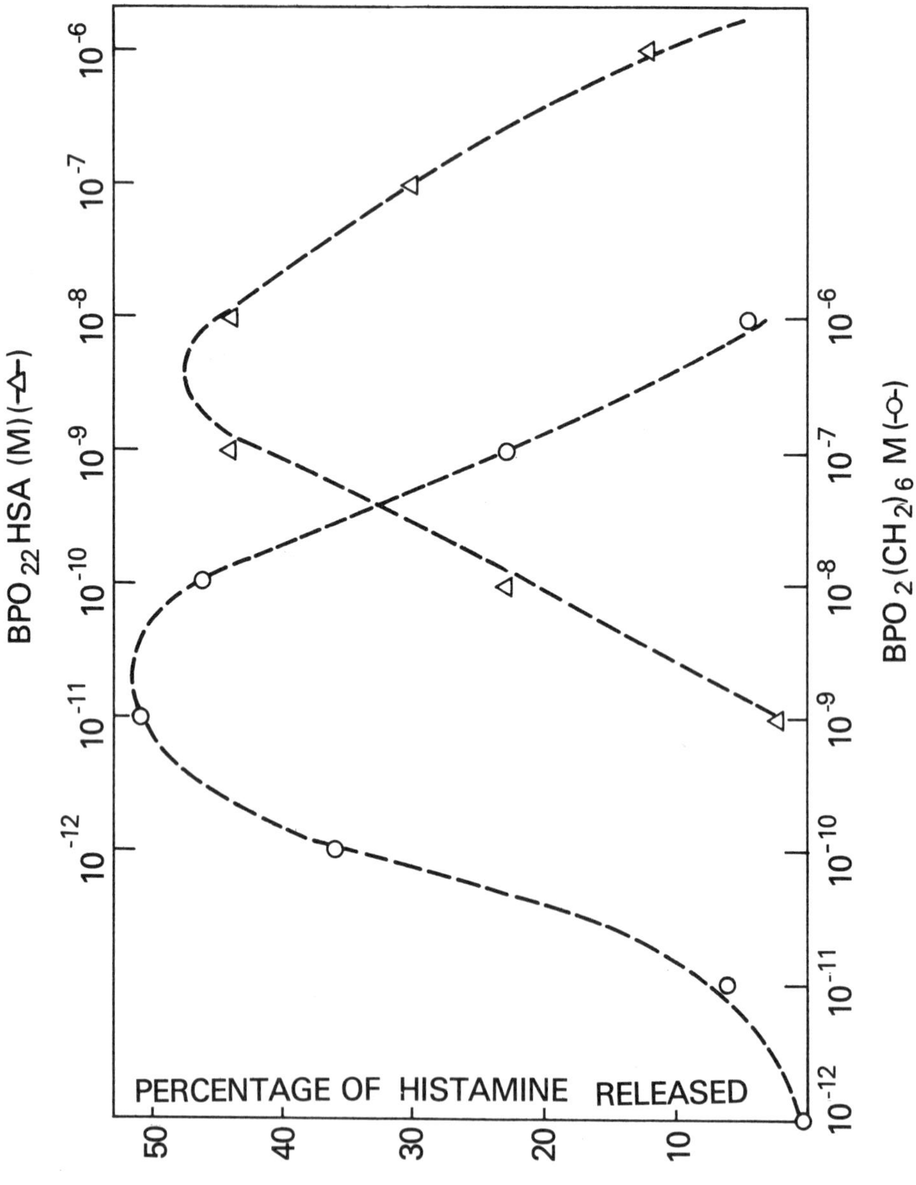

Fig. 8. Percentage of total histamine content released as a function of concentration at the steady state calculated (dashed line) using eq 18 with $K_2m_5=1.5$, $\alpha/\lambda = 0.27\pm.01$, $a = 1.4\pm0.2$ and $m_5A = 80$ (m_5A was invariant over a wide range of values). $K = 2.5\times10^8\ M^{-1}$ for $BPO(CH_2)_6$, and $10^8\ M^{-1}$ for $BPO_{22}HSA$.

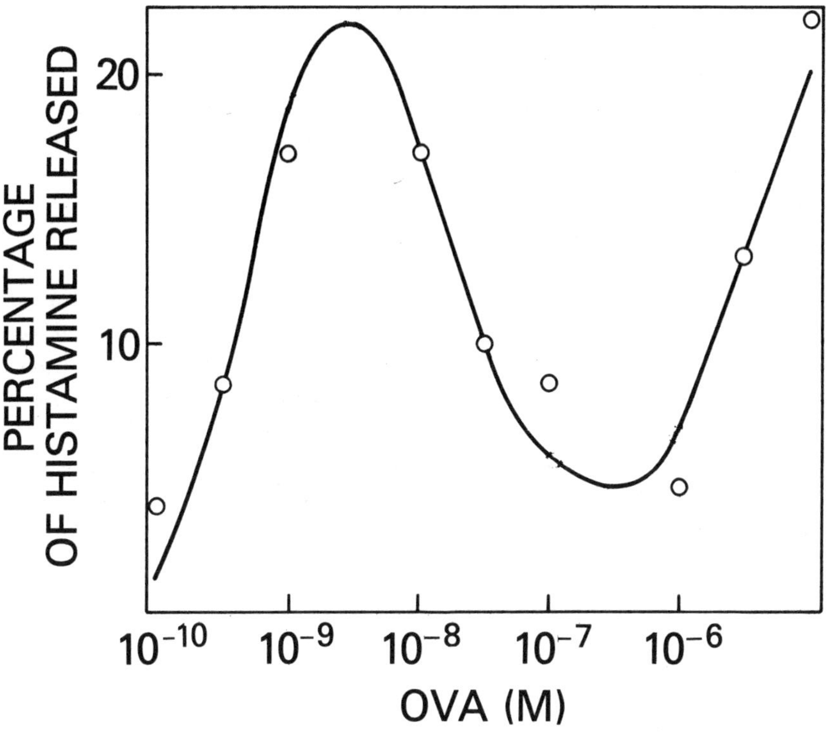

Fig. 9. Eq 18 fit to ovalbumin induced release with $\alpha/\lambda = 0.53 \pm 0.05$, $a = 3.1 \pm 0.4$, $K_2 m_5 = 200$, $m_5 A = 6$, and $K = 2 \times 10^6 M^{-1}$.

APPENDIX A. KINETIC EQUATIONS

The differential equations governing the time course of binding and cross-linking are (Fig. 1)

$$\frac{dm_1^*}{dt} = 2k_1 c^* m_4 - k_{-1} m_1^* - k_2 m_1^* m_4 + 2k_{-2} m_2^*$$

$$-k_3 m_1^* m_4 + 2k_{-3} m_3^* \qquad m_1^*(t=0) = m_1^*(0) \qquad \text{A.1}$$

$$\frac{dm_2^*}{dt} = k_2 m_1^* m_4 - 2k_{-2} m_2^* \qquad m_2^*(t=0) = m_2^*(0) \qquad \text{A.2}$$

$$\frac{dm_3^*}{dt} = k_3 m_1^* m_4 - 2k_{-3} m_3^* \qquad m_3^*(t=0) = m_2^*(0) \qquad \text{A.3}$$

with similar equations for m_1, m_2 and m_3.
These are to be solved subject to the constraint that the total receptor concentration be time independent

$$m_5 = m_1^* + m_1 + 2(m_2^* + m_2 + m_3^* + m_3) + m_4 \qquad A.4$$

The experiments discussed in this paper are carried out by allowing labeled ligand and cell membranes to equilibrate and then either diluting the system and adding cold ligand, or just diluting. In either case the initial condition on the free receptor site concentration, i.e. the concentration just prior to dilution, is:

$$m_4(0) = m_5 - m_1^*(0) - 2m_2^*(0) - 2m_3^*(0) \qquad A.5$$

It is also assumed that, in both situations, once the system has been "infinitely diluted", the binding of ligand from solution is negligible and the first term on the right hand side in eq A.1 can be neglected. The effect of cold ligand, which is not present when the system is only diluted, is to block all forward reactions, including those on the cell surface.

CASE I. Infinite Dilution with Excess Cold Ligand. The ligand concentration is supposed to be sufficiently high to rapidly saturate all free sites so that terms involving m_4 can be neglected. Then eq. A.1 integrates to

$$m_1^* = \sum_{i=1}^{3} a_i \exp(\lambda_i t_i) \qquad A.6$$

where

$$a_1 \equiv m_1^*(0) - a_2 - a_3$$

$$a_2 \equiv \frac{2k_{-2}m_2^*(0)}{k_{-1} - 2k_{-2}} \qquad k_{-1} \neq 2k_{-2}$$

$$a_3 \equiv \frac{2k_{-3}m_3^*(0)}{k_{-1} - 2k_{-3}} \qquad k_{-1} \neq 2k_{-3}$$

and $\lambda_1 \equiv -k_{-1}$, $\lambda_2 \equiv -2k_{-2}$, $\lambda_3 = -2k_{-3}$

Since the differential equation for the total bound label, B^*, can be written as

$$\frac{dB^*}{dt} = -k_{-1}m_1 \qquad A.7$$

then,

$$B^* = k_{-1} \sum_{i=1}^{3} (a_i/\lambda_i)\exp(\lambda_i t) \qquad A.8$$

CASE II. Dilution Alone. The equations are non linear and must generally be solved numerically. However when the cells are equilibrated at a dose of ligand small enough to leave most receptors free, the free receptor concentration (m_4) can be approximated by the total receptor concentration, m_5. Eqs. A.1-A.3 can then be approximated by three coupled linear differential equations whose solutions can be written explicitly as sums of three exponentials. The associated relaxation times are given by the negative reciprocals of the solutions to the cubic

$$\lambda^3 + \lambda^2(k_{-1}+k_{-2}+k_{-3}+k_2m_5+k_3m_5)$$

$$+\lambda\ [k_{-2}(k_{-1}+k_3m_5)+k_{-3}(k_{-1}+k_2m_5)+k_{-2}k_{-3}]+k_{-2}k_{-3}=0$$

The largest root to this equation (i.e. the value of λ closest to zero) is given approximately by

$$\lambda \cong -k_{-1}/(1+K_2m_5+K_3m_5)$$

The result assumes that the dissociation rate constant for singly bound antigen is small compared to the forward rate constants for multipoint attachments on the membrane. This is what one expects in those experimental situations which show a pronounced difference between dissociation kinetics with and without cold ligand.

The results may be very different for multivalent ligands (valence greater than two). Each of these can bind a number of receptors and large ligand-receptor lattices may form. Since several bonds must, in general, break before an antigen can dissociate, there will be more than two exponentials. Moreover there is a greater likelihood that some of the dissociations occurring on the membrane may in this case be diffusion limited, since if large massive networks form, they will be slow to diffuse apart subsequent to bond breakage. Dissociation kinetics in this case should be considerably more complex.

APPENDIX B. The Cluster Size Distribution Function with $K_3=0$.

Let $\hat{p}$ be the probability that a randomly picked receptor site is bound ($\hat{p}=(m_1+2m_2)/m_5$) and let p be the fraction of bound ligand sites that are on multiply bound ligand ($=\frac{2m_2}{2m_2+m_1}$). Then the concentration of receptor molecules in aggregates with N receptors and n singly bound ligands is

$$C_n(N,t) = \left(\frac{m_5}{2}\right)\binom{2}{n}(\hat{p}p)^{N-1}[\hat{p}(1-p)]^n(1-\hat{p})^{2-n} \qquad \text{B.1}$$

and the concentration of independent molecules with at least one crosslink is:

$$\sum_{n=o}^{2}\sum_{N=2}^{\infty} C_n(N,t) = m_2(1-m_2/m_5) \qquad \text{B.2}$$

Eqs B.1 and B.2 are valid for all time, so that $C_n(N,t)$ can be obtained as a function of time by solving the differential equations for m_1 and m_2[28]. Note that $C_n(N,t)$ is proportional to $(\hat{p}p)^{N-1} = (2m_2/m_5)^{N-1}$. Consequently, if the <u>intrinsic</u> rate for crosslinking (i.e. the rate when the centers of mass of the reacting units are appropriately positioned for reaction) is greater than the rate of separation by diffusion, any biological effect which depends upon large aggregate formation will be a very sensitive function of membrane viscosity.

REFERENCES

1. Siraganian, R.P., Hook, W.A. and Levine, B.B. (1975) Immunochem, 12, 149.

2. Ishizaka, T. and Ishizaka, K. (1978) J. Immunol., 120, 800.

3. Izersky, C., Taunog, J.D., Poy, G. and Metzger, H. J. Immunol. In press.

4. Schlessinger, J. This volume.

5. Bell, G.I. (1974) Nature, 248, 430.

6. Bell, G.I. (1975) Transplant. Rev., 23, 23.

7. DeLisi, C. and Perelson, A. (1976) J. Theor. Biol., 62, 159.

8. DeLisi C. and Thakur, A.K. (1977) Cell Immunol., 28, 416.

9. Dembo, M. and Goldstein, B. (1978) J. Immunol., 121, 345.

10. Wofsy, C., Goldstein, B. and Dembo, M. J. Immunol. In press.

11. DeLisi, C. and Siraganian, R.P. J. Immunol. In press.

12. Berman, M. and Weiss, M. (1978) SAAM27 Manual. U.S. Dept. of Health, Education and Welfare, NIH, Bethesda, Maryland, pp. 1-202.

13. Knott, G. (1976) MLAB Manual. Division of Computer Research and Technology, NIH, Bethesda, Maryland.

14. Fletcher, J.E., Spector, A.A. and Ashbrook, J.O. (1970) Biochem. 9, 4580.

15. DeMeyts, P. (1976) in Cell Membrane Receptors, Beers, R.F. and Bassett, F.G., eds., Raven Press, New York. p. 21.

16. Frazier, W.A., Boyd, L.F. and Bradshaw, R.A. (1974) J. Biol. Chem. 249, 5513.

17. Bell, G.I. (1974) in Proc. First Los Alamos Life Sciences Symposium, Mammalian Cells - Probes and Problems, Oak Ridge, DOE Technical Information Center.

18. DeMeyts, P., Van Obberghen, E., Roth J., et al (1978) Nature, 273, 504. and references cited therein.

19. DeLisi, C. and Metzger, H. (1976) Immunol. Commun., 5, 417.

20. For example, Lefkowitz, R.J., Caron, M.G., Mukherjee, C., Mickey, J. and Tate, R. (1976) in Cell Membrane Receptors for Viruses, Antigen and Antibodies, Polypeptide Hormones and Small Molecules, Beers, R.F. and Bassett, E.G. eds., Raven Press, New York.

21. Schild, H.O. (1949) J. Pharmacol. 4, 277.

22. DeLisi, C. Cell Biophys. In press.

23. DeLean A., Munson, P. and Rodbard, D. Molecular Pharmacol. In press.

24. DeLisi, C. and Siraganian, R.P. J. of Immunol. Submitted.

25. Becker, K.E., Ishizaka, T., Metzger, H., Ishizaka, K. and Grimley, P.M. (1973) J. Exp. Med. 138, 394.

26. Magro, A.M. and Alexander, A. (1974) J. Immunol. 112, 1762.

27. Siraganian, R.P. and Siraganian, P. (1975) J. Immunol. 114, 886.

28. DeLisi, C. and Perelson, A. Submitted.

29. Dembo, M., Goldstein, B., Sabotka, A. and Lichtenstein, L. (1978) J. Immunol., 121, 354.

30. Bretcher, P.A. (1975) Transplant. Rev., 23, 37.

31. Bown, F., Graham, J.D.P. and Taha, S.A. (1973) Eur. J. Pharmacol. 22, 64.

32. DeLisi, C. and Siraganian, R.P. Submitted.

33. Sips, R. (1948) J. Chem. Phys.,15, 490.

34. Tanford C. (1961) Physical Chemistry of Macromolecules, John Wiley and Sons, New York, p. 359.

35. See for example, Flory, P. (1968) Statistical Mechanics of Chain Molecules. John Wiley and Sons, New York.

36. Pecht, I., Givol., D. and Sela, M. (1972) J. Mol. Biol. 68, 241.

37. Dandlicker, W.B. and Levinson, S.A. (1967) Immunochem., 5, 171.

DISCUSSION

DR. DE MEYTS: Charles, first a very specific question. I would like to know what parameters you built in for simulating that analogs are displaced noncooperatively.

DR. DELISI: In order to have an asymmetric relation between the two displacement curves (Fig 3) not only must K_1 and K_2 for the cold analogues be different, but the ratio of K_1/K_2 must also be different. Thus the unlabelled ligands in the Figure have K_1M_5 = .06 and K_2M_5 = 19 (K_1/K_2 = $3.2\text{x}10^{-3}$) for the first analogue, and K_1M_5 = 0.006, K_2M_5 = 0.019 (K_1K_2 = .312) for the second. I expect the same sort of result for a heterogeneous system or for a system in which a conformational change occurs, so long as the ratios of the two equiibrium constants are sufficiently different for the two analogues.

Dr. Demeyts: I am not sure I understand correctly, but would this imply, then, that all low affinity analogs should displace asymmetrically in this system?

Dr. DeLisi: No. It would depend upon the extent to which this ratio changed. The ratio -- K1 to K2 again, let's call it, for the analog -- if the ratios are different than the displacement curves would not be parallel. If the ratios were not different, then I would expect the two displacement curves to be parallel.

DR. DEMEYTS: The second comment I would like to make is I think it is a very beautiful model. It probably applies to a large number of situations. As far as application to the insulin situation is concerned, I would like to say that we have been ourselves able to fit the scatchard plots, on one hand, and the accelerated dissociation kinetics, on the other hand, with a variety of different models, including two-step models, dissociating receptor models.

Where the problem arises is when you try to fit simultaneously the dissociation kinetics, the scatchard plot, and the association rates, where we can't so far get the fit with the simple models we have tested. And also you have to include the fact that in the insulin receptor system, the cooperative effect disappears at high concentrations of ligand, and I think we discussed this yesterday. I would like to know what you think of fitting that problem with this model.

DR. DELISI: I agree with everything you said. We haven't had a chance to do much. Ruth Chabay is now looking into this and trying to elaborate on it a bit. There is no doubt that the model as it stands in this very simple form does not include all your data. No very simple model will. The question is how much more extensive the model has to be, and where it breaks down. We do not know the answers yet.

DR. METZGER: Going back to the IgE system, I would like to ask both Micah and Charlie one question, and that is that it seems to me there is one fundamental piece of data which we don't have. We know that cross-linking is necessary, we know that dimerization is sufficient to give some sort of a response, but we really don't have any good information about what happens with trimers or tetramers.

In other words, does a trimer give you a single unit signal or two unit signals or what? Now, I would ask first of all Micah: do you have data which suggests that it is, in fact, the actual number of cross-links which determines the signal strength, so to speak, in which case one would have to say that a trimer is more effective than a dimer? Or are your data not precise for that? Because that is a fundamental piece of biochemistry that we just don't know at this point.

DR. DEMBO: In our system we can't answer your question because, unfortunately, the BPO binds to the antibody rather weakly, with an equilibrium constant on the order of $10^7 M^{-1}$, and you never get much more, according to our calculations, than a few percent in the form of trimer, even under optimal trimer-forming conditions.

So that it is a very subtle point. I mean one theoretically could distinguish various ideas as to exactly what the activity of the trimer should be as opposed to a dimer, but in practice it turns out that almost all the reasonable ideas give very similar results.

So, in one way it is good and it means that the question doesn't critically affect the model; and in another way it is bad because it means that modeling has a difficult time answerin the question.

DR. DELISI: The only data which we have that bears on this question involves the comparison between BPO_2 and BPO_{22} induced release (Fig 8). The fact that both curves can be fit with the same parameter set, except for affinity, seems to imply that to a good first approximation, only the total concentration of cross-links matters and not how they are distributed among the various sized aggregates.

Another aspect of our work which bears upon this, from a theoretical point of view, is that we can ask how sensitive our predictions are to the assumption that only the total concentration of cross-links matters. We can do this by developing a model for the opposite assumption; viz, that only the total concentration of aggregates matters, and not the number of crosslinks in the aggregate. Although the two models differ in their quantitative predictions; e.g. by how much must the number of IgE molecules per basophil be increased in order to convert the response in Fig 8 into one which is bimodal, the predictions are qualitatively the same. For the moment that is sufficient, because all we really care about testing right now is whether there is an increased desensitization when the number of IgE molecules is increased. However, we are clearly going to want more experimental information in addition to the BPO_2 VS. BPO_{22} experiments, which bears upon your question, and we are currently looking into ways to obtain it.

DR. PERELSON: Charlie, I noticed that in your model, the theoretical prediction for the $(BPO)_2$ curve was not symmetric I wonder if you would comment on that, compare it to Micah's. In the one slide that you showed comparing the histamine release curve for $(BPO)_2$ versus $(BPO)_{22}$, the $(BPO)_2$ theoretical curve was not symmetric.

DR. DELISI: It should have been symmetric.

DR. PERELSON: The second question is: I wonder if you could tell us what the four parameters were that you sort of obtained from curve fitting that go into the model.

DR. DELISI: One is equilibrium constant. One is the ratio of specific to nonspecific desensitization. I used, as Byron and Micah did, a Hill function. The other two parameters are the parameters of the Hill function.

DR. DEHAEN: I would like to raise an aspect that may bear on this model and how it applies to insulin. The German groups and some Japanese groups which I can't identify any more have described minimal peptide links that would give an insulin response, and they came up with peptides that, to my knowledge, if my recollection serves me right, were in that area in which you can cut off insulin and make the octapeptide, and it is active.

So I wonder if Dr. DeMeyts could comment?

DR. DEMEYTS: The group of Gunther Weizel has reported that small peptides of insulin have insulin-like effects on fat cells, which is quite possible, though a lot of people are skeptical about the purity of these peptides. But if you blow on a fat cell you get a biological effect. That doesn't prove anything.

We have tested some synthetic peptides on the binding, and they don't bind to the receptor at all. It is quite clear that as soon as you destroy the tertiary structure of insulin, just by cutting two amino acids, one at each end, you already reduce the biological potency to about one person. So the biological characteristics of these small peptides is really very questionable. In fact, they don't bind to the receptor when you study them directly.

I asked Weinzel himself for some peptides and he never answered, even when I wrote in German.

DR. GOLDSTEIN: I have a comment on the symmetry of the histamine release curve for $(BPO)_2$. We have emphasized that and we have seen it, and charles has seen it, and my question is: Why is it symmetric, as symmetric as it is? Because we did the calculation for a homogeneous set of IgE molecules, and now we have been able to do the theory for a heterogeneous IgE population. And when you introduce heterogeneity, you break the symmetry in the cross-linking curve.

Of course, one might say that we are lucky and we have a person whose IgE is relatively monoclonal, or it may be that the IgE response in general is not heterogeneous, but one would expect some heterogeneity and therefore some asymmetry.

One interesting thing about heterogeneity is that although it distorts the curve, it is very difficult to do anything dramatic. That is if you take two affinity distributions, take an extreme case, a person with two affinity distributions, that are widely distributed, the cross-linking curve will go up and come down. It won't do it symmetrically but it won't have a double hump.

DR. DELISI: I have been worried about something not quite identical but something similar, and that is, when you look at $(BPO)_{22}$, that is also symmetric. If you work out the theory for a multivalent, for a highly valent ligand, in fact the curve shouldn't be symmetric there either.

So with regard to the question of symmetry, I guess there are a couple of questions. One will have to look, I think, at the sorts of constraints on parameters which distinguish symmetric from asymmetric curves.

DR. WEBB: All of this discussion has been rather technical and I wanted to make a slightly different remark. Listening to these two last papers as an outsider in the field, I have the impression that the experimentalists in this area are extremely fortunate. It seems to me that these two modeling papers are just spectacularly successful. I have never seen anything in the variety of fields that I have worked in where there has been so much success in a complicated system like this by a chemical model.

DR. METZGER: Well, I think symmetrically -- and I don't feel immodest saying this because I wasn't involved -- but I think that the reason, of course, the theoreticians were attracted was by the success of the experimentalists who have really developed very nice, clean in vitro systems that are perturbable in simple ways.

Now, I am going to ask Dave Rodbard because he has something pertinent to comment on, and then Dave Segal has something somewhat related but not directly on this subject to present.

DR. RODBARD: Charles DeLisi has pointed out that the cross-linking model can give rise to a nonlinear scatchard, nonlinear dissociation, and accelerated dissociation in the presence of an increased amount of unlabeled ligand. I just wanted to point out or to emphasize, as Pierre DeMeyts has just said, that numerous other models will do this, including the subsite model developed by Andres de Leon and Peter Munson where the ligand binds in two steps or several conformations.

So at least here is another model, a closely-related model, which involves cross-linking within the molecule (i.e. intramolecular reaction) forming the active pool, again, rather than cross-linking between the molecules; and, of course, this can be tested by a variety of experimental approaches.

DR. DEMEYTS: For example?

DR. RODBARD: Well, this model doesn't help us with clustering. This doesn't give capping or things like aggregation; unless we superimpose that in this particular case.

I might just comment that there is another danger here that in order to do what you are doing with your model, Charles, I understand that you are assuming that the insulin forms dimers due to the very high concentration at the surface of the membrane. I might just point out that if you are forming dimers at the surface of the cell, at the surface of the membrane, this could also give rise to exactly the same phenomenon, that is, an unstirred layer, non linear scatchard, non linear dissociation, and apparent cooperativity.

One slide more, please.

DR. DEMEYTS: Still it is linear scatchard plots, I am sorry. It is a purely kinetic phenomenon.

DR. RODBARD: I concede. Slide, please. This is some data kindly provided to me, unpublished by Len Harrison, on the insulin receptor placenta solublized by triton, and in doing an experiment previously done by Pierre on a variety of cells showing the increase in the rate of dissociation of the labeled insulin as a function of the unlabeled insulin in the medium.

I just show this because of perhaps a superficial striking resemblance to the histamine release curves where we have a fairly symmetrical relationship covering seven or eight orders of magnitude. And at first you might say this looks very much like the percent of immunoglobulins cross-linked as a function of ligand concentration, but if that were true, then this leads to a rather puzzling question, and that is: If this is related to the percent cross-linking to the receptor with a peak at about this concentration, then we would expect a lower dissociation.

That is, if we have cross-linking of the ligand we might expect a lower rate of dissociation rather than a more rapid rate of dissociation. This might suggest that there is a combination of cross-linking and negative cooperativity operating simultaneously.

Thank you.

DR. DELISI: You said so many important things, I hardly know where to begin. But perhaps I could begin with that last picture. It is true even in the basophil system that even in those cases in which the curve peaks and then drops at high concentrations, it does not always come down due to lack of cross-linking. It often drops to zero well before the peak in the cross-linking curve.

In the basophil system the reason for that, or the reason we propose for that has to do with the presence of another signal -- viz, that cross-linking not only stimulates release, but desensitization as well. So it is in general more complicated than simply following cross-linking curves.

With respect to your multisubsite model, I doubt that it will fit Pierre's data without modification. The reason I say this is that Pierre equilibrates, his system at relatively low concentrations of insulin.

If you equilibrate at low concentrations, what you have got is mostly intramolecular complex in your model. You have got very little singlybound because that is less stable. Consequently, the amplitude of this component will be very small and you won't be able to fit his data.

I have the intramolecular complex in my model as well and the only reason I can fit his data is because I do have both that and the cross-linked state. That is the only reason it works. Your model is in fact a special case of the one I presented, in which cross-linking is ommitted.

There was something else you touched upon which is also very important, which probably bears on the philosophy of modeling. As Pierre has pointed out and you pointed out, there are numerous two-step models. It happens that this cross-linking model just seems natural to me because one knows that things are diffusing -- I mean that is what half this conference is about. There is a lot of mobility on the surface. And there is an additional constraint in this calculation, namely, that one predicts the rate of formation of cross-links on the surface, and that rate leads to a diffusion coefficient which is just in the range of what is measured. That is not a rigorous test but at least makes some sense and ties things together.

So there are additional constraints. One has to worry about not just choosing a model with any old parameters, but what the interpretation of parameters are that makes some sense in terms of what we know already.

DR. HECHTER: Can I make a comment?

DR. METZGER: Okay.

DR. HECHTER: This very ingenious modeling effort illustrates how alternative possibilities can be generated. When, however, one talks about a polypeptide-like insulin, on the one hand, and immunoglobulins, it is quite possible that you get divalent, trivalent, cross-reactions of this type working one way, and with insulin quite another thing.

There is a very clean case in polypeptide chemistry. We make radioactive AVP. We make it monomer and it is biologically active. If we isolate a dimer through disulfide thiol exchange, we get an inactive compound. If it goes to any higher order, it is absolutely inactive. It is a peptide. It is a clean peptide.

The story is the same with oxytocin. It is a classic rule of neurohypophyseal peptide chemistry that any polymerization leads to biological activity loss. We have also demonstrated that you lose binding activity.

So, what I am trying to say is that a mechanism that may be really attractive, fundamental and very important for one set, extrapolated across the board because we are all talking about receptors, may lose part of its power.

DR. METZGER: I am not sure that any of us are particularly extrapolating. I think each of the groups recognize that there are fundamental differences.

DeLisi/Blumenthal, eds. Physical Chemical Aspects of Cell Surface Events in Cellular Regulation

ON DEDUCING THE NATURE AND EFFECT OF ATTRACTANT-RECEPTOR BINDING FROM POPULATION MOVEMENTS OF CHEMOTACTIC BACTERIA

LEE A SEGEL
Department of Applied Mathematics, Weizmann Institute, Rehovot, Israel and Department of Mathematical Sciences, Rensselaer Polytechnic Institute, Troy, N. Y. 12181*

ABSTRACT

Conditions are derived for the validity of a relatively simple model of population chemotaxis. The principal application is the analysis of experiments in the literature concerning populations of chemotactic S. typhimurium moving in gradients of L-serine. It is concluded that the experiments provide evidence for negative cooperativity in, or two sites for, attractant-receptor binding, and evidence against a ratio mechanism for the probability of reversal.

INTRODUCTION

Motile bacteria typically swim in more or less straight lines or "runs" for an exponentially distributed length of time and then enter a relatively brief phase of "tumbling" after which they head in a new direction that differs from the old according to a certain probability distribution. When swimming up a gradient of attractant, chemotactic bacteria apparently sense an increasing number of attractant-receptor complexes and use this information to decrease their tumble frequency; in contrast, tumbling frequencies seem little altered when swimming is predominantly down-gradient. Longer persistence in the direction of the gradient results in a chemotactic drift that brings bacteria toward relatively high attractant concentrations. For references on the basic biological facts cited in this paper, see the reviews in Refs. 1-3.

In this communication we shall discuss something of what can be learned from experiments with populations of bacteria. In these experiments the bacterial density b (number per unit volume) is observed as a function of position and time. We shall center our attention on the fact that if the experimental conditions are properly arranged, the population observations can be used to determine the binding law that governs the substrate-receptor interaction.

THE PARAMETERS OF POPULATION MOTION

Which parameters govern population motions? In answering this question we begin with the case where there are no chemical gradients, so that the motion of

*On leave, 1978-9 academic year, at Division of Applied Sciences, Pierce Hall, Harvard University, Cambridge MA 02138.

the bacteria can be regarded as random. Under such circumstances, the population will appear to diffuse. Let us consider in outline how to formulate the equation that governs the population density b in such circumstances, assuming for simplicity that b depends only on the single spatial coordinate x and the time t. A fairly detailed derivation can be found in Ref. 4.

Consider situations where bacterial births and deaths can be neglected. Then the number of bacteria in a given region of space can change solely because of net bacterial flow across the boundaries of this region. The mathematical expression of this conservation law can be shown to take the form

$$\partial b/\partial t = - \partial J/\partial x \ . \tag{1}$$

Here the flux density $J(x,t)$ is the net rate at which bacteria cross a unit area placed perpendicular to the x-axis. The net rate is obtained by counting crossings in the direction of increasing x as positive and the opposite variety as negative.

The simplest reasonable assumption to make concerning J is that it is proportional to the gradient in bacterial density. This gives, for some proportionality coefficient μ,

$$J = - \mu(\partial b/\partial x) \ . \tag{2}$$

Combination of (1) and (2) yields the diffusion equation

$$\frac{\partial b}{\partial t} = \frac{\partial}{\partial x}[\mu\frac{\partial b}{\partial x}] \ . \tag{3}$$

If μ is constant, (3) takes the more familiar form

$$\frac{\partial b}{\partial t} = \mu \frac{\partial^2 b}{\partial x^2} \ . \tag{4}$$

The success of diffusion theory in accounting for random particle motions of all kinds gives confidence that random bacterial motion is completely determined by (3) together with appropriate initial and boundary conditions. This means that population observations can permit determination of μ. (A simple procedure is suggested in Ref. 5).

Aside from μ, there are other parameters associated with random bacterial motion. For example the motion may be idealizable as consisting of straight line runs of uniform length L, carried out at uniform speed v, with random changes of direction between runs. We emphasize that v and L cannot be separately determined from population observations; they influence population movements only through the combination that gives rise to μ, since $\mu = (1/6)v^2/L$ (as is shown in elementary treatments of random walk theory).

In the presence of chemical gradients, chemotaxis must bring about a modification of (2). A "drift" in the direction of the gradient will be added to the purely random flux of (2). This is expressed by

$$J = -\mu(\partial b/\partial x) + bV . \qquad (5)$$

Once again, there are many other parameters aside from V that are associated with chemotaxis. An example[6] is the "up-down ratio", the ratio of bacteria moving up the gradient at a particular point to those moving down the gradient. But only V enters the determination of population motion, so that only V can be fully determined by observations of this motion.

THE AIM OF THE PRESENT WORK

Good motivation for the mathematical study[7] on which the present exposition is based is provided by the fact that this study shows that an "intuitively obvious" and widely used assumption that underlies (5) is not generally correct. In many situations (5) must be replaced by

$$K(\partial J/\partial t) + J = -\mu(\partial b/\partial x) + bV \qquad (6)$$

where K is some coefficient. The presence of $\partial J/\partial t$ is an expression of a usually-neglected history dependence or "inertia" in the system. That is, generally the flux J depends not only on the present state of the system, but also (with a "fading memory") on its past history.

If the state of the system is changing sufficiently slowly, the effect of history dependence should be negligible. But how slowly is "sufficiently slowly"? More generally we must determine conditions under which the simple equations (1) and (5) can be expected to be valid. Our strategy will be to write equations corresponding to a fairly general view of chemotaxis and then to simplify these equations. By checking that omitted terms are indeed negligible, we will be able to derive conditions under which our simplifications are permissible. We will omit all mathematical details. These can be found in Ref. 7.

A MODEL FOR CHEMOTAXIS

We shall base our model on the experimental findings outlined in the introduction. In our one-dimensional view of chemotaxis, the x-axis is directed along the gradient, and the velocity along this axis is actually a projection of the true velocity. Bacteria move either to the right (in the direction of increasing x) or to the left. A tumble corresponds to a reversal in the direction of movement.

It is convenient to distinguish between unbound receptor sites that are associated, respectively, with left- and right-moving bacteria. The number per unit volume of these sites, in the neighborhood of point x at time t, will be denoted by $E^+(x,t)$ and $E^-(x,t)$. The corresponding densities of receptor-attractant complexes will be denoted by $C^+(x,t)$ and $C^-(x,t)$.

Our model will embody the observation that the probability of bacterial tumbling obeys a Poisson distribution. In the present context, this means that the probability of a right moving bacterium at (x,t) reversing during a short time interval t, $t + \Delta t$ is given by $\sigma^+(x,t)\Delta t$, for some function σ^+. The corresponding function $\sigma^-(x,t)$ gives the probability per unit time of a left moving bacterium reversing.

Ignoring births and deaths, we now postulate that the temporal change of E^+, the density of unbound receptor sites on right moving bacteria, comes from (a) the net flow of right moving bacteria into the spatial region under examination, (b) an increase in unbound sites due to a break-up of complex, (c) a decrease due to the binding of attractant (concentration s) with complex, (d) an increase owing to the reversal of left-moving bacteria, (e) a decrease owing to the reversal of right-moving bacteria. (The last two effects arise simply because when a right-moving bacterium reverses, for example, its unbound receptors must be subtracted from the right-moving pool and added to the left-moving pool.) The five effects just cited are responsible, in order, for the five terms on the right-hand side of the following equation:

$$\frac{\partial E^+}{\partial t} = - \frac{\partial}{\partial x}(vE^+) + k_{-1}C^+ - k_1 sE^+ + \sigma^- E^- - \sigma^+ E^+ . \tag{7}$$

The model description continues with similar equations for the time derivatives of C^+, E^-, C^-. There is an equation stating that attractant density s changes by binding and unbinding with receptor, by possible bacterial consumption and by diffusion.

In postulating the final fundamental equation of the model, let us assume that the attractant concentration increases as x increases. Bacteria moving rightwards (up the gradient) will then be associated with a reversal probability that is an increasing function of the temporal rate of change in the number of their occupied receptor sites. If $n^+(x,t)$ denotes the density of right moving bacteria, then C^+/n^+ is the average number of receptor sites on right moving bacteria. It is convenient and causes no loss of generality to normalize the number of bound sites by the total number N of sites on an average bacterium. With all this, our key assumption on reversal probability is that for some increasing function f

$$\sigma^+ = f\left\{\left(\frac{\partial}{\partial t} + v\frac{\partial}{\partial x}\right)\left(\frac{C^+}{Nn^+}\right)\right\} . \tag{8a}$$

Here we have taken account of the fact that a bacterium senses a temporal change in receptor occupancy both because of a local change of conditions at its instantaneous position $(\partial/\partial t)$ and because the bacterium is moving at speed v into an adjacent region where conditions in general will be different $(v\partial/\partial x)$.

The reversal probability of bacteria moving down the gradient remains constant to first approximation, so that

$$\sigma^- = f(0) . \tag{8b}$$

The total bacterial density b is obtained by summing right- and left-moving bacteria

$$b = n^+ + n^- , \tag{9}$$

and the flux density J is obtained by subtracting left- from right-moving flux densities

$$J = v(n^+ - n^-) . \tag{10}$$

With these relations, manipulations of the basic equations of our model can be shown[7] to imply the conservation law (1) as well as to yield

$$\frac{\partial}{\partial t}\left(\frac{J}{v}\right) + \frac{J}{v}(\sigma^+ + \sigma^-) = -\frac{\partial}{\partial x}(vb) - b(\sigma^+ - \sigma^-) . \tag{11}$$

Equation (11) has the form of the history effect equation (6) that we discussed above.

SIMPLIFICATION FOR SMALL TEMPORAL AND SPATIAL GRADIENTS

Let us now discuss simplification of our basic model under the assumption that there is little difference between left- and right-moving bacteria, i.e. that the differences $E^+ - E^-$, $C^+ - C^-$, and $n^+ - n^-$ are small. It can be shown[7] that for consistency this assumption requires that spatial derivatives are proportional to ε and temporal derivatives to ε^2, where $\varepsilon << 1$ is a measure of the common magnitudes of $E^+ - E^-$, $C^+ - C^-$, etc. Then the receptor-attractant chemical reaction can be regarded as at equilibrium at the local attractant concentration. Deleting small terms, one <u>deduces</u> the basic flux equation (5). If v is assumed to be constant, μ and V are found to be given by the following expressions:

$$\mu = \frac{v^2}{2f(0)} \quad , \quad V = -\frac{v^2 f'(0)}{2f(0)} \frac{d\hat{C}}{ds} \frac{\partial s}{\partial x} . \tag{12a,b}$$

Here $\hat{C} \equiv (2C/Nb)$ is the total number of receptor molecules divided by the total number of sites, i.e. the average number of bound receptors.

Let us consider the more general case where v is allowed to vary. A rather general assumption is that for some function g

$$v = g(s, \partial s/\partial x) . \tag{13}$$

It is quite reasonable to assume a dependence on s, for a change in substrate concentration might well enhance or depress the speed of swimming. The necessity of assuming a dependence on $\partial s/\partial x$ follows from the fact that v is a projection on the direction of the attractant gradient of the true velocity. The bacterium is expected to be more aligned along the x-axis when gradients are large, so that one anticipates that $|v|$ will increase with $|\partial v/\partial x|$.

Extension of our analysis shows that when gradients are small, as we are assuming, the dependence of v on $\partial s/\partial x$ (and, even more, on higher derivative terms) makes a negligible correction to (12b). If $v = v(s)$, one must add the term $(v/2\sigma)dv/ds$ to the right side of (12b). In most cases, it is found experimentally that the linear speed does not in fact noticeably depend on substrate concentration, so that (12b) can generally be taken as it stands.

Since the gradient dependence of v has been found to be negligible, the true average speed v_{av} can be obtained by assuming that it has equal projections on all three axes. Under such circumstances, Pythagorus' theorem yields

$$v_{av} = \sqrt{3}v . \tag{14}$$

We have seen that the key formulas (5), (12) and (13) are valid approximations when the bacteria experience sufficiently slow changes in their environment. Further analysis[7] shows that changes will indeed be sufficiently slow, i.e. $\varepsilon << 1$, if the drift speed $|V|$ is small compared to the linear speed $|v|$, or, equivalently, if the length of a typical run is small compared to the distance over which attractant concentrations change significantly.

Another assumption whose validity must be scrutinized is that wherein we took the reversal probability σ^+ to be a function of the time derivative of the concentration of receptor-attractant complex C. As Spudich and Koshland[8] showed in essence, the dependence of σ^+ on the rate $\partial C/\partial t$ is based on the fact that the number of tumbles suppressed is a function of the change in occupancy ΔC. As is shown formally in Ref. 7, the rate dependence of σ^+ then follows if the concentration change Δs experienced by the bacteria in all small time intervals Δt is small enough so that most tumble suppression by an instantaneous concentration change ΔC will be completed within Δt time units. Let us take as typical the result that tumble suppression from a 0.02mM jump in serine concentration is completed after about a minute, i.e. that a 3×10^{-4}mM concentration change can be accommodated each second. Then a bacterium moving at a typical speed of 15μ/sec fulfills the requirement on concentration change if it faces gradients of less than 2×10^{-5}mM/μ .

We note here that a formal analysis along the lines laid out above[9] shows that the hypothesis that the tumble probability σ^+ depends on the receptor occupancy C^+ (not its derivative) is not tenable. This hypothesis requires that the average tumble probability in positive chemotaxis be an increasing function of attractant concentration. But experiments show that this function is either constant or decreasing.

ANALYSIS OF SOME EXPERIMENTS WITH SALMONELLA TYPHIMURIUM

The usefulness of our findings can be seen by examining experiments and their analysis as considered by Dahlquist, Elwell, and Lovely[6]. These authors first considered the average velocity of a population of Salmonella moving up various gradients of the attractant serine. Observations were all taken at the same serine concentration, 10^{-3}M. The length scale of the gradient $s/(ds/dx)$ varied from $4\times10^{4}\mu$ to $4\times10^{3}\mu$. In contrast to our assumption that V is proportional to s', Dahlquist et al found that the data was well fit by the saturating curve

$$\bar{V} = V_{max} \frac{s'/s}{\gamma_0(s) + s'/s} \quad , \quad s'(x) = \frac{ds(x)}{dx} . \tag{15}$$

if $V_{max} = 7\mu/sec$ and $\gamma_0 = 0.25\ mm^{-1}$.

The length scale in these experiments is satisfactorily long compared to a typical observed run length of 100μ, so that the simplifications we made under the hypothesis that $\varepsilon << 1$ should be appropriate. On the other hand, knowing ds/dx and s, one can compute that ds/dx varies from 2.5×10^{-5}mM/μ to ten times this value. The first value is about at the boundary of our estimate of gradients that permit the assumption that σ^+ depends on $\partial C^+/\partial t$, while the second value exceeds this boundary. It is consistent with our estimates, then, that the drift velocity is observed to saturate. According to our point of view, the reason for the saturation is that the greater tumble suppression expected from steeper gradients is not fully realized because the bacteria experience the increase in attractant concentration too quickly to permit this increase to suppress its full quota of tumbles.

Dahlquist et al[6] analyzed a second experiment in which serine was maintained with the following spatial distribution

$$s(x) = s_0 \ , \ x < 0 \ ; \ s = s_0 \exp(-\gamma x) \ , \ x > 0 .$$

By observing the accumulation of bacteria at the "knee" in the attractant profile, $x = 0$, the drift velocity was measured as a function of the "plateau concentration" s_0. The length scale of attractant variation was kept at the constant value γ throughout the experiment. The experimental points are shown on Fig. 1.

Dahlquist et al[6] suggested that V_{max} could be regarded as independent of the concentration s_0, since the swimming speed was observed not to depend on s_0. They thus fit the data by choosing γ_0 to be a function of s_0 that has the value of 3.5 at $s_0 = 10^{-6}$M, has a minimum of 0.25 at 10^{-3}M, and then increases slightly at higher concentrations.

With Michaelean binding, our central result (12b) can be written

$$V = \chi_M \frac{d\hat{C}}{ds}\frac{\partial s}{\partial x} \quad , \quad \hat{C} = \frac{s}{K + s} , \tag{16a,b}$$

where χ_M is a constant.

Segel[7] chose the parameters in (16) to obtain the dashed curve in Fig. 1. The better fit of the solid curve was obtained by taking $n = 1/2$ after replacing (16b) by

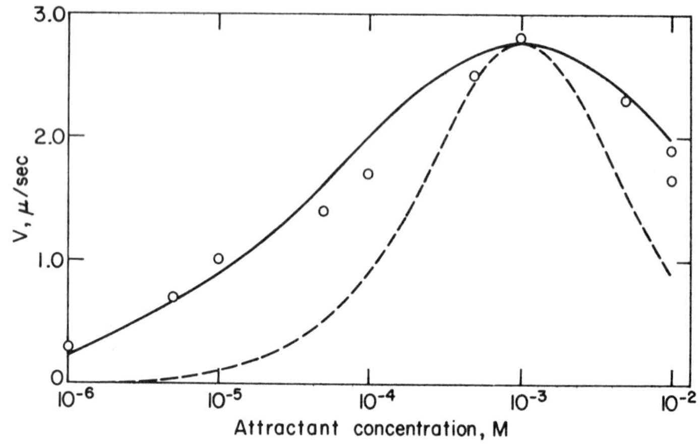

Fig. 1. Experimental[10] values of drift velocity V (small circles) and theoretical values with the Michaelean kinetics of (16) (dashed line) and the negatively cooperative or two-site kinetics of (17) with n = 1/2 (dashed line). From Ref. 7.

$$\hat{C} = s^n/(K^n+s^n) \tag{17}$$

and keeping the same values of V_{max} and K. Segel noted that a value less than unity like $n = 1/2$ can be accounted for by the existence of negative cooperativity in binding. He should also have pointed out the well known fact[4] that a two-site binding model gives the same results. Dahlquist et al[6] had earlier observed that the data does not seem to be well fit if Michaelean binding is assumed, and that a two-site hypothesis is made more likely by the unsuccessful attempts to isolate serine receptor mutants for E. Coli.

The gradients in the experiment under discussion range from 1.5×10^{-7}mM/μ to 1.5×10^{-3}mM/μ. (In this range $\gamma_0(s) >> s'/s$, and (15) is consistent with (16).) Only up to a plateau concentration of 10^{-4}M, then, are the gradients less than the 2×10^{-5}mM/μ estimate for the largest gradient permitting the hypothesis that tumbling probability has a simple rate dependence on $\partial c^+/\partial t$. There thus seems little justification for Segel's attempt to explain the observed maximum in Fig. 1 on the basis of the approximate formula (5) and (16a): at the very large gradients that correspond to the right side of the figure, the approximations are unlikely to be valid.

The fact that the approximations of (5) and (16a) are only reasonable for the left portion of Fig. 1 also weakens the case for the necessity to invoke either negative cooperativity or two sites to fit the experimental points. Nonetheless, the case remains reasonably strong even if one confines the comparison of the curves for $n = 1$ and $n = 1/2$ to the left half of Fig. 1, for the former curve seems clearly not to have the general shape required by the data.

SUMMARY AND CONCLUSIONS

We have been concerned with the flux J of chemotactic bacteria with density b. The simplest reasonable view is that J results from a chemotactic drift V that is superimposed on random diffusion. In mathematical terms

$$J = -\mu(\partial b/\partial x) + bV . \qquad (5)$$

By simplifying a model that seems to contain the essence of the fundamental observations concerning bacterial chemotaxis we have been able to show that (5) is valid when spatial changes are slow and temporal changes even slower [$\partial/\partial x = O(\varepsilon)$ and $\partial/\partial t = O(\varepsilon^2)$ respectively, where ε is the ratio of linear bacterial speed to the chemotactic drift speed and $\varepsilon \ll 1$.] Under these circumstances the formulae in (12) relate the effective diffusion coefficient μ and the drift velocity V to various factors, notably $\hat{C}(s)$, the average number of bound receptors when receptors are in chemical equilibrium with the local attractant concentration s.

If spatial and temporal changes are not sufficiently slow to warrant the simplifications we have made, it is difficult to extract significant molecular information for population observations. There are at least three major complications when spatial and temporal gradients are large. (a) The flux depends not only on the present chemical environment of the organism but also on its past history. (b) There is no longer adequate justification for passing from the occupancy law that governs tumble suppression to a rate law for the tumble probability per unit time. (c) The results depend strongly on the function f that represents how information about the state of the receptors is transduced into instructions concerning tumbling.

All the above-mentioned complications are absent in many experimental situations where, as we have seen, conditions do indeed permit the simplifications leading to (5) and (12). In particular, the transduction function f enters only in the form of the constants $f(0)$ and $f'(0)$ and so does not affect the qualitative nature of the results. The major consequence of this fact is that in small gradients, the manner in which the chemotactic drift depends on attractant concentration provides complete information on the equilibrium binding relation between attractant and receptor. Let us again stress that in large gradients the drift velocity results from a tangled combination of chemical history and the details of transduction.

We have discussed how our general conclusions apply to the analysis of experiments involving the chemotaxis of *Salmonella* to serine. Taking this discussion a little further, we note that Dahlquist *et al* showed that a "ratio mechanism" for what we call σ^+ could be ruled out as a way to explain the small gradient observations on the left half of Fig. 1. They pointed out that "most chemical mechanisms which one might imagine to be important for chemotaxis are ratio taking in nature" and concluded their paper with the remark that "if a ratio-taking mechanism is involved, some other steps in the data-processing machinery must be involved which modify the observed concentration dependence of the response." It is our principal point that the data processing machinery is only involved via the constants $f(0)$ and $f'(0)$, so that the experiments under consideration seem definitively to disprove hypotheses based on a ratio mechanism.

ACKNOWLEDGEMENT

This work was supported by National Science Foundation Grant MCS76-07492.

REFERENCES

1. Adler, J. (1975) Ann. Rev. Biochem., 44, 341.
2. Berg, H. (1975) Ann. Rev. Biophys. Bioeng., 4, 119.
3. Koshland, D. E. Jr. (1974) FEBS Letters, 40, Supplement, S3.
4. Segel, L. A. (ed) (1979) Mathematical Models in Molecular and Cellular Biology, Cambridge University Press.
5. Segel, L. A., Chet, I. and Henis, Y. (1977) J. Gen. Microbiol., 98, 329.
6. Dahlquist, F. W., Elwell, R. A. and Lovely, P. S. (1976) J. Supramol. Structure, 4, 329.
7. Segel, L. A. (1977) SIAM J. Appl. Math., 32, 653.
8. Spudich, J. L. and Koshland, D. E. Jr. (1975) Proc. Nat. Acad. Sci. (USA), 72, 710.
9. Segel, L. A. (1976) J. Theoret. Biol., 57, 23.
10. Dahlquist, F. W., Lovely, P. and Koshland, D. E. Jr. (1972) Nature New Biol., 236, 120.

V

PHYSICAL CHEMICAL ASPECTS OF CELL SURFACE EVENTS IN THE REGULATION OF THE IMMUNE RESPONSE

CHAIRMAN'S REMARKS

Cell surface receptors have two principal functions - to recognise and bind a specific ligand and to transduce this binding event into an intracellular signal. While a great deal is known about the recognition properties of lymphocyte receptors, almost nothing is known about their transducing function. In a way, this is surprising, especially in the case of antigen-specific receptors on B lymphocytes which we know are antibody molecules. Far more is known about the structure and genetics of these receptors than is known about any other vertebrate cell surface receptors. In addition, for 10 years we have been able to visualise these receptors by the use of labelled anti-immunoglobulin antibodies or antigens and thus follow the fate of ligand after it binds to receptor, in a way that endocrinologists are only now beginning to be able to do with labelled hormones. And yet, we still know almost nothing about the signalling process at the molecular level. Why should this be so?

One important reason is that almost all lymphocyte responses are complex and involve the interactions of two or more cell types. For example, even the simple proliferative response of T lymphocytes induced by the lectin Concanavalin A appears to involve at least three cell types. In addition, this lectin binds to a large number of different membrane glycoproteins, only some of which presumably play a role in signalling the response being measured. Thus, trying to draw conclusions about relevant molecular events following ligand binding is a pretty fruitless enterprise. This is why the study of basophils and mast cells has been so much more rewarding than the study of lymphocytes in terms of understanding receptor events in cell signalling. The response of these cells is remarkably simple and involves one ligand, one receptor and one cell type.

One might ask why lymphocyte responses are so complex? Why are so many cells involved at every stage? You might find the answers if you tried designing an immune system from scratch. You would be faced with three different kinds of problems - all of them problems of recognition and regulation. Firstly, you would have to find a way of having the system recognise a virtually infinite universe of foreign antigens and be able to respond in a highly specific way to any one of them. Secondly, since this responses must be designed to inactivate and/or eliminate foreign invaders, and therefore must be destructive, you would have to find a way of avoiding responses to self antigens. Thirdly, you would want to assure that the type of response was appropriate to the particular invader; for example, the response to a bacterial toxin would probably have to be different from the response to certain viruses.

The immune system has evolved elegant solutions to these three problems. It has solved the first problem by inventing a unique class of proteins, antibodies, which can exist in a million different forms - each with a unique recognition site for antigen. By distributing these antibodies on the surface of a million different lymphocyte clones so that each clone recognises a specific antigen, and by letting antigen select the appropriate clones to become activated, the specificity of the immune reponse is assured. This is the clonal selection hypothesis for which there is now compelling evidence.

The immune system has solved the second problem by devising ways of eliminating or suppressing those lymphocyte clones which have receptors for self antigens. We still do not understand the regulatory mechanisms involved in such 'self tolerance' but there is increasing evidence that more than one mechanism is involved.

The immune system has solved the third problem by evolving a variety of different types of responses - some involving the secretion of various classes of antibodies and others mediated by a variety of different subclasses of T lymphocytes. But how does the immune system know which types of response to mount against a particular antigenic challenge? It seems likely that this will turn out to be the principal role of the major histocompatibility complex glycoproteins (H-2 or HLA and Ia): by associating with foreign antigens in specific ways they may direct each class of antigen to the appropriate part of the immune system.

In this session, we are going to ignore much of the complexity of the immune system and concentrate on B lymphocytes. The only known function of these cells is to make antibodies. Early in their lives they make antibody and insert it into their plasma membranes as receptor for antigen. When these cells bind antigen, under the appropriate circumstances (which involves complex interactions with macrophages and various subclasses of T cells) they are stimulated to divide and differentiate into effector cells which make more of the same antibody and secrete it into the blood.

We shall begin with Bill Paul who will describe the proliferative response of B cells induced by anti-immunoglobulin antibodies. He will be followed by Greg Siskind who will talk about the control of antibody affinity and we shall finish with George Bell who will put it all into a theoretical framework for us.

CELL SURFACE RECEPTORS FOR IMMUNOGLOBULIN G

DAVID M. SEGAL, JULIE A. TITUS, AND JOYE F. JONES
Immunology Branch, National Cancer Institute, NIH, Bethesda, Maryland 20014

INTRODUCTION

The immune system responds to challenges from a vast number of antigens by synthesizing an equally impressive variety of antibodies, each able to bind specifically to a particular immunogen. These antigen-antibody complexes are subsequently removed or neutralized by various effectors[1] including phagocytic or cytolytic leukocytes. Immune complexes are recognized by these effector cells via surface receptors for immunoglobulin[2,3].

Antibody molecules contain variable regions, which bind antigens and constant regions, which interact with immune effectors[4]. Based on constant region differences, immunoglobulins fall into five major classes. In this article we will consider how immunoglobulin G (IgG), the major class in serum, interacts with receptor-bearing cells. The structure of the IgG molecule shown schematically in Fig. 1, reflects the dual function of this class of proteins.

The antigen binding functions are restricted to the N-terminal (Fv) portions of the two Fab fragments of each molecule, and these vary widely within the total immunoglobulin pool. In contrast, the cell-binding portions of IgG molecules reside near the C-terminal end, on the Fc fragments, and these segments are nearly identical in all IgG molecules. Therefore, although the concentration of any given antibody is typically low, the overall concentration of ligand for the "Fc receptors" in serum is high (10^{-4}-10^{-5}M). Receptors for IgG differ from others (e.g., hormone receptors) in that they must recognize very low amounts of antigen-associated Fc fragments in the presence of huge excesses of free immunoglobulin.

Effects of forming multivalent immunoglobulin complexes

The formation of immunoglobulin aggregates or complexes greatly enhances the affinity with which IgG binds to Fc receptors on cells[1,5]. Such complexes would form in nature, for example, when multideterminant antigens (viruses, bacteria, macromolecules, etc.) bind several antibody molecules. The avid binding of complexes compared with monomeric IgG is most certainly due to an increase in the total binding energy when several Ig-receptor interactions take place, in contrast to a single interaction when one IgG molecule binds to one receptor. Conformational changes within the antibody molecule as a result of binding antigen, if they occur, do not seem to have any effect on the affinity of IgG for its receptor; cross-linking or aggregating IgG by methods not involving

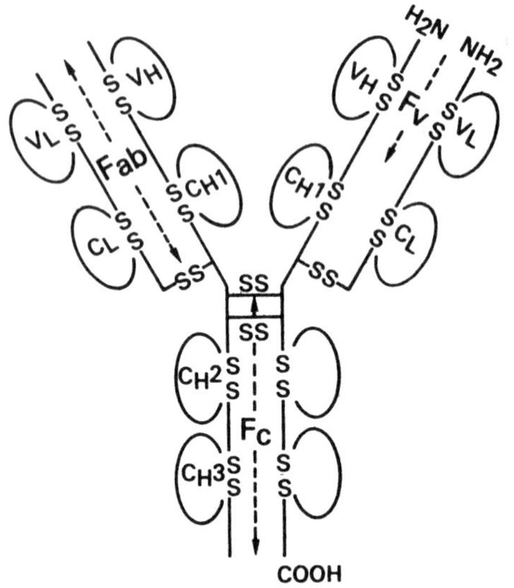

Fig. 1. Schematic representation of IgG molecule. The whole IgG molecule has a molecular weight of 150,000, and consists of two disulfide-linked heavy chains, of 50,000 Daltons each and two 25,000 Dalton light chains, disulfide-linked to the heavy chains. The antigen binding portions of the molecule reside in the N terminal parts of the two Fab's. The part which binds to cells is thought to be located at the C-terminus of the Fc portion.

the combining site also enhance the binding affinity, and monovalent hapten does not effect the binding of antibody to Fc receptors.

Quantitative measurements of the binding of IgG to cells bearing Fc receptors has recently been made possible by the development of complexes of well defined size formed by cross-linking anti-Dnp antibodies covalently at their combining sites with bivalent affinity labelling reagents[6,7]. Purified monomer, dimer, and trimer fractions and mixtures of larger oligomers were radioiodinated and used in cell binding studies[8].

In general, monomeric IgG binds too weakly to Fc-receptors to measure directly, although dimers and trimers bind with measurable affinity. One problem incurred with cell binding assays is the rapid rate at which small IgG oligomers elute from Fc receptors when cells are washed in aqueous media. A convenient way to circumvent this problem is to incubate cells with radiolabelled oligomer, and to sediment the cells through a mixture of water-immisible oils with density less than the cells, but greater than the medium[8]. Using this procedure, cells and bound immunoglobulin are separated from medium containing free oligomer, effecting about a 500 fold purification. Corrections are made for non-specific binding by subtracting the radioactivity sedimenting with cells incubated with labelled oligomer in the presence of a large excess of unlabelled protein.

With this binding assay we are able to quantitate the numbers and affinities of Fc receptors on any population of cells which forms a single-cell suspension. Fig. 2 shows how oligomers formed from rabbit IgG anti-Dnp antibodies bind to cells of the macrophage line, P388D_1. Clearly

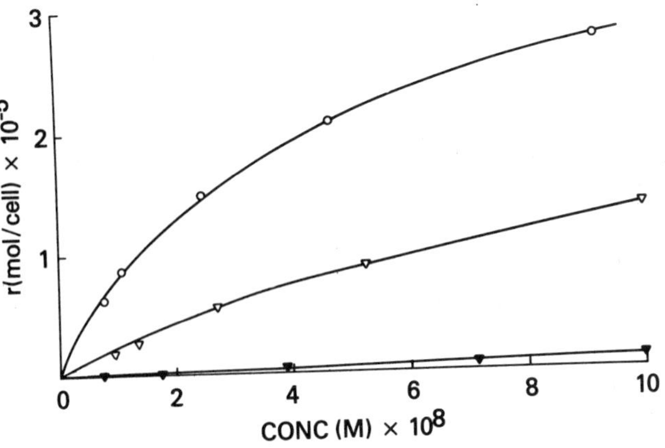

Fig. 2. The binding of affinity cross-linked oligomers of rabbit IgG antibodies to P388D_1 cells at 30°. Increasing concentrations of oligomers were equilibrated with the macrophage-line, P388D_1, and the amount of bound protein determined as described in the text. r is the number of bound 150,000 Dalton subunits/cell and Conc is the molar concentration of free subunit in the incubation mixture. -▼- monomer, -▽- dimer, and -○- trimer binding. Taken from ref. 8.

binding increases with oligomer size; no binding of monomer can be detected, while dimer and trimer bind with measurable and increasing affinities. The interaction of oligomer with cells is both reversible and time dependent (Fig. 3). The binding reaction approaches equilibrium with time,

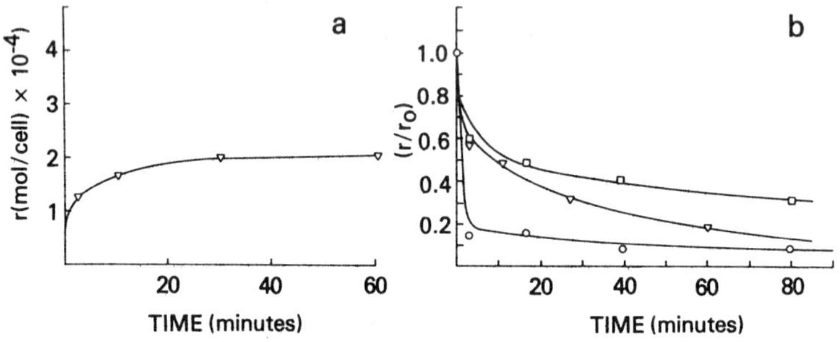

Fig. 3. Kinetics of oligomer binding to P388D$_1$ cells at 30°. a) Uptake of trimer at a concentration of free trimer of 9.0x10^{-9}M. b) Release of radio-labelled (O) dimer, (▽) trimer, and (□) a mixture of trimer and tetrameter, after the addition of a large excess of unlabelled oligomer. r_o is the amount of bound oligomer prior to addition of unlabelled protein. Taken from ref. 8.

and the bound oligomer can then either be chased off with cold oligomer, or washed off. The release of bound oligomer is size dependent, large oligomers being released more slowly than smaller ones.

Fc receptors on various cell populations

The largest oligomer which we are able to isolate as a homogeneous preparation in reasonable yields is the IgG trimer. This oligomer binds to cells with an affinity which is easily measured, and we routinely use it to compare binding properties of various types of cells.

The binding of trimer to human splenocytes from several donors is shown in Fig. 4. The binding curves of Fig. 4a are hyperbolic and give linear Scatchard plots (Fig 4b).

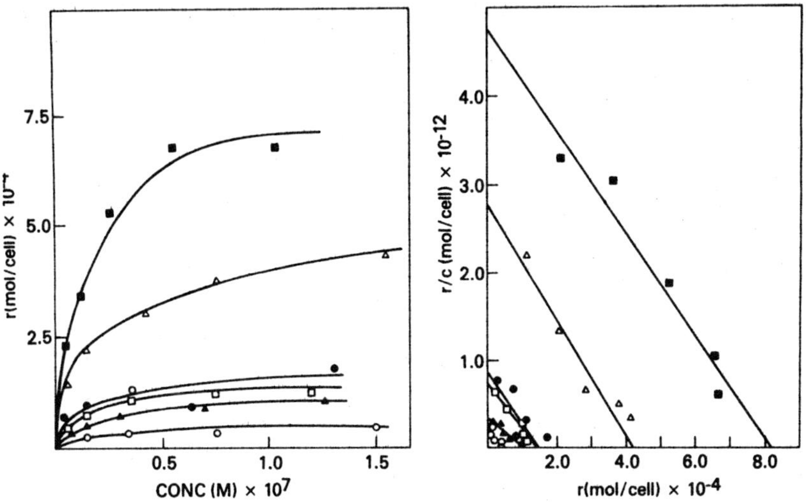

Fig. 4. The binding of rabbit IgG trimers to human spleen cells at 0°. a) Each curve represents trimer binding to cells from different donors. See Fig. 2 for definition of axes. b) Scatchard plots of the data presented in a).

These plots show that the average number of receptors per cell (from intercepts with the abscissa) in the populations vary considerably, while the binding constants (slopes) are similar. The populations tested in Fig. 4 contain many cell types and it is therefore impossible to say from these data that a particular cell type has a given receptor number. The linearity of the Scatchard plots, however, implies that the various cell types have receptors for the rabbit trimer which do not differ widely in affinity.

We have also examined the Fc receptors on relatively purified populations of human leukocytes[9]. A summary of the data is given in Table 1.

TABLE 1

IgG Receptors on Various Cell Types[a]

Population	n[b]	Kax10^{-7} (M^{-1})	Receptors Per Cellx10^{-5}	Percent[c] FcR$^+$	Receptors Per FcR$^+$ Cellx10^{-5}
Human Cells					
PMN Leukocytes[d]	3	5.8 (2.7-9.4)	4.1 (2.8-5.0)	96.0 (94.5-97.5)	4.3 (3.0-5.2)
Mononuclear Leukocytes[d]	5	4.7 (1.4-7.9)	.63 (.20-1.3)	35.0 (18.5-47.5)	1.7 (.92-2.7)
Lymphocytes[d]	7	2.7 (.83-5.7)	.21 (.05-.43)	13.9 (7.0-17.5)	2.1 (.49-3.4)
Monocytes[d]	2	4.6 (3.4-5.7)	0.31 (.26-.36)	94.5 (91.0-98.0)	.33 (.29-.37)
Splenocytes[d]	6	5.1 (2.9-6.3)	.29 (0-.83)	17.8 (5.2-36.5)	1.1 (0-1.9)
Murine Cells					
Splenocytes	3	5.6 (4.9-6.7)	.17 (.10-.23)	--	--
P388D$_1$ (macrophage line)[e]	-	4.2	3.6	--	--
PL-1 (plasmacytic lymphosarcoma)	2	2.1 (1.6-2.5)	.17 (1.6-1.8)	--	--

[a] Binding of rabbit trimer at 0°. Average and range.
[b] Number of samples assayed.
[c] Cells staining for Fc-receptors by immunofluorescence (11).
[d] From reference 9.
[e] From reference 8.

As with the unfractionated cells, the purified populations seem to bind with similar affinities, while the average numbers of receptors per cell vary considerably between populations. When binding studies are made in conjunction with fluorescent labelling, the average number of receptors per receptor-bearing cell can be estimated. Based upon these calculations, we have found that human polymorphonuclear (PMN) leukocytes, and a subpopulation of lymphocytes (11.8%) have the highest receptor density ($\sim 10^5$ receptors per cell) while circulating monocytes have fewer receptors. As a comparison we have included the binding parameters for several types of murine cells. The affinities of mouse Fc receptors for the rabbit trimer are similar to those observed in human cells. Comparisons of receptor properties using IgG from one species and cells from another are possible because both receptors and immunoglobulin appear to have been conserved during the course of evolution. Thus the binding of rabbit trimer to human cells is inhibitable by both human and murine IgG[9].

The binding studies summarized in Table 1 pertain to receptors which bind rabbit trimer with measureable affinity. We have no data implicating a weaker type of receptor (which might bind larger oligomers with measurable affinity), but it has been reported that human[10] and murine[11] lymphocytes have a subpopulation of cells which bind immune complexes only at high protein concentrations.

Class and Subclass Specificities of Fc Receptors.

In addition to IgG receptors, cells often have receptors for other classes of immunoglobulins. For example mast cells and basophils have extremely high affinity receptors for IgE [at least 5 orders of magnitude higher in Ka than the IgG receptor[12]]. Other receptors for IgM[13-17] and IgA[18] on various cell types are clearly different from those binding IgG, as shown by inhibition studies.

Since immunoglobulins of the IgG class in many species have structural differences within the constant regions which give rise to small numbers of subclasses, it is natural to question whether differences exist in the way in which proteins of the various subclasses interact with IgG Fc receptors. Indeed several investigators have studied this problem using a variety of techniques. Generally immunoglobulins of a known class or subclass are aggregated by artificial means (by cross-linking chemically or by heating) or by complexing with antigen. Aggregates are incubated with the receptor-bearing cells and the amount of bound immunoglobulin is determined by fluorescence microscopy, rosette formation (if the immune complex is an antibody-coated red cell) or bound radioactivity. Proteins of different subclasses are assumed to interact at the same site if the binding of immunoglobulins of one subclass is inhibited by the addition of protein of another subclass.

Various laboratories have reported subclass dependent differences in binding and antibody-mediated responses. Reports in the guinea pig which has two IgG subclasses (IgG1 and IgG2) show an enhanced affinity for IgG2 on macrophages[19-22]. Studies in humans, which have four subclasses indicate that IgG1 and IgG3 bind with higher affinities than IgG2 and IgG4 to several cell types: lymphocytes[5,23,24], neutrophils[23,25] monocytes[6,27], and placental cells[31]. Reports in the human also indicate that IgG1 and IgG3 are more effective in cell mediated cytolysis[24].

In our laboratory we have studied the subclass specificity of IgG receptors in the mouse. The availability of large quantities of homogeneous immunoglobulins (myeloma proteins) of the four IgG subclasses (IgG1, IgG2a, IgG2b and IgG3), the existence of several receptor-bearing cell lines, and the ease of obtaining large numbers of normal hematopoietic cells has made the mouse especially convenient for studies of this type. In our experiments[32] myeloma proteins of different subclasses were tested for their abilities to inhibit the binding of rabbit trimer to various cell types. Fig. 5 shows, for example, that trimer binding is completely inhibitable by a whole IgG2a myeloma protein or its Fc fragment, but not by the Fab.

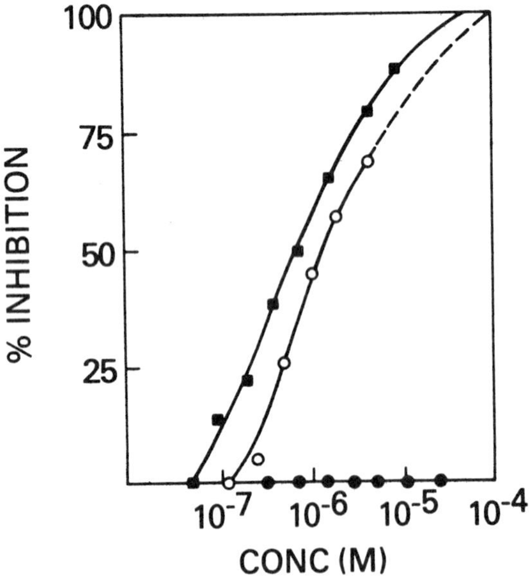

Fig. 5. Inhibition of binding of trimeric rabbit IgG to P388D$_1$ cells by (-■-), HOPC1, a murine γ 2a myeloma protein, (-○-) the Fc fragment of HOPC1 and (-●-) the Fab fragments of HOPC1. Trimer was incubated with increasing molar concentrations of inhibitor and the percent inhibition was determined relative to the amount of trimer bound in the absence of inhibitor. Taken from ref. 32.

In similar studies we observed that the binding of rabbit trimer to three cell types (a macrophage line, a lymphocytic line and normal spleen cells) was totally inhibitable by monomeric IgG1, 2a and 2b, but not by IgG3, IgM or IgA myeloma proteins.

Reverse experiments showed that oligomers of IgG1, 2a, and 2b bound to the three cell types and could be completely displaced by IgG from the rabbit, which has a single subclass. Using data from these experiments, one of which is illustrated in Fig. 6, we concluded that all of the IgG receptors on the three cell types examined, bound proteins of the IgG1, 2a and 2b subclasses, but not IgG3, IgM or IgA. Moreover, by determining the concentration of

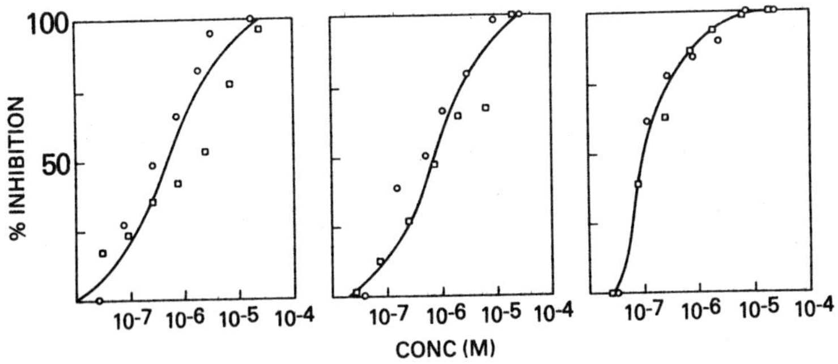

Fig. 6. Competition between rabbit IgG and murine IgG2a (HOPC1) for binding sites on three cell types. Left, P388D$_1$ cells (macrophage line), center, TC-1 cells (murine lymphocytic cell line) and right, normal C57Bl/6 spleen cells. ○ ^{125}I rabbit trimer inhibited by untreated HOPC1 monomer. □, ^{125}I HOPC1 trimer binding inhibited by monomeric rabbit IgG. Taken from ref. 32.

monomeric myeloma protein required to inhibit the binding of rabbit trimer by 50%, the affinities for the myeloma proteins appeared to be the same within experimental error (Table II).

TABLE 2

RELATIVE AFFINITIES* OF IgG1, IgG2a, AND IgG2b MYELOMA PROTEINS FOR THREE CELL TYPES AS DETERMINED BY THEIR INHIBITION OF BINDING OF ^{125}I RABBIT IgG TRIMER

Protein	Subclass	Cell Type		
		$P388D_1$	AKTB-1	Normal spleen
MOPC 300	IgG1(κ)	5.4	5.6	6.1
FLOPC 1	IgG1(κ)	5.5	6.0	6.2
avg ± S.E.		(5.5 ± .1)	(5.8 ± .2)	(6.2 ± .1)
HOPC 1	IgG2a(λ)	5.2	5.4	6.1
LPC 1	IgG2a(κ)	5.3	5.6	6.2
avg ± S.E.		(5.3 ± .1)	(5.5 ± .1)	(6.2 ± .1)
MOPC 141	IgG2b(κ)	5.5	6.1	6.0
UPC 120	IgG2b(κ)	5.5	5.6	5.9
avg ± S.E.		(5.5 ± 0)	(5.9 ± .3)	(6.0 ± .1)
Total Avg ± S.E.		5.4 ± .1	5.7 ± .1	6.1 ± .05

*pK_d (negative log of dissociation constant) as determined by Sips analysis.

In addition to binding oligomer via the major IgG receptors shown in Fig. 6, $P388D_1$ cells can bind small amounts of monomeric IgG2a (Fig. 7). This binding is not observed after trypsin treatment of the cells, and therefore many investigators believe that a second, trypsin sensitive receptor exists on the mouse macrophage and related cell lines[3,33-39]

There is some controversy in the literature about this receptor. In our experiments the binding of monomeric IgG2a can be inhibited by proteins of the IgG1 and 2b subclasses, at sub-physiological levels. However, others have presented data showing a more restricted specificity for the "trypsin sensitive receptor". We wish merely to point out here that the number of types and the specificities of IgG receptors is controversial, and refer the reader to references[3,33-39].

<u>Characteristics of IgG binding</u>

The binding affinities of oligomers of increasing size for receptor-bearing cells do not reflect simple additivities of free energy of the receptor-immunoglobulin interaction[8]. If energies were strictly additive, for example, the binding constant of the dimer should be the square and the trimer, the cube, of the monomeric binding constant. The Ka for rabbit monomeric IgG has been estimated to be $5 \times 10^{5}\ M^{-1}$ from the concentration required to inhibit the binding of labelled trimer. The dimer and trimer bind with affinities of $1 \times 10^{7}\ M^{-1}$ and $3 \times 10^{7}\ M^{-1}$ respectively, far lower than the affinities expected assuming an additive model.

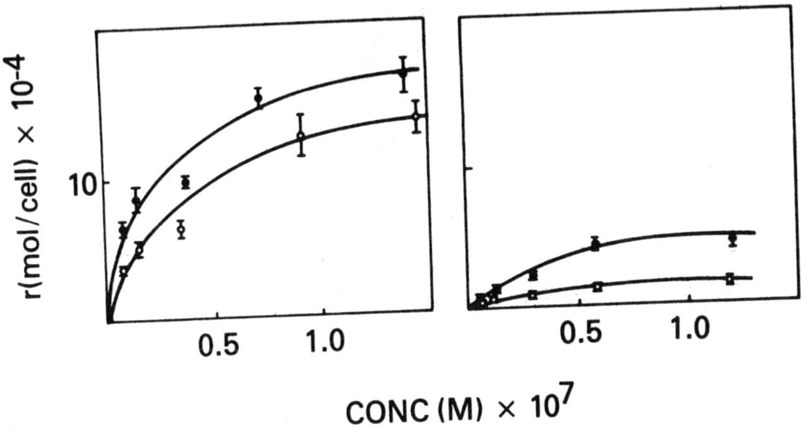

Fig. 7. Binding of the γ2a myeloma protein, HOPC-1, to $P388D_1$ cells at 0°. Left, binding of trimeric HOPC1, right, binding of the monomeric protein. Closed symbols, binding to untreated $P388D_1$ cells, open symbols; binding to trypsin-treated $P388D_1$ cells. Taken from ref. 32.

One reason for the deviation from additivity could be that the oligomers are linked in configurations not conducive to efficient simultaneous binding of all Fc portions of an oligomer. It is therefore important to measure affinities of oligomers cross-linked in different ways. Our data show that IgG trimers prepared with three different cross-linking reagents bind with similar affinities to receptor-bearing cells; a tightly binding oligomer has not yet been found.

The oligomers seem to bind multivalently. If the dimer bound monovalently, the interaction would be too weak to detect in our system (Fig. 2). The trimer appears to bind by all three Fc portions since the number of subunits bound per cell determined at saturation for the trimer is similar to that for the dimer. If the trimer bound dimerically then the saturation value should be $\frac{3}{2}$ that of the dimer (Fig. 2).

It also appears that oligomer binding fluctuates between intermediates in which, for example, the trimer binds trimerically, dimerically and monomerically. This is shown by the ability of monomeric IgG to increase the rate of release of dimer from the cell surface (Fig. 8). The most obvious explanation for

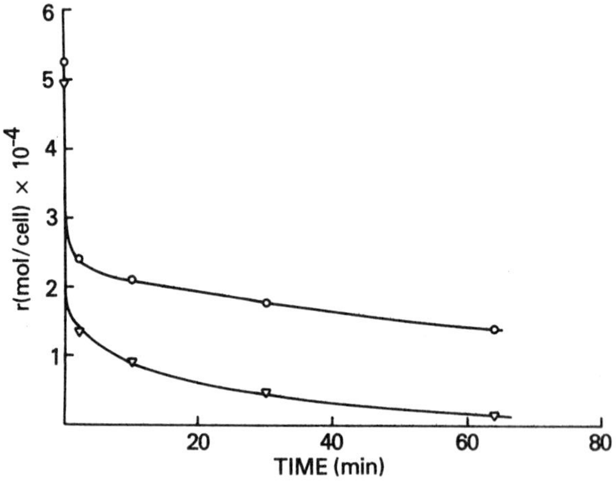

Fig. 8. The effect of monomeric IgG on the loss of ^{125}I rabbit trimer from P388D_1 cells at 30°. Duplicate samples of cells were equilibrated with 2.4×10^{-8}M trimer (subunit concentration) and at zero time, each was made 3.7×10^{-7}M in unlabelled trimer. -○- no further immunoglobulin added, -▽- sample was brought to 8.5×10^{-6}M in monomeric rabbit IgG. Taken from ref. 8.

this phenomenon is that when dimer binds to the cell surface it fluctuates between states in which both immunoglobulin subunits are bound and others in which only one binds. Monovalently-bound dimer can then either exchange off from the cell, or form a second interaction with free receptor on the cell surface. In the presence of a large excess of monomer, the concentration of free receptor on the cell surface decreases as the monomeric protein competes for receptor with the dimer. The dimer thus is pushed into states where it binds monomerically, and is finally released from the cell surface. Given the complexity of interactions of oligomeric IgG with cell-surface receptors, it is not surprising that a simple additivity of free energies of interaction does not describe the binding behavior. Oligomer binding probably occurs by a mechanism somewhere in between two extremes: one where free energies are additive and the other where IgG subunits of an oligomer bind independently of each other.

The binding of oligomers and high molecular weight aggregates[40] to cells is inhibitable by monomeric IgG at concentrations below physiological levels (Fig. 6). It is therefore clear that the establishment of meaningful receptor-ligand interactions _in vivo_ must involve processes other than simple binding. One such process could be cell-cell contact, which could exclude free IgG from the area where receptors and antibody came into contact. In this way an effector cell could recognize an antibody-coated target cell in an _in vivo_ environment.

Fc Receptors in Cytolysis

The function of IgG Fc receptors in biological systems is not completely understood. However, the role of these receptors in lysis or phagocytosis has been well documented. A particularly convenient assay for receptor function which we have employed in our laboratory is the measurement of antibody-dependent cell mediated cytotoxicity (ADCC). We have used a system originally devised by Perlmann and Perlmann[41], in which effector cells are incubated with ^{51}Cr-labelled chicken red blood cells (CRBC), which have been coated with antibody. Lysis of the target cells (CRBC) is followed by measuring ^{51}Cr release.

The role of the Fc receptor in ADCC has been established by two main lines of evidence, in our laboratory as well as in many others. First, only cells which have Fc receptors can serve as effectors of ADCC. Thus cells of the ABLS-5 or AKTB1-lymph node lines, which have no detectable Fc receptors, are unable to lyse antibody-coated CRBC, whereas other related lines which are Fc^+ can mediate ADCC. Secondly, ADCC can be inhibited by oligomeric IgG (Fig. 9) or by various aggregated IgG preparations[42,43], but not by other proteins which do not bind to Fc receptors, including $F(ab)'_2$ fragments, reduced and akylated IgG, or oligomers made from them and IgM[44,45].

We have investigated the role of antibody in ADCC by using oligomers of defined size to coat Tnp-treated CRBC and have compared the ability of cells and complement to lyse the oligomer-sensitized targets[45]. In these studies the oligomers were cross-linked with the bivalent affinity-labelling reagent, bis-Dnp-pimelic ester[7]. This reagent cross-links anti-Dnp antibodies at their combining site regions, but leaves the sites unoccupied, so that the oligomers are free to bind other Dnp or Tnp-containing antigens. Oligomers prepared in this way bind to Tnp-CRBC, and the number of oligomer molecules bound per cell can be determined by using radioiodinated protein.

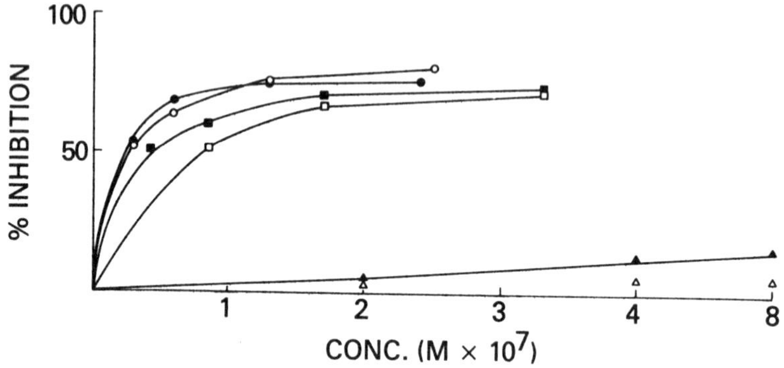

Fig. 9. Inhibition of ADCC mediated by $P388D_1$ with oligomers of rabbit anti-Dnp antibodies. Effector cells and oligomer were mixed and antibody-coated, ^{51}Cr-labelled CRBC were added. Open symbols, lysis was allowed to procede for 40 hr with an effector to target ratio of 5:1. Closed symbols, 24hr incubation with a 10:1 effector:target ratio. The abscissa represents the molar subunit concentration of oligomer in the incubation mixture, the ordinate is the percent inhibition of ADCC, relative to the amount of lysis observed in the absence of inhibitor. -D-, monomer; , dimer; 0, trimer. Taken from ref. 42.

The data in Fig. 10 show the lysis of target cells as a function of oligomer size and amount of bound oligomer using complement and three types of cells as effectors. With all effectors, the degree of lysis increased with the number of subunits (i.e., IgG molecules) bound per cell. However, with complement as effector, lysis increased as the size of the bound oligomer increased; each subunit when bound as part of a cluster was more efficient at initiating lysis than when it was bound monomerically. In sharp contrast, ADCC was not influenced by oligomer size. Monomers and small oligomers were just as efficient at initiating lysis as large ones.

Antibody dose response curves are sigmoidal when complement is the effector, indicating that a threshold density of antibody is required for lysis. In contrast, the dose response curves for ADCC are hyperbolic indicating that, instead of a threshold response, lysis increases steadily as the number of antibody molecules increases.

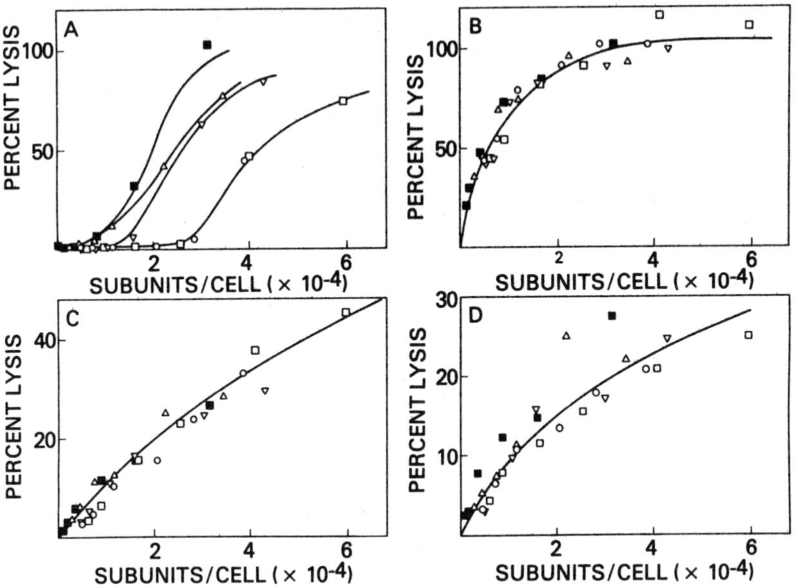

Fig. 10. Lysis of oligomer-coated, Tnp-CRBC by complement and effector cells. Effectors were (A) Dnp-absorbed guinea pig complement at a three-fold dilution. (B) Normal C57Bl/10 Sn spleen cells at a 10:1 effector:target ratio, (C) $P388D_1$ cells (murine macrophage line) at a 10:1 effector:target ratio, and (D) PL-1 cells (murine lymphosarcoma line) at a 10:1 effector:target ratio. The ordinates are the percentages of lysis relative to the amount of ^{51}Cr released by target cells incubated with 2M HCl, and the abscissae are the average numbers of molecules of 150,000 Dalton subunits bound per target cell. Oligomers used were -○- monomers, -□- dimers, -▽- trimers, -△- heavy fraction 1 (mainly tetramers-hexamers) and -■- heavy fraction 2 (pentamers and larger). Taken from ref. 45.

The differential effect of antibody oligomers on the two effector systems has a reasonable molecular explanation. Both systems require binding of many antibody molecules in order to elicit lysis, implying a multi-valent effector system which, in the case of complement, is known to be true[46]. In ADCC, since effector cells have thousands of Fc receptors on their surfaces, multivalency is also a reasonable assumption.

Effectors of complement mediated lysis are molecules of limited size, and in order for multivalent interactions to occur between antibody and complement, antibody molecules must be located very close to each other, within a few hundred angstroms. The necessity for clusters is reflected in the enhanced lysis with increasing oligomer size. In addition, the sigmoidal shapes of the antibody dose response curves probably arise from the formation of clusters on the target cell surfaces with increasing antibody densities. Cells, however, with Fc receptors which are freely mobile over an area of several square microns would be able to recognize target-bound antibody at any point on the target-cell surface. Thus clustering of antibody molecules would not be required for ADCC. The data also suggest that while the binding of several Fc receptors leads to enhanced lysis, there is no requirement for the cross-linking of neighboring Fc receptors to "trigger" the ADCC response. If this were so, cluster formation on the target would enhance lysis, and this is not observed.

REFERENCES

1. Metzger, H. (1978) Contemp. Top. Mol. Immunol. In press.
2. Cerottini, J. C., and Brunner, K. T. (1974) Adv. Immunol. 18:67.
3. Silverstein, S. C., Steinman, R. M. and Cohn Z. A. (1977) Ann. Rev. Biochem. 46:669.
4. Cathou, R. E., and Dorrington, K. J. (1975) in Subunits in Biological Systems, Timasheff, S. M. and Fasman, G. D. eds., Marcel Dekker, New York. Vol. 7, part C.
5. Dickler, H. B. (1976) Adv. Immunol. 24:167.
6. Segal, D. M. and Hurwitz, E. (1976) Biochemistry 15:5253.
7. Plotz, P. H., Kimberly, R. P., Guyer, R. L., and Segal, D. M. (1978) Immunochemistry. In press.
8. Segal, D. M. and Hurwitz. (1977) J. Immunol. 4:1338.
9. Alexander, E. A., Titus, J. A., and Segal, D. M. (1978) In preparation.
10. Arbeit, R. D., Henkart, P. A., and Dickler, H. B. (1977). Scand. J. Immunol. 6:873-878.
11. Arbeit, R. D., Henkart, P. A., and Dickler, H. B. (1976) In In Vitro Methods in Cell-Mediated and Tumor Immunity, Bloom, B. R. and David, J. eds., Academic Press, New York, pp. 143-154.
12. Kulczycki, A. and Metzger, H. (1974) J. Exp. Med. 140:1676-1695.
13. Ramasamy, R., Richardson, N. E. and Feinstein, A. (1976) Immunology 30:851.
14. Moretta, L., Ferrarini, M., Durante, M. L., and Mingari, M. C. (1975) Eur. J. Immunol. 5:565.
15. Lamon, E. W., Andersson, B., Whitten, H. D., Hurst, M. M. and Ghanta, V. (1976) J. Immunol. 116:1199-1203.
16. McConnel, I. and Hurd, C. M. (1976) Immunology 30:835-839.
17. Santana, V. (1977) Immunology 32:273-278.

18. Stroker, W., Hague, N. E., Lum, L. G., and Henkart, P. A. J. Immunol. In press.

19. Leslie, R. G. Q. and Cohen, S. (1974) Immunology 27:577.

20. Berken, A., and Benacerraf, B. (1966) J. Exp. Med. 123:119.

21. Guercio, P. Del., Tolonc, G., Andrape, F. B., Biozzi, G., and Binaghi, R. A. (1969) Immunology 16:361.

22. Liew, F. Y. (1971) Immunology 20:817.

23. Lawrence, D. A., Weigle, W. O., and Spiegelberg, H. L. (1975) J. Clin. Invest. 55:368.

24. Spiegelberg, H. L, Perlmann, H. and Perlmann, P. (1976) J. Immunol. 116:1464.

25. Henson, P. M., Johnson, H. B., and Spiegelberg, H. L. (1972) J. Immunol. 109:1182.

26. Huber H., and Fudenberg, H. H. (1968) Int. Arch. Allergy 34:18.

27. Okafor, G. O. Turner, M. W., and Hay, F. C. (1974) Nature 248:228.

28. Hay, F. C., Torrigiani, G., and Roitt, I. M. (1972) Eur. J. Immunol. 2:257.

29. Huber, H., Douglas, S. D., Musbacher, J., Kochwa, S., and Rosenfield, R. E. (1971) Nature 229:419.

30. Abramson, N., Gelfand, E. W., Jandl, J. H., and Rosen, F. S. (1970) J. Exp. Med. 132:1207.

31. McNubb, T., Koh, T. Y., Dorrington, K. J. and Painter, R. H. (1976) J. Immunol. 117:882.

32. Segal, D. M. and Titus, J. A. (1978) J. Immunol. 12:1395.

33. Unkeless, J. C., and Eisen, H. N. (1975) J. Exp. Med. 142:1520.

34. Unkeless, J. C. (1977). J. Exp. Med. 145:931.

35. Heusser, C. H., Anderson, C. L., and Grey, H. M. (1977) J. Exp. Med. 145:1316.

36. Anderson, C. L., and Grey, H. M. (1974) J. Exp. Med. 139:1175.

37. Walker, W. S. (1976) J. Immunol. 116:911.

38. Walker, W. S. (1977) J. Immunol. 119:367.

39. Diamond, B., Bloom, B. R., and Schorff, M. D. (19) J. Immunol. In press.

40. Knutson, D. W., Kijlstra, A. and VanEs, L. A. (1977) J. Exp. Med. 145:1368.

41. Perlmann, P. and Perlmann, H. (1970) Cell. Immunol. 1:300.

42. Hurwitz, E., Zatz, M. M., and Segal, D. M. (1977). J. Immunol. 118:1348.

43. MacLennan, I. C. M. (1972) Clin. Exp. Immunol. 10:275.

44. Wisloff, F., Michaelsen, T. E., and Froland, S. S. (1974) Scand. J. Immunol. 3:29.

45. Jones, J. F., Plotz, P. H. and Segal, D. M. (1978) In preparation.

46. Reid, K. B. M., and Porter, R. R. (1975) Contemp. Top. Mol. Immunol. 4:1.

B LYMPHOCYTE ACTIVATION BY ANTI-IMMUNOGLOBULIN ANTIBODIES

DONNA G. SIECKMANN, IRWIN SCHER AND WILLIAM E. PAUL
Laboratory of Immunology, National Institute of Allergy and Infectious Diseases, National Institutes of Health; Department of Immunology, Naval Medical Research Institute; and Department of Medicine, Uniformed Services University of the Health Sciences, Bethesda, Maryland 20014

The immune system is the principal defense system of the body. The understanding of its organization and regulation and of the mechanisms involved in its activation is of considerable importance, both in a practical sense and as a model for the delineation of other complex systems. One major problem encountered in attempting to dissect the workings of the immune system is the diversity of the cellular elements which make up the system. The immune system consists principally of three families of cells, namely: 1) thymus-dependent (T) lymphocytes, which are derived from the thymus during ontogeny and which serve effector and regulatory functions; 2) thymus-independent (B) lymphocytes, which are derived from the bone marrow and which are the precursors of antibody-secreting cells; and 3) monocytes and macrophages, which play accessory roles in T and B lymphocyte activation and which are involved in effector functions of the immune system. This paper will be limited to a discussion of B lymphocytes and will focus on the diversity of this family of cells and the physiology of their activation at the plasma membrane level.

B lymphocytes are known to have between 10^4-10^5 antibody molecules on their plasma membrane,[1] which serve as receptors for antigen.[2] These surface membrane immunoglobulin (mIg) receptors are generally 8S IgM monomers[3,4] or 7S IgD molecules;[5,6] however, other immunoglobulin isotypes also occur.[7-9] The structure of these receptors appears to be very similar to that of secreted antibody; in particular, the receptor has the same V region or idiotype (antigen combining region) as the antibody which the cell will eventually secrete after differentiation to the plasma cell stage.[10,11] It is known that in resting, non-activated B lymphocytes, 80-90% of the Ig made in the cell is located on the cell surface membrane.[12] This mIg turns over very slowly, with a half-life of 20 hours.[13] Possible differences in the structure of mIg, which make it a membrane bound (integral)[14] protein, rather than being transported out of the cell in secretion, are currently under investigation.[15,16] If they are found, biochemical differences between mIg and secreted Ig may offer some information as to the mechanism by which mIg functions as a receptor molecule within the membrane.

B Cell Diversity

One of the major problems in studying B lymphocyte activation is the ontogenetic or developmental diversity of an adult B cell population. B lymphocytes originate from stem cells within the hematopoietic tissue of the embryo.[17] In avian species, these stem cells originate in the bursa of Fabricius,[18,19] and this organ appears to play an important role in the development of the B cell system. A mammalian equivalent of the bursa, however, has not been found. In mammalian species, stem cells for the B cell pool are found in fetal liver, blood, and in the yolk sac.[20] The first cell type of the B cell lineage can be detected in mouse fetal liver at 12-13 days of gestation. This cell, called a pre-B cell,[21,22] is a large, rapidly dividing cell, which contains IgM in the cytoplasm, but does not appear to express Ig on its plasma membrane. Pre-B cells have none of the other B lymphocyte cell surface markers, such as Fc receptors or receptors for complement. However, both allotypic[23] and idiotypic[24] markers have been identified within these cells, suggesting that pre-B cells already manifest clonal diversity. This would also suggest that there is a linear relationship between this set of cells and the mature B cell population.

The current view of B cell development is that the large pre-B cells develop into small pre-B cells, which then differentiate into immature B lymphocytes.[25] The latter appear around 17 days of gestation in the fetal liver and spleen;[26] they are identified by the presence of IgM on their cell surface. The immune potential of these cells is unknown; however, it is thought that antigen recognition by immature B lymphocytes results in an unresponsive state.[27]

Immature B lymphocytes undergo further differentiation steps in which they acquire Ia antigens,[28] whose expression is controlled by the major histocompatibility complex, as well as receptors for the third component of complement.[29] Other B lymphocyte specific membrane antigens (the Lyb antigens) are expressed as the B lymphocyte matures.[30,31] It is not certain whether these alloantigens are present on all maturing B lymphocytes, or whether they are markers for sublines of B cells.

The maturing B lymphocyte also acquires a second immunoglobulin, IgD,[32,33] on its membrane, which expresses the same antigen-binding specificity as the already present IgM receptor.[34-36] Acquisition of mIgD occurs very late in development, however, since adult levels of IgD^+ lymphocytes are not reached until 3-4 weeks of age.[37] Thus, within a population of mature B lymphocytes there exist B cells, which have two different types of antigen receptors, with identical antigen-binding specificities. In addition, it has been shown that the number of IgM and IgD molecules on an individual cell can vary;[38] thus, any population of B cells exists as a continuum of cells: ranging from cells which have large amounts of mIgM, with little or no mIgD, to cells which possess large amounts of mIgD, with little or no IgM being present.[39] This diversity in expression of isotypes of mIg suggests distinct functional roles for mIgM and mIgD. Furthermore, it provides an example of the extreme heterogeneity of B lymphocytes and presents a challenge in experimentally testing for the function and activation requirements of B cells, since any population will be a diverse mixture of different maturational states.

The difficulties presented by the developmental diversity of the B cell population are compounded by a diversity of B cells in terms of their antigenic specificity. The major function of the immune system is to recognize and to respond to any antigen which is presented to the system. This is achieved by the existence, in each individual, of a large array of distinct clones of B lymphocytes, each of which is able to recognize one related family of antigens and to respond to it.[40] Thus, even in a population of lymphocytes that is developmentally homogeneous, a second element of diversity exists, which is due to the distinctive antigen-binding capacity of each lymphocyte. Consequently, only a very limited pool of cells exists which would possess receptors for any one antigen. Such antigen-binding lymphocytes may be identified by their ability to bind radioactively labeled antigens, using radioautography as a detection system. It has been shown, for example, that ∿0.1% of a normal lymphoid cell population will bind the dinitrophenyl hapten.[41]

The existence of B lymphocyte diversity in both antigen specificity and in development presents a major problem in terms of examining reactions of a defined population of B cells, since any defined population constitutes a very small fraction of the entire B cell population.

B Lymphocyte Activation

One central issue which is addressed in most theoretical models of B lymphocyte activation is the role of membrane bound Ig in activation of the B cell. The principal question is whether or not mIg acts as a receptor in having both the function of recognition (of antigen) and of the generation of a signal for some cellular function (antibody formation). The "one-non-specific signal" theory of B cell activation proposed by Coutinho and Möller,[42,43] suggests that Ig only serves to bind antigen to the cell and that no activation signal is delivered through mIg. They propose that some other portion of the antigen possesses an intrinsic stimulant (polyclonal activator) which binds to a distinct nonimmunoglobulin receptor. The main function of mIg is viewed as focusing this polyclonal activator onto the surface of appropriate cells. Antigens which serve to support this hypothesis include molecules of lipopolysaccharide (LPS) to which haptenic determinants, such as trinitrophenyl (TNP), have been coupled. The response to TNP-LPS is thought to be the result of TNP reacting with mIg and thus focusing the mitogenic portion of LPS onto the membrane. This model requires that all antigens possess some intrinsic mitogenic element or that a mechanism exists, such as T cell function, to endow antigens with such activation elements. One difficulty presented by this hypothesis is its failure to explain the need for two types of Ig receptors, IgM and IgD, on the B cell membrane. It is also difficult to account for the differential interaction of the mIg receptors with antigen which generally leads to antibody formation, but may in certain instances lead to immunologic tolerance.

Other models for B cell activation suggest that two signals are required,[44] one signal being delivered through the Ig receptor and the other being given by accessory cells. According to these notions, signals delivered through the Ig receptor alone would lead to tolerance induction. Others have suggested that a differential recognition of antigen by the mIgM and mIgD receptors may determine whether interaction of the cell with antigen results in the induction of immunity or unresponsiveness.[45]

Activation by Anti-Ig Antibodies

Lymphocyte diversity, as discussed above, presents technical problems for study of the immediate consequences of the interaction of an antigen with membrane bound Ig specific for that antigen. An alternative approach has been to study the cellular consequences of interaction of anti-immunoglobulin (anti-Ig) antibodies with mIg. This method has made it possible to stimulate large numbers of B lymphocytes regardless of their antigenic specificity, and also provides a tool for study of the function of the different isotypic specificities of mIg.

Previous studies by Sell and Gell[46] have shown that anti-Ig antibodies, specific for allotypic determinants on rabbit Ig, are able to cause the proliferation of rabbit peripheral blood cells in vitro. Since this initial observation, it has been reported that anti-Ig antibodies of the appropriate species specificity are able to induce proliferative responses in cultures of lymphocytes from chickens,[47,48] pigs,[49] and humans.[50-53] Although it has previously been reported that mouse lymphocytes are unable to respond to anti-Ig in culture,[54,55] it has now become possible to demonstrate such a response using improved culture conditions and chromatographically purified anti-Ig reagents.[56-58] Our laboratory, in particular, has been involved in studying the induction of a proliferative response by normal mouse lymphocytes when they are cultured with anti-Ig specific for either mouse μ heavy chain or κ light chain determinants.[58]

The stimulatory anti-μ antibody is prepared from the serum of goats previously immunized with mouse IgM myeloma proteins. The antiserum causes modest proliferation of mouse lymphocytes before absorption and purification on affinity columns bearing myeloma proteins of various specificities. The anti-μ is absorbed from the serum by passage over a column of insolubilized IgMκ myeloma protein (TEPC 183), but not by passage over columns of $IgG_{2\alpha}\lambda$ (HOPC 1), $IgG_{2\alpha}\kappa$ (MOPC 173), or germ-free mouse serum. The stimulatory anti-μ activity can be recovered from the TEPC 183 column by elution with a low pH buffer. These absorption and elution studies demonstrate the specificity of the anti-μ preparation, which has very substantial mitogenic activity. In contrast, goat anti-ferritin antibodies prepared on an affinity column of ferritin have been shown to be non-stimulatory.

When purified in such manner, the anti-μ antibodies eluted from the IgM column are shown to precipitate only μ chains from a preparation of surface membrane proteins of mouse spleen cells, thus confirming the antigen-binding specificity of the preparation.

Anti-μ as prepared by affinity chromatography, is able to cause greater proliferation of adult mouse spleen cells than the original serum, possibly because of the removal of inhibitory serum components. Concentrations as low as 5 μg/ml are able to cause a detectable response, as measured by the uptake of ^{3}H-thymidine (^{3}H-TdR) between 48 and 64 hours of culture. However, an increase in the response is obtained as the anti-μ concentration is increased up to 250 μg/ml. It is interesting to note that much higher concentrations of anti-μ are needed than would be expected for the binding of all free μ determinants on the membrane. This might suggest that other factors, such as receptor aggregation, may be needed for cell activation, rather than a mere receptor-liqand interaction. On the other hand, proliferation may be due to only a minor component in the anti-μ preparation. This issue has not yet been resolved.

Cultures of spleen cells stimulated with anti-μ begin to proliferate after 48 hours and continue to show ^{3}H-TdR uptake for the next 48 hours. In contrast, such cultures do not proceed into a phase of antibody formation (Table 1). When these cultures are assayed for IgM-secreting cells by the protein A plaque technique,[59] they appear to have even lower numbers of plaque-forming cells (PFC) than control non-stimulated cultures. In contrast, cultures stimulated with LPS show a peak proliferative response at 48 hours which quickly declines within the next 24 hours. LPS-stimulated cultures also show an optimum number of IgM-PFC on day 2 and continue for several days to show polyclonal B cell activation to IgM synthesis. This dichotomy between the ability of the mitogen LPS to cause a polyclonal response, and the inability of a mitogenic anti-μ to do so, has not yet been explained. It is known that anti-μ is a potent inhibitor of polyclonal activation.[60-62] It is perhaps this duality of anti-μ effects which prevents a polyclonal response following activation to a proliferating state.

TABLE 1

INABILITY OF ANTI-μ TO INDUCE A POLYCLONAL ANTIBODY RESPONSE

Stimulant	Conc. (μg/ml)	cpm per culture[a]		PFC per culture[b]	
		d.2	d.5	d.2	d.5
Medium		4,826	22,417	57	623
Anti-μ	100	81,882	86,530	25	12
LPS	50	118,040	23,279	2,953	1,778

[a] Cultures were given a 4 hour pulse with ^{3}H-TdR.
[b] Numbers represent IgM-secreting PFC as detected with a protein A plaque assay developed with a rabbit anti-μ antiserum.

The ability of anti-μ antibodies to induce a proliferative response has been shown to be independent of the T lymphocyte or macrophage population.[63] The response is not diminished by prior treatment of responding spleen cells with anti-Thy 1.2 antiserum and complement to remove T lymphocytes. Likewise, spleen cells from athymic (nude) mice are able to respond in the congenital absence of a thymic-derived lymphocyte population. To exclude a macrophage requirement, adherent macrophage populations were depleted either by treatment with carbonyl iron or by passage over columns of Sephadex G-10. Neither method was able to diminish the response to anti-Ig, although these cells were depleted of their ability to give a primary antibody response to sheep erythrocytes, which is a macrophage and T cell dependent response.

Recently, we have been interested in obtaining a more defined description of the B cells which respond to anti-Ig, particularly with respect to the type of Ig on the surface membrane. Studies on the ontogeny of responsiveness to anti-Ig have shown that cells from mice less than 4 weeks of age are unresponsive to anti-Ig.[63] It is generally agreed that mice acquire adult levels of Ig$^+$ cells by 2 weeks of age.[29] This would then suggest that the possession of mIg by a B cell is not the only factor involved in determining its responsiveness to anti-Ig and that additional maturational steps beyond that of mIg appearance are required prior to acquisition of responsiveness.

We have recently utilized microfluorimetry[64] as a technique to determine the Ig phenotype of cells responding to anti-Ig. A fluorescence activated cell sorter (FACS) was used to isolate populations of normal adult mouse spleen cells based on their density of mIg.[39] This technique thus allowed an examination of cells which were mIg^+ or mIg^- and also of cells with different amounts of mIgM or mIgD on their membrane. Spleen cells were generally first depleted of T cells by treatment with an anti-thymocyte serum (ATS) and complement (C). They were then labeled with a fluorescein-conjugated rabbit anti-Ig and sorted on the FACS into different pools based on the amount of fluorescence per cell. Figure 1 shows a normal fluorescence profile of anti-Ig-labeled spleen cells. The fluorescence intensity is displayed on the abcissa in units of 1-1000 (channel number). The number of cells appearing with each

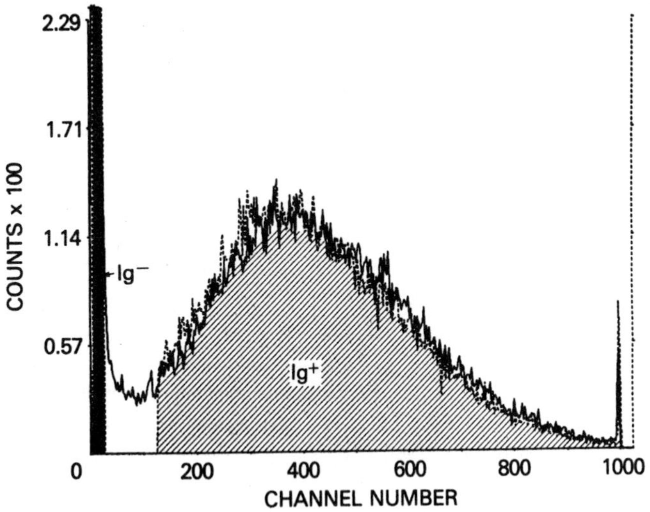

Fig. 1. FACS fluorescence profile of normal CBA/J spleen cells stained with a fluorescein-labeled $F(ab')_2$ anti-mouse Ig (——). A second profile (---) shows Ig^- and Ig^+ cells as they were detected during the sorting process.

fluorescence intensity is shown on the ordinate. It is assumed that the amount of anti-Ig which binds to the cell surface, and therefore the fluorescence intensity of the cell, is directly and linearly related to the number of Ig molecules on the surface membrane.

In the particular experiment presented here, cells in channels 1-30 were classed as Ig^-. These cells comprise a minor peak at the left end of the fluorescence profile. Cells in channels 50-1000 were classed as Ig^+ (89.1% of total). The Ig^+ cells were sorted into three fractions based on staining intensity and designated as either dull, intermediate, or brightly staining cells. It was found that anti-Ig responsiveness was a property of Ig^+ cells alone (Table 2). Ig^- cells were essentially unresponsive to anti-Ig and to LPS, and also to the T cell mitogen, Concanavalin A (Con A) because of prior removal of T cells. When responsiveness as a function of the fluorescense intensity was measured, it was found that the brightly staining Ig^+ cell fraction gave the greatest response to anti-Ig. Dull staining cells were much less responsive, although these cells still responded adequately to LPS.

TABLE 2

SEPARATION OF SPLEEN CELLS BY DENSITY OF THEIR MEMBRANE Ig[a]

Stimulant	Conc. (μg/ml)	ATS + C treated						Normal spleen
		Unsorted	Ig^+	Ig^-	Dull	Inter-mediate	Bright	
		(cpm per culture)						
Medium		1,488	763	558	1,160	814	1,829	1,718
Anti-μ	100	63,878	60,411	1,844	31,229	59,798	70,325	59,731
Anti-γκ	200	16,625	12,726	649	3,674	10,542	35,971	15,242
LPS	50	63,615	51,184	7,306	64,625	N.D.	61,281	63,724
Con A	2	992	415	448	800	485	1,274	184,043

[a] Cells were sorted on the FACS into three Ig^+ fractions, each 33% of the total Ig^+ cell population. Sort windows were as follows: dull, channels 50-204; intermediate, channels 205-325; bright, channels 326-1000.

These results, then, showed a differential responsiveness to anti-Ig, on the basis of mIg density, and thus suggested that sorting cells, on the basis of mIg density, also segregated Ig^+ cells into different functional groups. Correlation between mIg density and function has previously been shown for minor lymphocyte-stimulating (Mls) determinant and complement receptor (CR) bearing B cells,[65] which are included within the intermediate to bright Ig^+ cells, as isolated on the FACS. Although the Mls^+, CR^+ cell population falls into the same general area on the fluorescence profile as the anti-Ig responsive cells, it has not been determined whether they are in fact identical cell populations.

Since most adult splenic B lymphocytes carry both IgM and IgD on their membrane in varying amounts, it was of further interest to determine the surface isotype of the responsive cell. Spleen cells, depleted of T cells by treatment with ATS and C, were labeled with a fluorescein-conjugated rabbit anti-mouse μ

[F(ab')$_2$ fragments] and sorted into μ^+ or μ^- cell pools, or into individual μ^+ fractions, each consisting of 33% of total μ^+ cell population. The latter were termed dull, intermediate, or brightly staining cells. As shown in Table 3, anti-Ig responsiveness again appeared only in the Ig^+ cell fraction. When cells with varying densities of surface IgM were examined, it was found that responsiveness to anti-μ and to anti-γκ appeared predominantly in the dull μ^+ cell fraction. Bright μ^+ cells were almost unresponsive to anti-γκ and poorly responsive to anti-μ, although higher amounts of anti-μ in culture were able to stimulate the bright cells with increasing efficiency. A similar correlation has been found for Mls^+, CR^+ lymphocytes, in that the majority of these cells are found in the dull μ^+ fraction.[65] On the other hand, LPS seemed to stimulate all fractions very well; indeed, the bright staining cell fraction gave a greater response than the other fractions.

TABLE 3

SEPARATION OF SPLEEN CELLS BY DENSITY OF THEIR MEMBRANE IgM[a]

Stimulant	Conc. (μg/ml)	ATS + C treated						Normal spleen
		Unsorted	Ig^+	Ig^-	Dull	Inter-mediate	Bright	
		(cpm per culture)						
Medium		990	674	277	726	549	676	1,509
Anti-μ	10	48,293	49,574	1,307[b]	52,464	45,941	18,615	19,199
Anti-γκ	50	21,384	16,925	50[b]	30,356	6,686	3,985	8,959
LPS	50	62,402	52,524	30	34,397	41,578	72,214	61,128
Con A	2	1,200	595	226	837	347	557	182,591

[a] Cells were sorted on the FACS into three IgM^+ fractions, each 33% of the total IgM^+ cell population. Sort windows were as follows: dull, channels 40-138; intermediate, channels 139-290; bright, channels 291-1000.

[b] Anti-μ, 100 μg/ml; anti-γκ, 200 μg/ml.

These latter experiments suggested that, since the responding cell population was a high density Ig^+ cell with low density μ, another Ig isotype must be present on the membrane. In the experiment shown in Table 4, surface IgD was examined. Normal DBA/2 spleen cells were labeled with fluorescein-conjugated anti-δ (hybridoma 10-4.22)[66] and sorted into δ^+ and δ^- fractions and also into four δ^+ fractions, each consisting of 25% of the total δ^+ pool. When these cell pools were examined in culture for their response patterns, only IgD^+

cells responded to anti-Ig and to LPS. When surface density was varied, responsiveness to anti-μ and to anti-γκ extended from the intermediate to the brightly staining cell fraction, while the dull cell fraction responded poorly to anti-μ and was nearly unresponsive to anti-γκ. In contrast, all four fractions were equally responsive to LPS.

TABLE 4

SEPARATION OF SPLEEN CELLS BY DENSITY OF THEIR MEMBRANE IgD[a]

Stimulant	Conc. (μg/ml)	Unsorted	Total Ig^+	Ig^-	Dull	Dull to inter-mediate	Bright to inter-mediate	Bright
				(cpm per culture)				
Medium		626	428	164	404	404	428	309
Anti-μ	100	40,917	64,535	240	17,768	71,398	96,516	78,311
Anti-γκ	200	4,705	6,601	167	1,841	8,777	10,139	4,230
LPS	50	42,848	46,768	2,913	54,791	53,703	51,588	42,866

[a] Cells were sorted on the FACS into four IgD^+ fractions, each 25% of the total IgD^+ cell population. Sort windows were as follows: dull, channels 70-202; dull to intermediate, channels 203-304; bright to intermediate, channels 305-437; bright, channels 438-1000.

In summary, anti-Ig is able to induce proliferation of mouse B lymphocytes in culture, and it appears to preferentially stimulate B cells having a high density of mIg. These high density mIg^+ cells have a high density of mIgD and very little mIgM. This particular mIg phenotype is most characteristic of a class of more mature or later developing B lymphocytes. It has been shown that early B lymphocytes possess mIgM; however, as they mature, they acquire increasing amounts of mIgD while losing their mIgM.[67] Other mIg isotypes, such as IgG, may be acquired in further differentiation steps. B lymphocytes which have a predominance of mIgM and little or no mIgD are found in the neonate; and, as described above, cells from such mice are not stimulated to proliferate by anti-Ig. A similar relationship between anti-Ig responsiveness and the mIg phenotype has also been found using the CBA/N mouse strain.[63,68] This mutant mouse strain has been shown to lack a functional mature B lymphocyte population.[69-71] Although anti-μ antibodies are able to bind the surface of Ig^+ B lymphocytes from these mice, such cells cannot be stimulated to proliferation by anti-Ig.[63] This demonstrates that possession of mIg is apparently not the sole requirement for anti-Ig activation, but that a certain differentiation state of the cell is required.

The relationship of the immature B cell pool to the mature B cell pool is currently not well defined. The question of whether maturation represents an antigen driven event or a normal genetically programmed set of differentiation events occurring without antigenic intervention has not been resolved. Nevertheless, the ability of anti-μ to preferentially cause an activation to proliferation of only the mature B cell pool suggests that the function of mIgM in these cells is different than that of mIgM in less mature cells having predominantly IgM on their membrane. One way in which this difference might be expressed could be that interaction of anti-μ with mIgM on immature cells leads to inactivation (tolerance) of their antigen recognition system.

The significance of anti-Ig stimulation of B lymphocytes is yet to be determined. Although induction of Ig^+ B cells to DNA synthesis and the appearance of mitotic figures within such cultures has been demonstrated, the relevance of this proliferative phase to the scheme of total maturation of a resting B lymphocyte into an antibody-secreting plasma cell is unknown. As discussed above, anti-Ig stimulation alone does not lead to Ig synthesis. Additional signals induced through the membrane may be needed to cause the cell to proceed from proliferation to differentiation and to antibody synthesis.[72,73] On the other hand, the differentiation step may be controlled by the strength of the Ig receptor signal which is generated; this in turn might be regulated by the extent to which mIg is modulated by anti-Ig (or antigen). These possibilities are currently under investigation. Nevertheless, activation of B cells by anti-Ig demonstrates the importance of mIg in controlling the activities of the B lymphocyte. Furthermore, it suggests that mIg is an essential part of the antigen recognition mechanism of the cell, not only in determining the specificity of the cells stimulated by a certain antigen, but also in functioning as a receptor in providing a link between the cells' outer environment and the intracellular mechanisms involved in antibody synthesis.

ACKNOWLEDGMENTS

The authors wish to express their gratitude to Robert Habbersett for preparing the FACS sorted cells.

This work was supported in part by a Postdoctoral Fellowship from the Arthritis Foundation (to D.G.S.) and by the Naval Medical Research and Development Command Work Unit No. M0095-PN.001.1030 and Uniformed Services University of the Health Sciences Research No. C08310.

REFERENCES

1. Marchalonis, J.J., Atwell, J.L. and Haustein, D. (1974) Biochim. Biophys. Acta, 351, 99-112.
2. Sulitzeanu, D. (1972) Curr. Top. Microbiol. Immunol. 54, 1-75.
3. Vitetta, E.S., Baur, S. and Uhr, J.W. (1971) J. Exp. Med. 134, 242-264.
4. Marchalonis, J.J., Cone, R.E. and Atwell, J.L. (1972) J. Exp. Med. 135, 956-971.
5. Melcher, U., Vitetta, E.S., McWilliams, M., Phillips-Quagliata, J., Lamm, M.E. and Uhr, J.W. (1974) J. Exp. Med. 140, 1427-1431.
6. Abney, E. and Parkhouse, R.M.E. (1974) Nature (London), 252, 600-602.
7. Cooper, M.D., Kearny, J.F., Lawton, A.R., Abney, E.R., Parkhouse, R.M.E., Preud'-homme, J.L. and Seligmann, M. (1976) Ann. Immunol. (Inst. Pasteur), 127, 573.
8. Okumura, K., Julius, M.H., Tsu, T., Herzenberg, L.A. and Herzenberg, L.A. (1976) Eur. J. Immunol. 6, 467-472.
9. Yuan, D., Vitetta, E.S. and Kettman, J. (1977) J. Exp. Med. 145, 1421-1435.
10. Eichmann, K. and Rajewsky, K. (1975) Eur. J. Immunol. 5, 661-666.
11. Claflin, J.L., Lieberman, R. and Davie, J.M. (1974) J. Exp. Med. 139, 58-73.
12. Andersson, J., Lafleur, L. and Melchers, F. (1974) Eur. J. Immunol. 4, 170-180.
13. Melchers, F., Cone, R.E., von Boehmer, H. and Sprent, J. (1975) Eur. J. Immunol. 5, 382-388.
14. Melcher, U., Eidel, L. and Uhr, J.W. (1975) Nature (London), 258, 434-435.
15. McIhinney, R.A.T., Richardson, N.E. and Feinstein, A. (1978) Nature (London), 272, 555-557.
16. Williams, P.B., Kubo, R.T. and Grey, H.M. (1978) J. Immunol. 121, 2435-2439.
17. Miller, J.F.A.P. and Mitchell, G.F. (1969) Transplant. Rev. 1, 3-42.
18. Warner, N.L., Szenberg, A. and Burnet, F.M. (1962) Aust. J. Exp. Biol. Med. Sci. 40, 373-387.
19. Cooper, M.D., Peterson, R.D.A., South, M.A. and Good, R.A. (1966) J. Exp. Med. 123, 75-102.
20. Metcalf, D. and Moore, M.A.S. (1971) Haematopoietic Cells, North Holland, Amsterdam.
21. Raff, M.C., Megson, M., Owen, J.J.T. and Cooper, M.D. (1976) Nature (London), 259, 224-226.
22. Gathings, W.E., Lawton, A.R. and Cooper, M.D. (1977) Eur. J. Immunol. 7, 804-810.
23. Hayward, A.R., Simons, M.A., Lawton, A.R., Mage, R.G. and Cooper, M.D. (1978) J. Exp. Med. 148, 1367-1377.
24. Kubagawa, J., Vogler, L., Conrad, M., Lawton, A. and Cooper, M. (1978) Fed. Proc. 37, 1765.
25. Parkhouse, R.M.E. and Cooper, M.D. (1977) Immunol. Rev. 37, 105-126.
26. Owen, J.J.T., Cooper, M.D. and Raff, M.C. (1974) Nature (London), 249, 361-363.
27. Metcalf, E.S. amd Klinman, N.R. (1976) J. Exp. Med. 143, 1327-1340.
28. Hammerling, U., Chen, A.F. and Abbot, J. (1976) Proc. Natl. Acad. Sci. 73, 2008-2014.
29. Gelfand, M.C., Elfenbein, G.J., Frank, M.M. and Paul, W.E. (1974) J. Exp. Med. 139, 1125-1141.
30. Huber, B., Gershon, R.K. and Cantor, H. (1977) J. Exp. Med. 145, 10-20.
31. Ahmed, A., Scher, I., Sharrow, S.O., Smith, A.H., Paul, W.E., Sachs, D.H. and Sell, K.W. (1977) J. Exp. Med. 145, 101-110.
32. Rowe, D.S., Hug, K., Forni, L. and Pernis, B. (1973) J. Exp. Med. 138, 965-972.
33. Vitetta, E.S. and Uhr, J.W. (1975) Eur. J. Immunol. 6, 140-143.

34. Salsano, F., Froland, S.S., Natvig, J.B. and Michaelson, T.E. (1974) Scand. J. Immunol. 3, 841-846.
35. Pernis, B., Brouet, J.C. and Seligmann, M. (1974) Eur. J. Immunol. 4, 776-778.
36. Fu, S.M., Winchester, R.J. and Kunkel, H.G. (1975) J. Immunol. 114, 250-252.
37. Kearney, J.F., Cooper, M.D., Klein, J., Abney, E.R., Parkhouse, R.M.E. and Lawton, A.R. (1977) J. Exp. Med. 146, 297-301.
38. Vitetta, E.S. and Uhr, J.W. (1977) Immunol. Rev. 37, 50-88.
39. Scher, I., Sharrow, S.O., Wistar, R., Asofsky, R. and Paul, W.E. (1976) J. Exp. Med. 144, 494-506.
40. Paul, W.E. (1970) Transplant. Rev. 5, 130-166.
41. Davie, J.M. and Paul, W.E. (1971) J. Exp. Med. 134, 495-516.
42. Coutinho, A. and Möller, G. (1974) Scand. J. Immunol. 3, 133-146.
43. Coutinho, A. (1975) Transplant. Rev. 23, 49-69.
44. Bretscher, P. and Cohn, M. (1970) Science, 169, 1042-1049.
45. Mosier, D.E., Zitron, I.M., Mond, J.J., Ahmed, A., Scher, I. and Paul, W.E. (1977) Immunol. Rev. 37, 89-104.
46. Sell, S. and Gell, P.G.H. (1965) J. Exp. Med. 122, 423-439.
47. Skamene, E. and Ivanyi, J. (1969) Nature (London), 221, 681-682.
48. Kirchner, H. and Oppenheim, J.J. (1972) Cell. Immunol. 3, 695-699.
49. Maino, V.C., Hayman, M.J. and Crumpton, M.J. (1975) Biochem. J. 146, 247-252.
50. Adenolfi, M., Gardner, B., Giannelli, F. and McGuire, M. (1967) Experientia (Basel), 23, 271-272.
51. Daguillard, F., Heines, D.C., Richter, M. and Rose, B. (1969) Clin. Exp. Immunol. 4, 203-211.
52. Oppenheim, J.J., Rogentine, G.W. and Terry, W.D. (1969) Immunology, 16, 123-137.
53. Gausset, P., Delespesse, G., Hubert, C., Kennes, B. and Govaerts, A. (1976) J. Immunol. 116, 446-453.
54. Schreiner, G.F. and Unanue, E.R. (1976) Adv. Immunol. 24, 37-165.
55. Greaves, M., Janossy, G., Feldmann, M. and Doenhoff, M. (1974) in The Immune System, Sercarz, E.E., Williamson, A.R. and Fox, C.F. eds., Academic Press, Inc., New York, pp. 271-297.
56. Parker, D.C. (1975) Nature (London), 258, 361-363.
57. Weiner, H.L., Moorehead, J.W. and Claman, H. (1976) J. Immunol. 116, 1656-1661.
58. Sieckmann, D.G., Asofsky, R., Mosier, D.E., Zitron, I.M. and Paul, W.E. (1978) J. Exp. Med. 147, 814-829.
59. Gronowicz, E., Coutinho, A. and Melchers, F. (1976) Eur. J. Immunol. 6, 588-590.
60. Andersson, J., Bullock, W.W. and Melchers, F. (1974) Eur. J. Immunol. 4, 715-722.
61. Melchers, F., von Boehmer, H. and Phillips, R.A. (1975) Transplant. Rev. 25, 26-58.
62. Kearney, J.F., Cooper, M.D. and Lawton, A.R. (1976) J. Immunol. 116, 1664-1668.
63. Sieckmann, D.G., Scher, I., Asofsky, R., Mosier, D.E. and Paul, W.E. (1978) J. Exp. Med. 148, 1628-1643.
64. Loken, M.R. and Herzenberg, L.A. (1975) Ann. N.Y. Acad. Sci. 254, 163-171.
65. Scher, I., Ahmed, A. and Sharrow, S. (1977) J. Immunol. 119, 1938-1942.
66. Oi, V.T., Jones, P.P., Goding, J.W., Herzenberg, L.A. and Herzenberg, L.A. (1978) Curr. Top. Microbiol. Immunol. 81, 115-129.
67. Vitetta, E.S. and Uhr, J.W. (1975) Science, 189, 964-969.
68. Amsbaugh, D.F., Hansen, C.T., Prescott, B., Stashak, P.W., Barthold, D.R. and Baker, P.J. (1972) J. Exp. Med. 136, 931-949.
69. Scher, I., Steinberg, A.D. and Berning, A.K. (1975) J. Exp. Med. 142, 637-650.

70. Scher, I., Sharrow, S.O. and Paul, W.E. (1976) J. Exp. Med. 144, 507-518.
71. Mosier, D.E., Scher, I. and Paul, W.E. (1976) J. Immunol. 117, 1363-1369.
72. Dutton, R.W. (1975) Transplant. Rev. 23, 66-77.
73. Kishimoto, T. and Ishizaka, K. (1975) J. Immunol. 114, 585-591.

DISCUSSION

DR. RAFF: This paper is open for discussion.

DR. BELL: Bell, Los Alamos. Bill, you point to very peculiar sort of dose reponse effects, namely, that you think you use more anti-immunoglobulin antibodies than are necessary to saturate all the receptors and you still get increasing responses, the cells which have the maximum number of receptors don't respond best, and so forth.

I wonder if there are possibly some interactions between your B lymphocytes in this case. For example, your antibodies will not only cross-link receptors on the surface of one cell; they may stick two cells together, and one might entertain the hypothesis that only after cells have been stuck to each other will they be activated.

Do you have any evidence on whether individual cells will be responding or whether they might be aggregated? I wasn't quite clear about your culture conditions.

DR. PAUL: We don't have evidence. Our cultures are really very simple and straightforward. We put about from 50,000 to 500,000 cells in a microwell, so there is plenty of opportunity for the interaction of these cells. I suspect the only way we could test your ideas would be if we could develop a semisolid material so that we could distribute cells individually and be sure they weren't interacting.

That is not beyond the realm of possibility. It hasn't been done to these systems, but others have, using mitogenic agents, convinced B lymphocytes to grow in agar. It is conceivable one might be successful. We have not tried it, though, but it is clearly quite correct that we should be doing that.

DR. BELL: Well, one might try some experimental approach where you diluted the responsive B cells with something else so that opportunities for interaction between them would be very --

DR. PAUL: I think those are possible, but in the end, the other would be more decisive and indeed, I suspect, simpler.

DR. BLUMENTHAL: You have shown very beautifully how you isolate specific cells, identify their receptors, and go to the next step to find out what the transduction mechanism is. Could you comment a little bit on the more crude ways that people have taken of just perturbing cells with proteases, periodate, calcium ionophore, if something can be learned from those about lymphocyte activation?

DR. PAUL: Well, the answer is that of course a lot can be learned. Most of those crude ways you think of are mainly good for T cells, for example, activation by periodate. The calcium ionophore, as far as I know, principally activates T lymphocytes. And T lymphocyte activation is easy to study with such agents. I think we learn a lot about what cells have to do in order to divide when we study them.

Whether we are learning what cells do in the course of an immunological response, I am not sure. We may very well. The T cell is more difficult to study in this respect because we still have a very poor grasp on the nature of its receptor. For B cells, of course, LPS is the paradigm.

A good bit is known about early biochemical events. I am sure you know better than I, really. I know that there are all sorts of things that happen fast, and in these cells, or in B cells and T cells, responding to mitogenic agents has been studied for 15 years and more, and we know lots of things happen very fast. But we don't really have an understanding which of these events is central and which are simply, if you will, accompaniments of the activation process.

I must say that not being a cell biologist, I always feel very diffident in the face of people who are competent in this field. Perhaps Dr. Raff would like to comment.

DR. RAFF: I think you have to appreciate why the cell complexity makes those experiments almost impossible to interpret. For example, the cleanest experiment that people claim to have done is with, for example, a calcium ionophore such as A23187, which has stimulated a lot of thinking about whether calcium influx is related to stimulation.

But it has never been shown that this ionophore could stimulate lymphocytes directly. It is equally plausible and, I would suggest, probable that the ionophore is acting on macrophages to stimulate the production of the mitogen that now acts on a lymphocyte. In every T cell response that I am aware of, at least three cells are required. How are you going to know what you are actually studying if you are simply measuring the effect of a calcium ionophore, for example?

DR. METZGER: Bill, obviously one way that one could put a couple of things you mentioned together, which was the fact that the bright cells chosen with anti-IgD are most effective, plus the fact that you need a very high dose of your putatively specific anti-mu, is in fact that your anti-mu is contaminated with, let's say, anti-IgD. And I wonder whether you or Kimi could say something about the response of these cells to anti-delta.

DR. PAUL: I suspect that isn't the case because we have tried to activate cells with anti-delta, so far without success. We anticipated that anti-delta would actually inhibit proliferation since anti-kappa is much less stimulatory than is anti-mu.

Now, the problem with anti-delta is, that it is made against a membrane protein in the mouse. Thus one generally has a serum with anti-delta activity but no reliable estimate of the actual concentration of anti-delta. Relatively recently a hybridoma reagent has been produced which has specificity for delta. Even it, however, is not yet quantifiable in microgram terms because it is a mixture of different immunoglobulin molecules, including anti-delta but with some others. So we probably don't know its concentration better than with a four-fold or five-fold range.

Nonetheless, using as much of it as we can put in tissue culture without killing the cells, this reagent doesn't stimulate B cell activation. Thus, I do not think that the stimulatory effect of anti-mu can be due to contamination with anti-delta.

DR. ISHIZAKA: Not about this program, but I have one question, maybe too much technical, but when you use the cell sorter, it is very beautiful, but do you have any kind of data on treatment with anti-delta that did not affect or treatment with anti-mu that did not affect?

DR. PAUL: It is not a trivial question at all. It is a really critical question. Dr. Ishizaka's point is that after all, we are in the bind that we want to separate cells on the basis of the density of immunoglobulin; we use anti-immunoglobulin for the separation and then we use it again for stimulation. What confidence do we have that the first treatment hasn't really affected the outcome of the second treatment?

I am actually quite confident we are okay, for very curious reasons. We have spent a long time, over a year or more, in which we have tried to pulse lymphocytes with anti-immunoglobulin in the hope that we could learn something about early events in B lymphocyte activation by anti-immunoglobulin.

In fact, we have been terribly unsuccessful. Our results show that if you remove anti-immunoglobulin from the culture before 24 hours, there is no response whatever. Furthermore, if we do an experiment of the sort you describe, that is, add anti-Ig but do not sort the cells, such cells are no different from untreated cells in their subsequent response to anti-mu.

DR. ISHIZAKA: That means you have to have antibody for a long time to get the stimulation. How long do you have to have antibody, and what is going on?

DR. PAUL: Well, I don't know what is going on. If the anti-Ig is present for less than 24 hours, you get virtually no stimulation. In fact, we nearly have despaired of a pulsing technique; we then did what I regard as an experiment that could not fail. If we had the anti-immunoglobulin in for two days, then the cells respond at the end of that time.

So, we felt one experiment which would surely work is to add anti-immunoglobulin for 24 hours, wash the cells, and add it again for a second 24 hours. And indeed if you do that, they respond, but that is hardly surprising.

(Laughter)

So then we said let's now add the anti-immunoglobulin for 24 hours, 12 hours, or one hour, and then at the end of 24 hours we will put it back again. Now, when we do that, we can show that adding anti-Ig for a relatively short time, say from time zero to one hour, waiting 23 hours, then adding anti-immunoglobulin for a second 24 hours, in fact has an effect.

So the first hit does do something but it is very hard to reveal it. And we don't really understand what it does. It may well be just a simple cycling. One cycle of modulation may lead to some intracellular event which is critical. As I say, it has been exceptionally hard to show that.

DR. RAFF: You have a system that in some respects is comparable to the Goldstein and Brown LDL system. That is, you have a CBA/N mouse that as a normal number of IgM receptors, or at least they have a large number of IgM receptors, and yet they are not responding to the ligand.

So, if you label the ligand and then look at the fate of the ligand, can you distinguish the CBA/N?

DR. PAUL: Cells from the CBA/N mouse, I should say, behave like and look like, in many respects, neonatal cells. It has, in fact, not normal amounts of IgM but large amounts. It has about three times as much IgM on the average as a normal cell, and the same amount of IgD.

On the face of it, it would seem straightforward. You say let's examine the ways in which CBA/N cells and normal cells handle immunoglobulin. And I think it is a good experiment, but I don't think it is as simple as it appears, because these cells differ from normal cells in a variety of ways. It is unlikely that CBA/N cells will differ from normal cells in a single protein to which determination of responsiveness can be ascribed.

It is not quite as bad as studying two inbred strains, for example, but it is not as clean as it appears to be. Nonetheless, the experiments should be done. They are obviously the entering wedge. But I don't think when we get the answers, we will now have the answers that are really clean. They will just help us to frame somewhat better questions.

DR. GERSHON: I am Gershon. Do you think there might be any significance to the fact that cells, after coming out from the cell sorters, are not clean populations? If you would vary the different amounts of contaminants in the different populations, will you see a different type of responses?

DR. PAUL: You are suggesting we should intentionally mix. We haven't done that. I would be surprised to see much in the way of differences, but we haven't intentionally done that.

But you could argue with some validity that our populations aren't pure yet, and I would have to agree with that. We haven't done that.

DR. RAFF: Could I just finish this part of the session by giving an example where we actually do know something about immunoglobulin, signalling it. I thought Bill might talk about it. You just referred to it.

And that is what happens when you look at neonatal B cells, B cells that have just developed in the fetal liver or the bone marrow in the adult. These cells only have IgM on their surface, and Fc receptors for IgG, and no other markers, as Bill said. But they behave very differently when ligands such as anti-immunoglobulin and anti-mu bind to those receptors.

In these cells, within minutes, endocytosis of the complex occurs. And after several cycles of ingestion, ligand-induced ingestion and resynthesis or reappearance, then the cell shuts off. After 12 to 24 hours, it no longer will make IgM and put it on its surface, and such a cell is no longer able to respond to ligands or to antigen or to LPS, for example.

It has now been shown by a variety of people that antigen binding to the IgM receptors on these cells also shuts these cells off and induces them to become tolerant, if you like. Whereas, if you take a mature B cell and look at anti-mu, binding exactly to the same receptor, and there is no reason to expect that these receptors are different, then you get ingestion but it requires much higher concentration, takes much longer to occur, and it is always reversible up to six or seven days. It will always continue to put out more IgM, and this cell is not shut off in that way.

So here is the same receptor, presumably, on cells at different stages of maturation, where ligand binding induces a very different event. Now, as far as I know, it is the only good example where one has some molecular information on what a signal is actually doing through an immunoglobulin receptor.

There it is clear that a cycle of ingestion is somehow leading to the shut-off of synthesis in that cell.

DR. BELL: Does that take bivalent ligands?

DR. RAFF: Bivalent ligands do it much better, but monovalent ligands will do it. Monovalent anti-mu will do it. So it is very difficult to be sure there is no aggregation of that system, but certainly an anti-mu that will not induce visible aggregation or capping will shut these cells off. It requires much higher concentration.

DeLisi/Blumenthal, eds. Physical Chemical Aspects of Cell Surface Events in Cellular Regulation

THE INTERACTION OF SURFACE IMMUNOGLOBULINS OF LYMPHOCYTES WITH HIGHLY DEFINED SYNTHETIC ANTIGENS

HENRY KRAKAUER, JAMES S. PEACOCK, BENEDICT G. ARCHER AND TERESA KRAKAUER
Program in Biochemistry and Biophysics, Washington State University, Pullman, Washington 99164

I. INTRODUCTION

Synthetic antigens have long served as probes of the properties of immunoglobulins.(1) They proved very useful in the characterization of the binding sites of antibodies and are now being applied extensively in studies in cellular immunology whose objectives are 1) the definition of the molecular basis of immunogenicity, and 2) the clarfication of ontogenetic pathways of differentiation, particularly in the B lymphocyte line. Even the first objective has, so far, proved elusive. Though monovalent haptens are generally not immunogenic, certain exceptions, for example, the dinitrophenyl (DNP) derivatives of oligolysines and some conjugates of l-tyrosine, do exist. (2,3) Their mechanism of action is thought to involve interactions of receptor-hapten complexes on T and B cells, and may thus involve a conversion to functional polyvalency. The complication of T-B cell cooperation in the expression of immunologic activity is not, on the other hand, required in the case of thymus-independent (TI) antigens. These are principally, synthetic or naturally occurring polymeric materials bearing repeated determinants. (4) Thus, although they permit the avoidance of the complication of cell-cell interactions, they introduce that of molecular complexity.

The ideal antigen for the study of the molecular parameters of immunogenicity, that is of the constellation of effects on the cell that result from the binding of this antigen, should possess clearly identifiable portions for which the receptor is specific, epitopes, present in an exactly known multiplicity and at definable relative positions on the molecule. In addition, the carrier of these epitopes should be as nearly "invisible" as possible, i.e. should possess negligible affinity for macromolecular solution components or for cells. And this antigen should have a limited and identifiable target population. We present below candidates for this role.

2. THE ANTIGENS

The antigens are dinitrophenyl derivatives of polyethylene oxide (DNP_nPEO). Their synthesis and molecular characterization have already been described.(5) Briefly, they may be prepared from linear PEO with two -OH ends or with one -OH and one $-OCH_3$ end to yield strictly divalent or strictly monovalent derivatives respectively. Chain lengths of $-(CH_2CH_2-O-)_n$, n = 5 to n $\geq$ 1,500 are available for the divalent form. Alternatively, branched-chain PEO may be used to prepare multivalent antigens with an epitope density of up to 80 DNP groups per molecule, or one DNP per 5,000 D.

The attachment of the DNP moiety to the ends of the PEO is carried out in two steps in dry tetrahydrofuran, the first being the conversion of the -OH to an alkoxide,

$$\text{R-OH + n-butyl lithium} \xrightarrow{\text{THF}} \text{R-O}^-\text{Li}^+ + \text{butane}$$

and the second the displacement by the alkoxide of the fluorine of Sanger's reagent,

$$\text{R-O}^-\text{Li}^+ + \text{DNFB} \xrightarrow{\text{THF}} \text{R-O-DNP} + \text{Li}^+\text{F}^-$$

Both reactions go to completion. After transfer to an aqueous medium, the DNP-substituted polymer is purified by one passage over Biogel P-6 (Biorad) or Sephadex G-10 (Pharmacia) equilibrated with water or any suitable buffer. Fractionation according to chain length may be achieved by "sieving", for example on Sephadex G-200 for the larger antigens. Two steps suffice to approach homogeneity in molecular weight, as may be seen in Table I.

3. THE INTERACTIONS OF THE ANTIGEN WITH SOLUBLE RECEPTORS (Ig)

The binding of DNP_1PEO (mol. wt. 930 D) to DNP-specific antibody was characterized in studies employing calorimetry, quenching of fluorescence, and spectrophotometry, (6) and of DNP_2PEO, 25,000 D to 75,000 D, by light scattering photometry (7) . DNP_1-PEO behaves as a typical DNP hapten, binding to a series of antibody fractions with an affinity constant of about $3X10^6M^{-1}$ and a ΔH of -14 to -17 Kcal/mol.

The objective of the study of the interaction of DNP_2PEO with two subclasses of equine antibody, IgG and IgG(T), was the elucidation of the molecular basis for the failure of IgG(T) to precipitate multivalent antigens. At equilibrium,

Table I. Molecular Properties of DNP_2-PEO

$\overline{M}_w$ (a)	$\overline{M}_w/\overline{M}_n$ (b)	R_g (c)	$R_g/(\overline{M}_w)^{1/2}$ (d)
82,400±2400 D	1.13±0.06	12.2nm	0.42
65,200±1800	1.06±0.03	11.3	0.44
56,500±1800	1.00±0.04	8.4	0.35
42,800±1300	0.97±0.03	8.8	0.43
35,400±1300	1.03±0.03	-	-

(a) $\overline{M}_w$ - weight average molecular weight, determined by light scattering analysis. All uncertainties are standard errors of the mean.

(b) $\overline{M}_n$ - number-average molecular weight, determined by measurement of mass by differential refractometry and end-group analysis by spectrophotometry. A ratio of $\overline{M}_w/\overline{M}_n$ of 1 indicates homogeneity in size. The material of highest mol. wt. appeared near the void volume.

(c) R_g = root-mean-square radius of gyration, determined by light-scattering photometry, in 0.45 M K_2SO_4, 0.01 M phosphate buffered saline, at 25°C.

(d) This ratio should be constant for homogeneous polymer fractions.

a mixture of antibody (ab), divalent antigen (ag) and monovalent hapten (h) contains, in addition to these components, a variety of linear and cyclic complexes. The reactions and equilbrum constants connecting these various species are:

1) $ab_n\ ag_n \rightleftharpoons cyc_n$, $\quad \frac{1}{2n^{5/2}} Q \frac{V}{N} K$

2) $ag + ab_n\ ag_n \rightleftharpoons ab_n ag_{n+1}$, $\quad PK$

3) $ab + ab_n ag_n \rightleftharpoons ab_{n+1} ag$, $\quad PK$

4) $ag + ab_{n+1}\ ag_n \rightleftharpoons ab_{n+1}\ ag_{n+1}$, $\quad 4PK$

5) $ab + ab_n\ ag_{n+1} \rightleftharpoons ab_{n+1}\ ag_{n+1}$, $\quad 4PK$

6) $h + ab_n\, ag_n == h\, ab_n\, ag_n$, K

7) $h + ab_{n+1}\, ag_n == h\, ab_{n+1}\, ag_n$, 2K

8) $h + h\, ab_{n+1}\, ag_n == h_2\, ab_{n+1}\, ag_n$, (1/2)K

$ab_n\, ag_n$, $ab_{n+1}\, ag_n$ and $ab_n\, ag_{n+1}$ represent linear complexes with reactive ends, and cyc_n the closed or cyclic complexes containing n antibody and n antigen molecules. K is the equilibrium constant for the binding of the hapten to a receptor site on the antibody and P a statistical factor representing that fraction of configurations of the polymer chain of the antigen that remain accessible after it has bound to antibody to participate in a linear complex. (In fact, P accounts for all factors that cause the equilibrium constant for binding of DNP_2PEO to differ from 4K). The numerical factors result from the number of ways a given type of bond may form and be broken. Q is the ring closure factor or the probability that, with one end of the antigen bound to one antibody site, the other end of the antigen will be at the other site. For a ring of n antibody and n antigen molecules the factor is $Q_n = Q/n^{3/2}$. $\frac{V}{N}$ is the volume per molecule in the standard state chosen for K. (Q is in units of nm^{-3}, $\frac{V}{N}$ is 1.66 nm^3 for K in molar units. P is unitless (8)).

The intensity of light scattered by a solution containing complexes of conventionally behaving IgG or IgG(T) with DNP_2 PEO decreases with increasing concentrations of the hapten DNP_1PEO in a pattern shown in Figure 1. Analysis of such results by a procedure described elsewhere (7) yields estimates of Q and P, the two unknown quantities in the system of equations 1-8. (K, as noted above, can be obtained independently, and for the antibodies used in these studies, lay in the range 2-4 X $10^6 M^{-1}$.) (Table II)

Concentrations of the various molecular species present during the course of the titrations may be calculated in turn. These are shown for a representative experiment in Figure 2. It is evident that complexes containing more than one antibody and one antigen do not contribute greatly.

The most significant point immediately evident from Table II is that the ring closure probability Q is larger for all antigens in reactions with the non-precipitating antibody, IgG(T), than in reactions with the precipitating IgG. This confirms the supposition that the Fab arms are held closer together in IgG(T). The consequent lesser effectiveness in cross-linking macromolecular antigens could then indeed be the basis for its inability to precipitate them. (9)

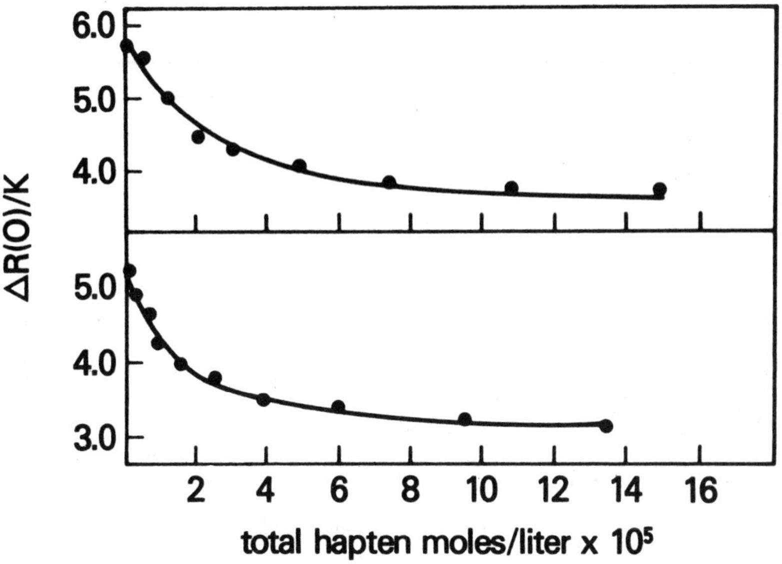

Figure 1. Light scattering measurements on equimolar mixtures of DNP_2PEO, mol. wt. 57,800 D, and IgG(T) (upper panel) or IgG (lower panel). The concentrations of antibody and antigen were $5X10^{-6}M$.

Table II. Estimates of the Ring Closure (Q) and Steric Interference (P) Factors

Antigen [(a)] Mol. Wt.	IgG		IgG(T)	
	Q	P	Q	P
74,800 D	$0.68 \times 10^{-5} nm^{-3}$	0.27	$1.21 \times 10^{-5} nm^{-3}$	0.41
63,000	1.09	0.36	1.70	0.41
57,800	1.00	0.27	1.72	0.39
45,400	0.98	0.16	2.61	0.35
35,600	1.06	0.12	2.87	0.28
25,000	-	-	8.64	0.18

(a) These are number-average molecular weights.

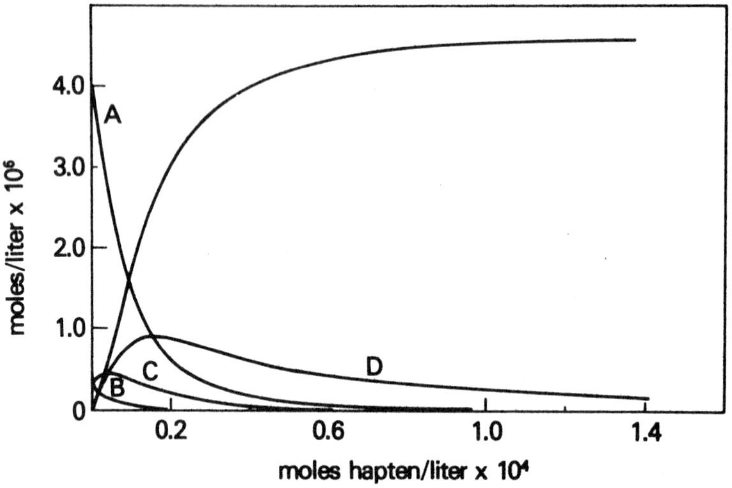

Figure 2. Concentrations of the predominant species of complexes containing IgG, DNP_2PEO (mol. wt. 57,000 D) and DNP_1PEO. Concentrations of antibody and antigen were $5X10^{-6}$M. Curve A: ab=ag, Curve B: cyclic complexes larger than ab=ag, Curve C: linear complexes other than habag or h_2ab. Curve D: habag, Curve E: h_2ab.

A quantitative estimate of the distance (R) between the binding sites on the antibodies may also be made. The simplest model on whose basis Q and R may be related assumes that the antigen DNP_2PEO behaves as an ideal polymer and that the Fab arms are held fixed at an average intersite distance R.

$$9) \quad Q = \left(\frac{3}{2\pi\beta^2 M}\right)^{3/2} \exp\left(\frac{3R^2}{2\pi\beta^2 M}\right)$$

M = molecular weight of the antigen, $\beta^2 M = R_g^{\,2}$ (see Table I, from which $\beta^2 = 0.01\ nm^2$ mol/gm)

Alternatively, free flexibility may also be allowed about the hinge in the antibody so that R may vary from R = 0 to $R = R_m$, some maximum distance. In that case

$$10)\qquad Q = \left(\frac{3}{2\pi\beta^2 M}\right)^{3/2}\left(\frac{3}{R_m^{\ 3}}\right)\int_{R_m}^{0} R^2 \exp\left(\frac{-3R^2}{2\beta^2 M}\right) dR$$

and the average intersite distance is

$$11)\qquad \langle R^2 \rangle = \frac{3}{5} R_m^2$$

In addition, account may be taken of the perturbation by the antibody molecule of the polymer segment distribution, in the manner that led to the formulation of the parameter P. In this model, the antibody is represented by an infinite plane beyond which the polymer cannot penetrate(10), placed at a distance d below the binding sites. These are held apart at a fixed distance R.

$$12)\quad Q = \left[1-\exp\frac{6d^2}{\beta^2 M}\right]\left[\frac{3}{2\pi\beta^2 M}\right]^{-3/2}\left[\exp\left(\frac{-3R^2}{2\beta^2 M}\right)\right]/\mathrm{erf}\left(\frac{3d^2}{2\beta^2 M}\right)^{1/2}$$

Figure 3 presents the analysis of Q in terms of the 3 models just described, and Table III the estimates of R (or $\langle R^2\rangle^{1/2}$ for the flexible hinge model).

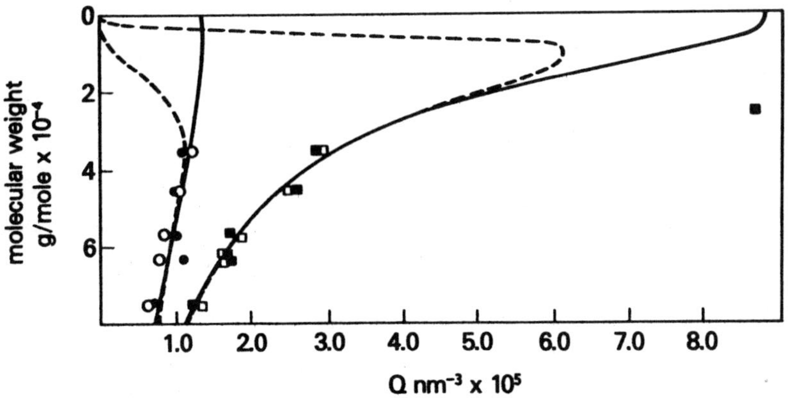

Figure 3. The ring closure probability, Q, as a function of antigen molecular weight. Filled circles and squares are experimental data for IgG and IgG(T) respectively. Dashed lines were calculated by means of eq. (9)(ideal polymer, fixed hinge), solid lines by means of eq. (10)(ideal polymer, flexible hinge), and open symbols by means of eq. (12)(perturbed polymer, fixed hinge).

Table III. Estimates of Intersite Distance in IgG and IgG(T).

	Model		
	Ideal Polymer Fixed Hinge (a)	Ideal Polymer Flexible Hinge (b)	Perturbed Polymer Fixed Hinge (c)
IgG	18.9 ± 0.7 nm (d)	20.0 ± 0.9 nm	16.0 ± 0.7 nm
IgG(T)	10.5 ± 0.5 nm	10.6 + 0.5 nm	11.3 ± 0.9 nm

(a) Equation 9 (b) Equations 11 & 12 (c) Equation 12

(d) uncertainties are standard errors

The models are indistinguishable within the range of antigen molecular weights useable in light scattering photometry. Nevertheless, they do consistently verify the qualitative conclusion that the Fab arms are held substanyially closer together in IgG(T) than in IgG. In addition, the estimates of of intersite distances are also reasonably consistent with other estimates of Fab dimensions obtained by measurements on solutions of antibodies.(11)

With the utility of DNP_2PEO as probes of properties of immune receptors established, it becomes possible to employ them in studies on membrane-bound receptors, the cell surface immunoglobulins. Such studies fall into two classes, those dealing with physical interactions and amenable to analysis by physical-chemical methodology, and those dealing with the biological consequences of these interactions. Preliminary data are available on both aspects.

4. THE BINDING OF DNP_2PEO TO IMMUNE LYMPHOCYTES

The peripheral blood lymphocytes (PBL) of ponies heavily immunized with DNP-BSA, such as those used as sources of anti-DNP IgG and IgG(T) used in the previously described studies, were purified in the usual way, by centrifugation over a Ficoll-Hypaque pad.(11) The cells were further selected for specificity for DNP by affinity chromatography. They were passed slowly at 4°C over gelatin beads derivatized with DNP (13) and retained by a fine nylon mesh in a 2.5 cm diameter jacketed column. After extensive washing of the column to remove cells not specifically adherent, collagenase was added and the temperature raised to 37°C to melt and hydrolyze the gelatin. The cells were then filtered through the nylon mesh, washed free of enzyme and hydrolyzed gelatin by pelleting and resuspension in fresh medium (L-15 supplemented with homologous serum) and incubated at 37°C for about 24 hours to permit regeneration of normal surface structures. PBL treated in this way remain inactive, do not lose responsiveness to con A, and show a very markedly enhanced DNA synthesis when stimulated with DNP_nBSA. In addition, whereas about 5% of crude immune PBL form rosettes with dinitrophenylated homologous red blood cells, (DNP-RBC), up to 47%, depending on the details of the purification procedures, do so with the enriched population. The results described below were obtained with the products of two separate enrichment procedures, one yielding cells 22% of which formed rosettes with DNP-RBC and the other 47%. Antigens were prepared in the manner described above, but with ^{3}H-DNFB, to yield DNP_2 PEO with a specific activity corresponding to 2.45 x 10^{15} cpm per mol of DNP groups under the conditions of the measurements. The binding assay was performed in the following manner. Cells that had been recovered from the DNP gelatin columns and had been preincubated for 24 hours at 37°C were pelleted and resuspended in phosphate-buffered (pH 7.4) saline at 4°. To aliquots of these cells, various amounts of DNP_2 PEO, fractionated according to chain length , were added. After a further incubation for one hour at 4°C, the cells were resuspended gently and layered over pads of Ficoll-Hypaque in small test tubes. They were then pelleted rapidly at 4000g. The upper portions of the tubes were washed thoroughly and the Ficoll-Hypaque layer containing the pellet was dried under reduced pressure. The cells were solubilized with Protosol and counted in Omnifluor (New England Nuclear). Parallel determinations of binding to cells that had failed to adhere to the DNP-gelatin column were performed. From these it was calculated that 10-15% of the antigen molecules bound non-specifically to DNP-specific cells. Table IV presents the results of the measurements in terms of the numbers of DNP groups bound specifically (i.e. to the receptors) per DNP-specific cell. Figure 4 presents the data in Table IV in the form of a Scatchard plot.

It is evident from Table IV and Figure 4 that the interaction of DNP_2 PEO with the cell surface DNP receptors can be adequately described in terms of binding to a lattice. The near linearity of the plot for the antigen of lowest

TABLE IV. The Specific Binding of DNP_2 PEO to DNP-Specific Cells.

Antigen: Concentration – Mol. Wt.	$4x10^{-7}N$ A	$4x10^{-7}N$ B	$1x10^{-6}N$ A	$1x10^{-6}N$ B	$2x10^{-6}N$ A	$2x10^{-6}N$ B	$5x10^{-6}N$ A	$5x10^{-6}N$ B	$1x10^{-5}N$ A	$1x10^{-5}N$ B
	DNP groups bound per cell x 10^{-5}									
$1.1x10^5D$	1.13	1.21	2.16	2.02	3.90	3.80	5.69	6.62	8.96	8.52
$6.0x10^4D$	0.71	0.81	1.38	1.66	2.10	2.45	3.29	3.84	4.71	5.40
$3.3x10^4D$	-	0.87	-	1.55	-	2.69	-	5.90	-	7.09
$1.2x10^4D$	0.91	0.93	2.19	2.03	3.81	3.87	8.16	7.71	12.5	11.6

Antigen concentration is given in terms of moles of DNP groups per liter (normality). A = cell preparation containing 3.30 x 10^6 DNP-specific lymphocytes per 7.0 x 10^6 lymphocytes in the 1 ml assay sample. B = cell preparation containing 1.75 x 10^6 DNP-specific lyphocytes per 8.0 x 10^6 lymphocytes in the 1 ml assay samples.

molecular weight is somewhat surprising. Curvature is observed only in its initial portions. This implies both very limited heterogeneity in the affinity of the receptors for the DNP goups and predominantly monovalent binding, except at very low extents of binding. The receptor homogeneity might have resulted from the very prolonged course of immunization undergone by the animals and may correspond to the limited range of affinities for DNP_1 PEO of antibodies (about a factor of 2) measured in connection with studies on IgG and IgG(T) described above. The predominantly monovalent binding of DNP_2 PEO of 12,000 D implies a low ring closure probability Q. This may be consistent with the dashed curves of Figure 3, implying further that the arms of the receptors are held relatively rigidly in an open position on the cell surface. Such a conclusion is at present highly speculative. A much more thorough quantitative analysis, currently underway, is clearly called for. The linearity of the plot does, however, permit extrapolation to the intercept on the abscissa and an estimate of about 6 x 10^5 receptor Ig molecules per DNP specific cell. The larger antigens with a higher average end-to-end distance have apparently a sufficient reach to permit ring closure on the same receptor or binding to adjacent receptors to occur. This aspect of potential cross-linking has major biological imlications (14) and also requires more detailed analysis.

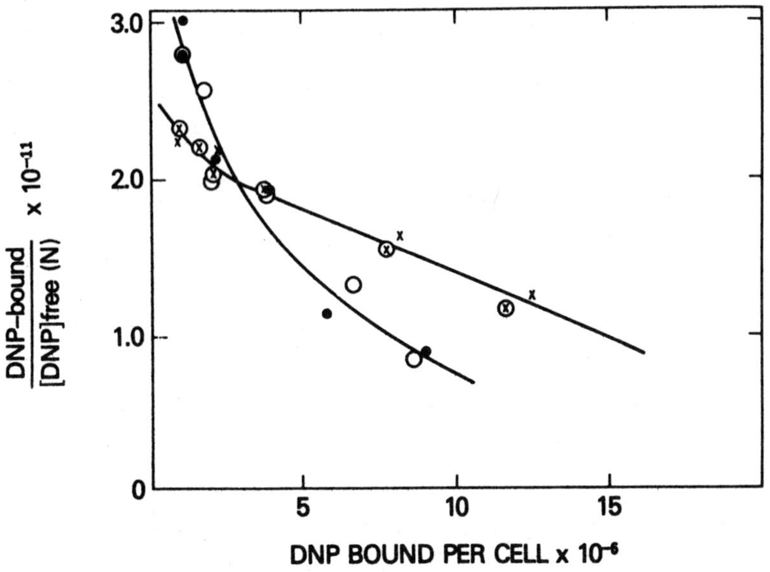

Figure 4. Scatchard plot of the binding of DNP_2PEO to DNP-specific equine lymphocytes. X, ⊠ : DNP_2PEO mol. wt. 12,000 D, columns A and B respectively of Table III: •, O : DNP_2PEO mol. wt. 110,000 D, columns A and B respectively.

5. THE IMMUNOGENICITY OF DNP_nPEO ANTIGENS

The immunogenicity of mono, di and multivalent DNP derivatives of PEO was investigated in New Zealand white rabbits. These materials were administered into the foot pads in complete Freund's adjuvant for priming and in incomplete Freund's adjuvant for boosting. The humoral response, consisting of anti-DNP antibody precipitated by dinitrophenylated horse gamma globulin, is displayed in Figure 5. Clearly, although di and multivalent DNP_n PEO are not as effective as DNP_n BSA, a well-known highly potent immunogen, they elicit quite substantial anti-DNP antibody levels, approaching 2 mg/ml. On the other hand, no precipitating anti PEO antibody was detectable. Intradermal injections of the antigens failed to elicit delayed hypersensitivity reactions. Figure 5 also suggests that DNP_1 PEO is not immunogenic, although in this in vivo test, in contrast to the in vitro studies to be shown below, a wide range of concentrations was not tested.

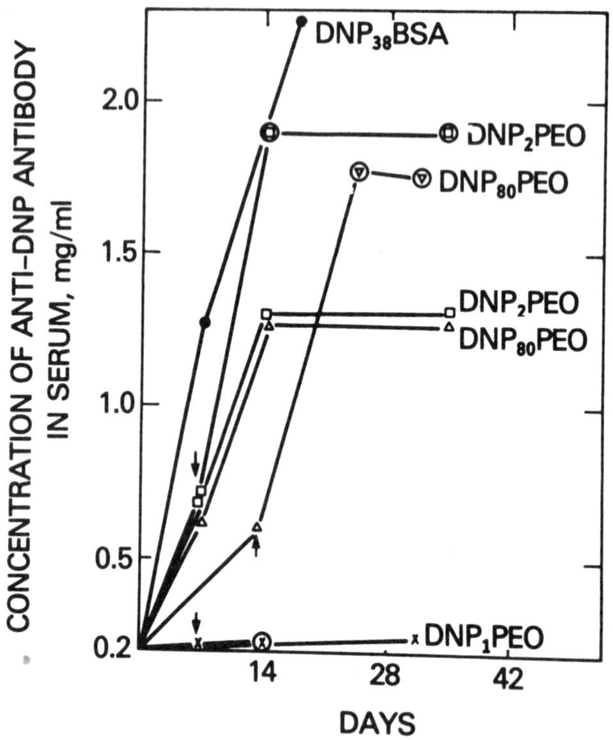

Figure 5. The humoral response of New Zealand white rabbits to immunization with DNP-antigens. Arrows indicate the times at which booster immunization was administered, and circled symbols refer to secondary responses. The mol. wt. of DNP_2PEO was 35,000 D, those of the others are given in the text.

The reactivity of peripheral blood lymphocytes, in whole blood cultures, and of splenic lymphocytes of immunized rabbits was tested in vitro. Whole blood cultures consisted of arterial blood diluted 1:30 with RPM1-1640 buffered with HEPES or with HCO_3^-/H_2CO_3 in a 95% air, 5%CO_2 atmosphere at pH7.4. The medium was supplemented with 7% heat-inactivated serum obtained from unimmunized rabbits. Spleen cells were cultured at a concentration of 10^6 cells per ml in the same medium, at a total volume of 2 ml. Incubations were carried out at 37°C for 120 hours in the presence of the antigens prior to addition of ^{3}H-thymidine and for 72 hours when stimulation was effected with concanavalin

A. Incubation with ^{3}H-thymidine was then continued for a further 24 hours. Various specific activities of the radiolabel were used in the different series of experiments.

Table V. Responses of Rabbit Spleen Cells to Stimulation by DNP-Antigens

In vitro stimulant	In vivo immunogens		
	None	$DNP_{40}BSA$	DNP_2PEO
	Incorporation of ^{3}H-Thymidine		
None	10,800 ± 1,000	3,100 ± 200	8,700 ± 1,300
$DNP_{40}BSA$-0.1 mg/ml	-	37,800 ± 2,700	-
$DNP_{40}BSA$-0.15 mg/ml	11,600± 700	41,400 ± 4,100	20,800 ± 3,8000
$DNP_{40}BSA$-0.2 mg/ml	-	20,000 ± 600	-
$DNP_{80}PEO$ 1 x 10^{-8}N	-	2,800 ± 500	-
" 1 x 10^{-7}N	-	2,400 ± 300	-
" 5 x 16^{-7}N	-	-	18,600 ± 1,500
" 1 x 10^{-6}N	-	3,500 ± 300	12,500 ± 600
" 1 x 10^{-5}N	-	3,700 ± 200	11,300 ± 1,500
" 4 x 10^{-5}N	-	2,800 ± 600	-
DNP_2PEO 1 x 10^{-8}N	-	-	8,000 ± 1,400
" 1 x 10^{-7}N	-	-	13,200 ± 300
" 1 x 10^{-6}N	-	-	18,200 ± 1,200
" 1 x 10^{-5}N	11,200 ± 1,700	-	34,600 ± 3,200
PEO,[(a)] 10 μg/ml	11,000 ± 1,600	-	-
", 200 μg/ml	5,000 ± 1,500	-	-
", 1 mg/ml	9,400 ± 400	-	-
", 7 mg/ml	6,400 ± 1,300	-	-

The data given here and in the following Tables are CPM ± SEM.
[(a)] Average molecular weight = 35,000.

Table V presents the in vitro responses of immune rabbit spleen cells to various doses of DNP_nPEO, and for comparison to $DNP_{40}BSA$. It is interesting to note that cells of rabbits immunized with DNP_2 PEO respond to $DNP_{40}BSA$, as well as to $DNP_{80}PEO$. The cross-reactivity with $DNP_{40}BSA$ suggests that DNP_2PEO is thymus independent (cf. also the failure to elicit delayed hypersensitivity). Once again, although DNP_2PEO is less effective than $DNP_{40}BSA$, it clearly does elicit a substantial proliferative response in primed lymphocytes. Neither it nor the carrier, the linear PEO from which is was synthesized,is a non-specific activator. This is evident from the failure of DNP_2PEO to stimulate non-immune lymphocytes, or of the failure of PEO to stimulate at similar or much higher doses. Thus PEO itself is not a polyclonal B cell activator. Neither does PEO stimulate immune cells (Table VI).

Table VI. Responses of Peripheral Blood Lymphocytes to the Carrier, PEO, and to the Monovalent DNP_1 PEO.

In Vitro Stimulant	In Vivo Immunogen: DNP_2PEO
	Incorporation of ^{3}H-Thymidine
None	620 ± 60
DNP_2PEO, 10^{-6} N	2380 ± 200
PEO, 0.017 mg/ml	700 ± 100
PEO, 0.17 mg/ml	480 ± 50
None	550 ± 80
DNP_2PEO, 10^{-7} N	850 ± 150
" , 2×10^{-6} N	1570 ± 160
" , 10^{-5} N	1090 ± 180
DNP_1PEO, 2×10^{-6} N	660 ± 150
" , 1×10^{-5} N	600 ± 80
" , 5×10^{-4} N	280 ± 200

The additional objective of Table VI is the demonstration that DNP_1PEO is not stimulatory to immune lymphocytes, again at doses at which DNP_2PEO is effective, and higher. It must be noted that the comparison in this table is between DNP_1PEO with a mol. wt. of about 930 D with DNP_2PEO with a mol. wt. of 35,000 D. That the failure to respond to DNP_1PEO is not due to its small size is shown in

Table VII in which the dependence of the immunogenicity of DNP_2PEO on mol. wt. (carrier chain length) is explored. The optimal antigen size is 10,000 D, but a clear response is still observed with DNP_2PEO of 1,000 D, material the same size as DNP_1PEO. A very similar pattern is obtained with whole blood cultures.

Table VII. The Dependence of the Mitogenicity of DNP_2PEO on Carrier Chain Length.

Antigen Mol. Wt..[(a)]	Antigen Concentration[(b)]	Spleen Cell Cultures	Whole Blood Cultures
		Incorporation of ^{3}H-Thymidine	
Background		6,800± 700	460± 90
6.9×10^4D	1×10^{-5}N	21,000± 900	1,400± 220
3.5×10^4D	"	23,000± 1,600	1,800± 180
1.0×10^4D	"	39,000± 200	3,300± 390
3.0×10^3D	"	18,000± 1,900	1,400± 360
1.0×10^3D	"	15,000± 600	900± 180

(a) This is a number average mol. wt.
(b) This concentration, given in terms of DNP groups, was determined to be the optimal or nearly so for each fraction in experiments not shown.

Table VIII. The Dependence of the Mitogenicity of DNP_nPEO on Epitope Density

Antigen Valence[(a)]	Antigen Concentration[(b)]	Spleen Cell Cultures
		Incorporation of ^{3}H-Thymidine
Background		6,800± 700
80	$1x10^{-7}$N	18,500±1,400
63	$1x10^{-7}$N	17,000±1,600
38	$3x10^{-7}$N	25,000±2,300
20	$1x10^{-6}$N	40,000±3,300
9	$1x10^{-6}$N	16,700±1,400
4	$3x10^{-6}$N	13,000±1,200

(a) Number of DNP goups per branched-chain PEO molecule of $4x10^5$ mol. wt.
(b) These concentrations, in terms of DNP groups, were found to be optimal in experiments not shown.

The effect of valence of multivalent DNP_nPEO on immunogenicity is shown in Table VIII. The carrier is a branched chain PEO of mol. wt. of 400,000 capable of bearing 80 DNP groups at saturation. The optimum epitope density is 20, and at a valency of 4 it is substantially less effective. Interpretation of the results in this table is rendered difficult, as is true of all results with materials only whose average valency is known, by the uncertainty in the location and accessibility of the DNP groups in partially derivatized molecules.

Table IX. The Effect of Anti-Thymocyte Serum on the Response of Immune Peripheral Blood and Spleen Lymphocytes to DNP_2PEO

Treatment	Stimulant	Spleen Cell Cultures	Whole Blood Cultures
		Incorporation of ^{3}H-Thymidine	
None	None[a]	23,000 ± 2,300	400 ± 160
	Con A, 60 μg/ml[a]	234,000 ± 16,300	10,200 ± 1,100
	None	46,800 ± 1,000	600 ± 80
	DNP_2PEO, 10^{-5}N	118,000 ± 16,000	------
	DNP_2PEO, 10^{-6}N	--------	2,200 ± 300
ATS[b]	None[a]	15,700 ± 2,000	270 ± 20
	Con A, 60 μg/ml[a]	20,800 ± 3,000	300 ± 10
ATS + 10% untreated cells	Con A, 60 μg/ml[a]	61,000 ± 8,200	-------
ATS	None	22,500 ± 5,000	500 ± 50
	DNP_2PEO, 10^{-5}N	76,200 ±13,500	------
	DNP_2PEO, 10^{-6}N	--------	1,600 ± 190

(a) These incubations were carried out for 72 hours prior to addition of ^{3}H-T, the optimaltime for Con A stimulation. The others were carried out for 120 hours, the optimal time for antigen stimulation.
(b) ATS was used at a dilution of 1:50 (see text)

The potential thymus-independence of the DNP_2PEO antigens was investigated further in cultures from which T-cells were removed. Anti-rabbit thymocyte serum was produced in Hartley strain guinea pigs injected intraperitoneally twice, two weeks apart, with rabbit thymus cells. The resulting serum, after absorption with rabbit red blood cells, was found to cause, at a dilution of 1:50 and in the presence of fresh rabbit serum as a source of complement, the lysis of over 95% of rabbit thymocytes and about 50% of rabbit spleen cells. The surviving spleen cells failed to respond to stimulation by con A, a T-cell specific mitogen (15). On the other hand, the response of neither peripheral blood lymphocytes, nor of spleen cells (Table IX) to DNP_2PEO was significantly impaired by treatment with the antithymocyte serum. The experiment in which a small number of untreated cells was restored to the treated cells, shown in Table IX, was carried out to demonstrate that the abolition of the response to con A was not the result of the generation by anti-thymocyte serum of an inhibitory activity specific against con A but rather, consequently, of an actual loss of responsive cells. We have not rigorously demonstrated the total T-cell independence of DNP_2PEO because of the possible survival of a few. It is clear, however, that the help of a substantial number of T cells is not required. Overall, a presumption of T-cell independence appears to be adequately supported. One additional useful property of DNP_2PEO antigens should be noted: the ready reversibility of the mitogenic stimulus that it delivers (Table X). This reversibility appears to be a characteristic of divalent mitogens and a consequence of much greater ease of dissociation compared with such multivalent mitogens as con A and DNP_n BSA (16).

Table X. The Reversiblity of Stimulation of Rabbit Lymphocytes with DNP_2PEO

Treatment	^{3}H-Thymidine Incorporation
None	1,800 ± 100
DNP_2PEO, 10^{-5}N	11,000 ± 2,000
DNP_2PEO, 10^{-5}N, removed by replacement of medium after 22 hours	2,200 ± 900

6. CONCLUSION

The principal attributes of the antigens described above as probes of cellular function are their ability to activate lymphocytes through normal physiologic processes, their simplicity, the range over which their structures can be varied and the precision with which they can be defined. Thus not only can their interactions with receptors be studied within the framework of conventional thermodynamics and kinetics, but their effects on the molecular properties of the cell surface can be approached in the context of a reasonably defined physical state. For example, mobilities of immunoglobulin receptors as well as of other membrane components can be characterized in terms of meaningful saturation fractions. Most interesting will be the correlations of these surface events with the intracellular events associated with the blast transformation triggered by the antigens. The objective of such studies is the dissection of the pathways that lead to cellular proliferation and differentiation. These studies are currently underway.

ACKNOWLEDGMENTS

This work was supported in part by USPHS NIH grants AI-107471, GM-07160, and fellowship AI-50496(to B.G.A.). H. K. is a Research Career Development Awardee of the NIH (GM-70441).

REFERENCES

1. Kabat, E.A., Structural Concepts in Immunology and Immunochemistry, 2nd Edition, Chapter 2, Holt, Rhinehart and Winston, N.Y. 1976.
2. Yaron A., Dunham, E.K. and Schlossman, S.F., Biochem. 1974 13: 347.
3. Prange, C.A., Green, C., Nitecky, D.E., and Bellone, C.J., J. of Immunol. 1977. 118: 1311.
4. Kontiainen, S., Makela, O., and Hurme, M. Quart. Rev. of Biophysics. 1975. 8: 507.
5. Cooke, M.P., Archer, B.G., and Krakauer, H., Biochem. Biophys. Res. Comm. 1974, 57: 1032.
6. Archer, B.G., and Krakauer, H. Biochem. 1977. 16: 615.
7. Archer, B.G., and Krakauer, H. Biochem. 1977. 16: 618.
8. The formulation of equilibrium constants for inter- and intramolecular reactions is discussed by Crothers, D.M. and Metzger, H., Immunochem. 1972. 9. 341, whom we follow.
9. Klinman, N.R. and Karush, F., Immunochem. 1967. 4: 387.
10. Di Marzio, E.A., J. Chem. Phys. 1965. 42, 2101.
11. Pilz, I., Puchwein, G., Krattky, O., Haager, O., Gail, W.E., and Edelman, G.M., Biochem. 1970. 9:211.
12. Krakauer, T. and Krakauer, H., Cell. Immunol. 1976, 26, 242.
13. Gold, E.F., Kleinman, R. and Ben Efraim, S., J. Immunol. Methods. 1974. 6: 31.
14. DeLisi, C., these PROCEEDINGS.
15. Sheppard, H.W., Redelman, D., and Sell, S. Cell Immunol 1976. 24: 34.
16. Krakauer, T. Ku, G.S.B., Peacock, J.S., and Krakauer, H. Biochem. Biophys. Res. Comm. 1978. 82: 1251.

REGULATION OF ANTIBODY AFFINITY AND HETEROGENEITY*

GREGORY W. SISKIND

From the Division of Allergy and Immunology, Department of Medicine
Cornell University Medical College, New York, New York 10021

It is well known that the antibody produced by individual animals to most antigens is a highly heterogeneous population of molecules. A variety of types of heterogeneity can be distinguished. (a) With most complex antigens, antibody specific for a number of distinct antigenic determinants is produced. Even with relatively simple haptenic determinants, the hapten groups located at different positions in the molecule are presented in different orientations and with different near neighbors so that a variety of structurally different antigenic determinants probably exist. (b) If one considers the antibody response to a single haptenic determinant (e.g., the 2,4-dinitrophenyl hapten (DNP)), it is obvious that antibodies of a variety of different immunoglobulin classes and subclasses are produced. (c) The antibody, of a given class and subclass, which is specific for a particular determinant, itself consists of a highly heterogeneous population of molecules. This heterogeneity can be demonstrated in terms of electrophoretic mobility, isoelectric focusing, idiotype, or the binding properties (affinity) of the antibody for the antigenic determinants. This last type of heterogeneity, with which we will mainly deal in this paper, presumably reflects the diversity of B-cell clones which is stimulated to produce antibody. It is generally believed that individual B lymphocytes are only capable of producing one or at most a small number of different antibody molecules. It is further assumed by most workers that the catalogue of information, which is required to produce the complete array of antibody molecules that the animal is capable of producing, is present prior to antigen exposure. According to such a clonal selection type hypothesis (1) individual B cells bear on their surface

* Supported in part by a research grant from the National Institutes of Health, U.S.P.H.S., Number CA-20075.

immunoglobulin molecules (probably of the IgM and/or IgD classes) which have antigen binding properties identical to those of the antibody molecules which that cell and its clonal progeny will secrete following appropriate stimulation by antigen. It is assumed that antigen (probably after some preliminary processing events possibly involving macrophages) interacts with cell surface antigen receptors on B lymphocytes. This interaction, most likely in concert with signals delivered via T cells or macrophages, leads to the activation of the B cell followed by proliferation and differentiation into antibody secreting cells and B-memory cells. Thus, it is the interaction of antigen with cell surface, immunoglobulin, antigen receptors which results in selection of a set of B cells to proliferate and differentiate into antibody secreting cells and B-memory cells. Presumably, this interaction is responsible for imparting specificity to the immune response.

If, as is generally believed the binding properties of secreted antibody are identical to those of the antigen binding sites on the precursors of the B cells which secreted the antibody, then the affinities of serum antibodies will reflect the process of selection and activation of B cells to become antibody secreting cells. It is well known that early after immunization, a heterogeneous population of antibody molecules is produced which is of low average affinity for the antigenic determinant. With time after immunization, there is a progressive increase in average affinity (2-8). The rate of increase in affinity is faster after lower doses of antigen. This shift in affinity has been interpreted (5) in terms of a microevolutionary selection theory. B cells bearing high affinity antigen receptors preferentially capture antigen (especially when antigen concentration is low) and are thereby preferentially stimulated to proliferate and secrete antibody. With time there is a shift in the distribution of antibody secreting cells so that higher affinity clones become predominant in the population. Regulation of affinity can thus be viewed in terms of antigen mediated selection of high affinity antibody producing cells. This interpretation dominated theorization about the regulation of the selection for high affinity antibody production through the 1960's. In addition, it was known that circulating antibody specifically depressed antibody production. It had been shown (9,10) that antibody mediated suppression operated in vivo in the regulation of the normal immune response. As can be predicted on the basis of a selectional theory of the immune response, passive antibody preferentially suppresses low affinity antibody synthesis (3). That is, circulating antibody acts to decrease effective free antigen concentration and thus tends to increase the selective pressure for the production of high affinity antibodies (11).

A second factor involved in regulating the selection for high affinity antibody synthesis is the degree of cell proliferation which takes place during the course of the immune response. Clearly, for a shift in the population distribution of B cells with respect to affinity to take place cell proliferation is required Increased cell proliferation would be expected to be associated with efficient cell selection, while decreased cell proliferation would be expected to result in a reduced efficiency of selection for high affinity antibody production. Consistent with these predictions are the effects of adjuvants which increase the rate of selection for high affinity antibody production and nonspecific immunodepressants which depress the rate of increase in affinity (12).

In the late 1960's and early 1970's it became clear that T lymphocytes play an important role in regulating the magnitude of the immune response to most antigens. Helper T cells augment the magnitude of the response to most antigens (T-depedendent antigens) while suppressor T cells depress the magnitude of the immune response either by depressing the activity of helper T cells or by acting directly on B cells. At least some of the effects of suppressor T cells appear to be mediated through the secretion of antigen specific suppressor factors. Helper and suppressor T cells could influence selection for high affinity antibody secreting B cells through two general mechanisms. First, there may be a specific effect on B cells which directly influences their selection. Second by increasing or decreasing the degree of cell proliferation there may be an indirect effect on the efficiency of selection for high affinity antibody producing subpopulations. Both mechanisms are probably involved and it is difficult, a priori, to know which mechanism plays a more important role in any given experimental manipulation of the immune system. One would predict that if the binding of antigen to antigen receptors on B cells is involved, the results would be markedly influenced by the affinity of the cell surface antigen receptors. Under such circumstances, a small depression in magnitude (e.g., 25 to 50 percent) would be expected to be associated with a marked depression in affinity (i.e., a preferential effect on high affinity antibody producing cells). In contrast, where the effect on affinity is mainly due to alteration in the degree of proliferation, one would expect that a modest decrease in magnitude (e.g., 50 percent) would be associated with little or no change in affinity; only a marked depression in magnitude (e.g., 85 percent or greater) will be associated with a decrease in affinity. These predictions are consistent with the limited data available in the literature. (a) In B-cell tolerance, a slight decrease in magnitude is associated with a marked decrease in affinity (13,14). (b) Specific suppressor T-cell activity leads to a marked depression in affinity associated with a modest decrease in the magnitude of the response (Dekruyff and Siskind, unpublished observations). (c) Restriction of helper T-cell activity leads to a decrease in affinity (15). A moderate decrease in magnitude (ca. 30 percent) has no effect on affinity, while a marked decrease in magnitude is associated with a profound decrease in affinity (Dekruyff and Siskind, unpublished observations). (d) Specific T-cell tolerance results in a decrease in affinity only when the reduction in the magnitude of the response is very marked (14,16 and Dekruyff and Siskind, unpublished observations). (e) Non-specific immunodepression by cytotoxic drugs has little effect on affinity if there is only a 50 percent depression in magnitude, but results in a profound depression in affinity if there is a marked (85 percent) decrease in the magnitude of the response (12). (f) Similarly, nonspecific depression of the magnitude of the immune response as a consequence of antigenic competition is associated with a decrease in affinity only when the depression in magnitude is of a marked degree (ca. 85 percent) (17-19). Thus, it appears that in situations which involve an interaction between antigen and B-cell antigen receptors, a small decrease in magnitude is associated with a marked decrease in affinity. In contrast, in situations where the change in affinity is a secondary consequence of decreased proliferation, an effect on affinity is only seen when there is a very profound decrease in magnitude. These general observations indicate that when evaluating the effect of immunodepression on antibody affinity it is necessary to perform the studies under a variety of degrees of immunodepression if a reasonable interpretation is to be made.

Thus, up to the mid-1970's the regulation of antibody affinity could be viewed as mainly the consequence of antigen selection of B cells in an affinity dependent manner. Specific antibody produced during the response, mainly as a consequence of its interaction with antigen, suppresses the further stimulation of low affinity antibody synthesis and facilitates selection for high affinity antibody secreting cells. This selection system is modulated by specific and nonspecific enhancing and suppressing effects of different T-cell subpopulations and possibly other accessory cells.

In the mid-1970's a new concept was introduced by Jerne into considerations of regulation of the immune response (20). It was reported by a number of workers that one could produce antibodies specific for the antigen combining site of an antibody molecule. The antigen combining site, when viewed as an antigenic determinant, is referred to as an "idiotype" or "idiotope". Antibody specific for the antigen binding site of an antibody molecule is referred to as "anti-idiotype antibody". Anti-idiotype antibodies are capable of specifically inhibiting the capacity of an animal to produce antibody molecules bearing that idiotype. Evidence suggests that T and B cells share common idiotypes. The complexity of biological effects which could be brought about by anti-idiotype antibodies (or idiotype specific T cells) becomes self evident if one considers the possibility of inhibiting specific clones of B cells, of helper T cells or of suppressor T cells in varying proportions. The effects would of course, be idiotype specific and might involve numerous clones which have similar or identical idiotypes. Jerne proposed that interactions between idiotypes and auto-anti-idiotype antibodies (or cells) are involved in the normal regulation of the immune response. In effect, he proposed that antigen induced expansion of a set of antigen specific clones bearing a set idiotopes would stimulate an auto-anti-idiotype reaction leading to the expansion of an idiotype specific T-cell population and the production of anti-idiotype antibodies. These anti-idiotype antibodies and cells act to depress the expansion of clones of cells bearing that idiotype. Jerne described the immune system as being regulated by a network of idiotype specific interactions. According to such a view, the immune system normally exists in a state of equilibrium (or perhaps more accurately, in a steady state condition). Introduction of antigen into the system results in the expansion of a set of clones and thereby perturbes the steady state. Anti-idiotype specific cells are stimulated and a new steady state is achieved in which the elements (clones of cells) are distributed in a manner different from that of the original population. The immune response and its regulation can thus be viewed as an antigen induced perturbation, of an idiotype specific network, which normally exists in a steady state condition. Regulation of the immune response and immunologic memory are the consequence of the rearrangement of elements within the network so as to establish a new steady state with certain elements (clones) present in a higher incidence. Relatively little direct support for this type of network regulation exists at present. Recently, we have obtained some evidence for the production of auto-anti-idiotype antibodies during a normal immune response by mice (Goidl, Schrader, Thorbecke and Siskind, unpublished observations).

At present one can, therefore, view the regulation of antibody affinity and heterogeneity as basically an antigen driven, affinity dependent, selective process. Antibody mediated suppression probably operates to favor selection of high affinity antibody secreting cells. Selection basically takes place as a consequence of high affinity antigen binding cells (precursors of high affinity antibody secreting cells) preferentially capturing antigen and being, thereby, preferentially stimulated to proliferate and differentiate into antibody secreting cells and memory cells in a type of microevolutionary system. This process of cell selection is undoubtedly modulated as a consequence of specific and nonspecific augmenting and/or suppressing activities by different subpopulation of T cells and other accessory cells.

Finally, the various cellular elements probably exist in a steady state condition in a type of network of idiotype specific interacting elements.

REFERENCES

1. Burnet, F.M. (1959) The Clonal Selection Theory of Acquired Immunity, Vanderbilt University Press, Nashville, Tenn.
2. Eisen, H.N. and Siskind, G.W. (1964) Biochemistry, 3, 996.
3. Siskind, G.W., Dunn, P. and Walker, J.G. (1968) J. Exp. Med., 127, 55.
4. Goidl, E.A., Barondess, J.J. and Siskind, G.W. (1975) Immunology, 29, 629.
5. Siskind, G.W. and Benacerraf, B. (1969) Advan. Immunol., 10, 1.
6. Werblin, T.P. and Siskind, G.W. (1972) Transplant. Rev., 8, 104.
7. Werblin, T.P. and Siskind, G.W. (1972) Immunochemistry, 9, 987.
8. Werblin, T.P., Kim, Y.T., Quagliata, F. and Siskind, G.W. (1973) Immunology, 24, 477.
9. Graf, M.W. and Uhr, J.W. (1969) J. Exp. Med., 130, 1175.
10. Bystryn, J.-C., Graf, M.W. and Uhr, J.W. (1970) J. Exp. Med., 132, 1279.
11. Siskind, G.W. (1970) in Developmental Aspects of Antibody Formation and Structure. Edited by J. Sterzl and I. Riha, Academia Publishing House of the Czechoslovak Academy of Sciences, Prague.
12. Mond, J., Kim, Y.T. and Siskind, G.W. (1974) J. Immunol., 112, 1255.
13. Theis, G.A. and Siskind, G.W. (1968) J. Immunol., 100, 138.
14. Andersson, B. and Wigzell, H. (1971) Eur. J. Immunol., 1, 384.
15. Gershon, R.K. and Paul, W.E. (1971) J. Immunol., 106, 872.
16. Theis, G.A., Green, I., Benacerraf, B. and Siskind, G.W. (1969) J. Immunol., 102, 513.
17. Brody, N.I. and Siskind, G.W. (1969) J. Exp. Med., 130, 821.
18. Harel, S., Ben-Efraim, S. and Liacopoulos, P. (1970) Immunology, 19, 319.
19. Kim, Y.T., Merrifield, N., Zarchy, T., Brody, N.I. and Siskind, G.W. (1974) Immunology 26, 943.
20. Jerne, N.K. (1974-1975) The Harvey Lectures, 70, 93.

THEORETICAL MODELS FOR CELL-CELL INTERACTIONS IN IMMUNE RESPONSES

GEORGE I. BELL
Theoretical Division, Los Alamos Scientific Laboratory, University of California, P.O. Box 1663, Los Alamos, New Mexico 87545

ABSTRACT

Quantitative theoretical models are described for treating interactions between cells which are mediated by mutually complementary receptors or by ligands, such as antigens or antigen-antibody complexes, which can bind and form links between the cells. It is found that relatively few receptor-receptor or receptor-ligand bonds can establish firm adhesion between the cells. Conditions on receptor numbers, mobility, and binding constants are defined for rapid binding between cells. As mobile receptors diffuse into and become concentrated in regions of intercellular contact, the cell membranes will be thereby locally modified, possibly forming a contact cap. An analysis is made of the extent and rate of such receptor accumulation. Some applications of these considerations to problems in immunology, including dual recognition, are sketched.

INTRODUCTION

Interactions between cells appear to be critical for the initiation and regulation of immune responses. For example, interaction of lymphocytes with macrophages or other adherent cells is required for _in vitro_ initiation of lymphocyte activity[1]. Interactions between B lymphocytes and the various classes of T lymphocytes are required for initiation and regulation of _in vivo_ immune responses[2]. Interactions between recirculating lymphocytes and endothelial cells lining the blood vessels of lymphoid tissue are necessary for maintaining the functional patterns of lymphocyte migration[3,4]. Interactions between nascent T cells and epithelial cells of the thymus appear to determine which histocompatibility antigens are recognized as self by the T cells[5].

In many cases it is possible that these interactions are mediated by diffusible factors or macromolecular complexes[6-8] and that direct contact between the interacting cells is not required. In these cases, such theoretical methods as have been described by DeLisi and Dembo at this workshop could provide a quantitative approach to treating interactions between cells. When, however, the interactions require direct contact between the cells, new theoretical considerations become important. Several will be described in this paper.

Various arguments suggest that direct contact between cells is very important in the immune system. Electron micrographs of lymphoid tissue[9], or of macrophages and lymphocytes[10], or of killer lymphocytes and their target cells[11], all show large areas where the cells are in very close proximity. In those cases where the cellular receptors have known molecular specificity such as the immunoglobulin molecules on B lymphocytes or Fc receptors on macrophages or lymphocytes, it seems likely that intercellular molecules bridges must frequently form. Thus we would expect that B cells would bind to antigenic cells, and that B cells would become stuck to each other by mutually complementary soluble antigens or to macrophages by antigen antibody complexes. Indeed some of these forms of intercellular adhesion are exploited by immunologists in various rosette assays. In passing I suggest that the systematic variation of binding conditions in indirect rosette assays[12] may be a useful way to check some of the theoretical considerations presented below.

In this paper I will present three theoretical predictions concerning adhesion between cells which is mediated by specific molecular bonds and the resulting effects on cell behavior. The antigen-antibody bond is a good example of the kind of specific molecular interaction which I have in mind as forming an intercellular bond. It is reversible, rather weak yet specific. I will consider both direct binding between cells mediated by complementary receptors such as antigen and antibody, and indirect binding where an intercellular ligand or complex is required in order to form the bridge between two cells.

It is useful to have in view at the outset a molecular model of the cell membrane. For this I adopt the fluid mosaic model[13,25] and take the membrane to be a phospholipid bilayer in which various glycoproteins are dispersed and retained by virtue of the favorable free energy of interaction of their hydrophobic portions with the lipid environment. Some of these glycoprotein molecules may serve as receptors for adhesive contacts with other cells (or with surfaces) although glycolipids or other cell surface molecules could also serve. I will generally assume that the receptor molecules are more or less free to move in the plane of the membrane and that their motion is random, as in diffusion, rather than ordered.

It may be observed that this is a rather passive model of the cell membrane, in which turnover of membrane components, connections of the receptors with the cytoskeleton or each other, and ordered notion of the receptors (e.g. capping) or cells is not considered. As the predictions of this passive model become better understood and are compared with observed cell behavior, it may be possible to deduce more clearly which if any of these complications must be invoked to explain some aspect of cell behavior.

Now for the predictions.

1. Relatively few receptor-receptor or receptor-ligand bonds can establish firm adhesion between two cells. In order to arrive at this prediction and of the quantitative meanings of "relatively few" and "firm" we must be able to estimate the force which is required to break a bond and to compare this with other forces

to which the cell may be subject or which may act between cells. These subjects have been treated in a recent paper[14] and I shall here only summarize the main ideas.

Since we are considering reversible bonds such as those between antigen and antibody, no force at all is required to disrupt a single bond. The bond will in time spontaneously break due to thermal motion. However when two cells are stuck together by many bonds, these are most unlikely to all break simultaneously and force will have to be applied to separate the cells. The effect of the force on an individual bond is to shorten its lifetime and we may estimate the force required to reduce the bond lifetime to a very small value.

Let us consider how the free energy of a pair of binding molecules would vary if the molecules were displaced relative to each other along some direction of minimum work. A curve as shown in Figure 1 would be expected. Binding implies

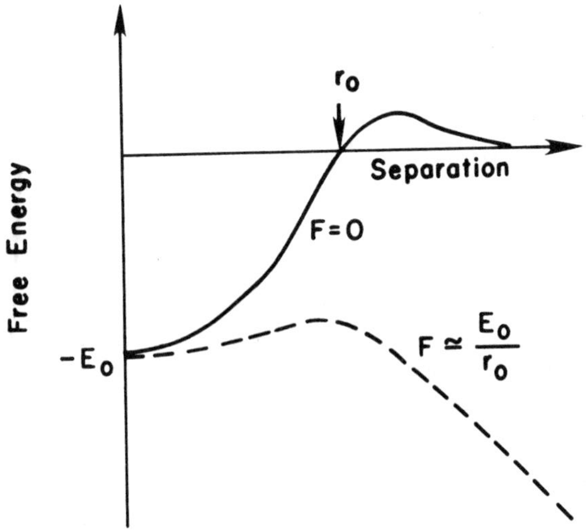

Fig. 1. Free energy plotted against receptor-ligand separation along some direction of minimum work. Zero separation corresponds to the bound state, of free energy $-E_o$ relative to the separated reactants. A force tending to separate the reactants will decrease the free energy linearly with distance, as shown.

a free energy minimum of $-E_o$, relative to the completely separated state where E_o is the free energy decrease in the reaction. In addition, let r_o be the range of the energy minimum. A force tending to separate the molecules will decrease the energy linearly with separation and a force of magnitude E_o/r_o will abolish the energy minimum and rapidly break the bond. Letting f_o denote this force and measuring E_o in keal/mole and r_o in Å,

$$f_o = 7 \times 10^{-6} \frac{E_o}{r_o} \frac{\text{dynes}}{\text{bond}} \qquad (1)$$

More careful analysis[14] shows that cells can be separated by forces smaller than f_o per bond. It is not necessary for the force to be strong enough to break every bond simultaneously. For a representative case, it was found that a critical force, $f_c \simeq 1/3\ f_o$ would suffice to separate the cells.

To obtain a numerical value for f_o, consider a typical antigen antobody bond having an equilibrium constant of $10^6\ M^{-1}$ and hence a free energy change of 8.5 keal/mole. For the range of the interaction, I believe it is appropriate to take a distance somewhat smaller than the size of the antibody binding site and have somewhat arbitrarily taken $r_o = 5Å$. With these values, $f_o = 1.2 \times 10^{-5}$ dynes/bond and the critical force, $f_c \simeq 4 \times 10^{-6}$ dynes/bond. I shall use this value of the critical force for comparison with other forces.

Note that the force to break a bond is proportional to the bond energy E_o, which is known or measurable for many potential bonds, and inversely proportional to the range r_o which is much less certain. It is to be expected that r_o will depend somewhat on which reactants form the bond and that for any particular reactants, r_o, will have a substantial uncertainty, perhaps of the order of a factor of two.

We thus expect that a force $f_c \simeq 4 \times 10^{-6}$ dynes will be sufficient to break a typical receptor-receptor or receptor-ligand bond. But might not the receptor molecule break or pull loose from the cell at smaller forces? The force f_c is most unlikely to break a covalent bond since such bonds are characterized by much larger values of E_o ($\sim$ 70 keal/mole) and smaller values of r_o ($\sim$ 1.4Å) than the antigen antobody bond. It is less clear that the force f_c will not pull the receptor molecule out of the membrane. I have estimated[14] that the force required to extract a glycolipid molecule is comparable to f_c and that the force required to extract a particular transmembrane protein (glycophorin) is only a few times larger. Other membrane proteins, especially those having small or no cytoplasmic tails might be considerably easier to pull out. It is thus unclear whether a receptor will be pulled out of the membrane before the receptor-ligand bond breaks. I expect that as a variety of ligands and receptors are examined, examples of both kinds of behavior will be found.

How does the force f_c compare with other forces to which a cell may be subject, such as hydrodynamic forces or forces which may be exerted on the membrane by a cell's cytoskeleton? If a stationary spherical cell of radius r is located in a fluid stream of velocity v and viscosity η, the force upon it will be $F = 6\pi\eta rv$. For a lymphocyte having $r = 4\ \mu m$ in a medium having the viscosity of water ($\eta = 7 \times 10^{-3}$ g/cm sec at 37° C), $F = 5.3 \times 10^{-5}$ v dynes where v is the velocity in cm/sec. If we divide this force by f_c we obtain an estimate of the number of bonds required to hold the cell in a fluid stream of velocity v as

$$N_v = 5.3 \times 10^{-5}\, v/f_c = 13v \quad . \qquad (2)$$

If for example we are considering adhesion of a recirculating lymphocyte to an endothelial cell in the venule of a lymph node[4] where v = 0.3 cm/sec, equation (2) tells us that the adhesion could be mediated by about four of our typical bonds. Of course we don't know what the receptor molecules are in this case but the essential conclusion is that if such adhesion were mediated by specific molecular bonds, then relative few bonds ($\lesssim$ 10) could suffice. This result also suggests that a little antigen on the endothelial cells might enhance the adhesion of specifically reactive lymphocytes. Similar calculations could be done for the adhesion of granulocytes to blood vessel walls.

Cells are expected to have attractive van der Waals forces between them as well as repulsive electrostatic forces[16,17]. The net result is predicted to be a long range attractive force $\sim 10^{-5}$ dynes per μm^2 of area when the cell membranes are separated by an optimal distance ($\sim$ 100Å). This means that such a force would be required to further separate the cells. Note that a similar force would be required to break two or three specific bonds per μm^2. It follows that a few specific bonds per μm^2 will be more important than the expected non-specific van der Waals attraction between cells.

Finally, let us consider the forces which contractile elements of the cytoskeleton may exert on the cell membrane. Such contractile elements are made of actin, myosin, and possibly other proteins[18] and in order to exert a force on the membrane they are presumably anchored to "receptors" on the cytoplasmic surface of the membrane. It is reasonable to conjecture that such anchoring interactions are similar to the bonds we have been considering. If so, then when contractile elements attempt to pull loose a cell from some point of adhesion, they can succeed, to first approximation, only if there are more bonds per unit area of receptors with contractile elements, than bonds of receptors with extracellular ligands. Of course various mechanical constraints may have to be also satisfied. Contractile elements can exert a tension between two points of adhesion, but they can only pull the cell loose from that point in which it is more weakly bound.

Comparisons with other forces in biology are made in ref (14).

On to the second prediction.

2. Bonds between complementary receptors on cells can form very rapidly provided the receptors are sufficiently numerous, mobile, and accesible.

I have described elsewhere[14] the calculation of the rate of binding between complementary receptors on adjacent cells. Two assumptions were necessary. The first is that the cells are (locally) separated by a distance and medium such that the receptors are accessible to one another and the second is that something is known about the rates of reaction when at least one of the receptors or ligands is in solution. Ideally, forward and reverse rate constants for the reactants in solution should be known. In addition, the diffusion coefficients of each receptor in the plane of the membrane should be known.

Theory can then be used[14,19] to go from the reaction rates for reactants in solution to those for membrane bound reactants. Two important effects must be taken into account. First, the concentration of a membrane bound reactant will be very high next to the membrane. That is, locally, the number of molecules per unit volume will be the number per unit area on the membrane divided by the thickness of the region ($\sim$10-100Å) next to the membrane within which the molecules are confined. If, for example I take as receptor 10^5 immunoglobulin molecules on a B cell of radius 4μm the number of molecules per unit area will be $10^5/4\pi r^2$ = $5 \times 10^{10}/cm^2$. If their binding sites are located within a 20 Å band of the surface, the local concentration is $\sim 5 \times 10^{10} \div 2 \times 10^{-7}$ cm^{-3} = 2.5×10^{17} molecules $/cm^3$ = 0.4×10^{-3} M, a remarkably high concentration of specific antibody. Because of these high concentrations, it will be difficult to break up intercellular binding by soluble inhibitors unless very high concentrations can be used.

The second important effect is that the diminished mobility of membrane bound reactants, as compared to those in solution, may decrease both forward and reverse rate constants. Theory of the encounter complex[14] predicts that the diffusion coefficient will not alter the equilibrium constant but may affect both reaction rates equally. In addition, steric considerations will be somewhat different for the membrane bound reactants[19].

Consider two adjacent cells having complementary receptors. Then the maximum rate of bond formation per unit area between the cells will be the product of a forward rate constant for membrane bound reactants and the number of receptors per unit area, N_1 and N_2, on each cell. If the diffusion constants for translation of the receptors in the membrane are D_1 and D_2 and N_b is the number of bonds per unit area, we have

$$\frac{dN_b}{dt} \leqslant 2\pi\varepsilon\ (D_1 + D_2)\ N_1 N_2 \tag{3}$$

Here ε is a factor by which the forward rate constant is less than the diffusion limit. For receptors such as immunoglobulin molecules on B cells[20], $D \sim 10^{-10}$ cm^2/sec, and for a particular case I estimated[14] $\varepsilon \simeq 0.1$. With these parameters and for N_b, N_1, and N_2 in molecules per μm^2,

$$\frac{dN_b}{dt} \lesssim 0.013\ N_1 N_2 \qquad \mu m^{-2}\ sec^{-1}$$

Hence if the receptors were as abundant as Ig molecules on B cells ($N_1 = N_2$ = 500 μm^{-2}), $dN_b/dt \lesssim 3000\ \mu m^{-2}\ sec^{-1}$. This indicates that under appropriate conditions, the rate of intercellular bond formation can be very high. Note that for rapid bond formation, mobility of one of the receptors will suffice. The other could be fixed to a surface.

Cells may also be stuck together by soluble multivalent ligands such as lectins, antobodies, or antibody-antigen complexes. Rates of bond formation can be estimated as for membrane bound receptors with a few additional complications[21]. For one thing, an intercellular bond can form only between a ligand which is bound to one cell and a free receptor on the other. Hence bond formation will be hindered by ligand excess as well as deficiency. In addition, if the ligand binds to the same kind of receptor on both cells, it can also cross-link receptors on each cell as analyzed by Dembo and DeLisi in this volume. Such crosslinking will render receptor and ligand molecules unavailable for intercellular bonding. Finally there is the intriguing possibility that such ligands will cause portions of a cell membrane to stick to rather distant portions of the same cell. For example the tip of a microvillus might become stuck to a portion of the cell membrane. What would then happen if the cell tried to retract or further extend its microvillus and might this have anything to do with endocytosis, capping, or other cell activities? I shall consider one possibility in the final section.

3. Membranes in a local region of cell contact can become rapidly modified as additional receptors diffuse into the area and stick there. Consider two cells which have come into contact with each other and have become stuck together by the formation of intercellular bonds over some local area as shown in Fig. 2. Additional receptors will by chance diffuse into the contact area and become stuck there by binding to complementary receptors on the other cell. Therefore receptors will accumulate in the contact area, possibly reaching much higher concentrations than in the rest of the membrane and grossly altering the local properties of the membrane. Thus a cap may form in the contact area, simply due to diffusion and reactions of receptors[22,25]. I shall call this a "contact cap," to distinguish it from an accumlation of multivalent ligand and crosslinked receptors on a single cell.

To what extent will receptors accumulate in the contact area? An estimate may be made as follows. When the binding has reached equilibrium there will be no net flux of receptors into or out of the contact area. Hence the number of free receptors per unit area will be the same in the contact area and outside of it. If most of the cell's receptors are still free, then

$$N_b = KN_1N_2 \tag{4}$$

where K is the equilibrium constant[14]. This tells us that if $KN_1 >> 1$, there will be a marked accumulation of the second cell's receptors in the contact area, while if $KN_1 \lesssim 1$, there will not be. Similarly if $KN_2 >> 1$, the first cell's receptors

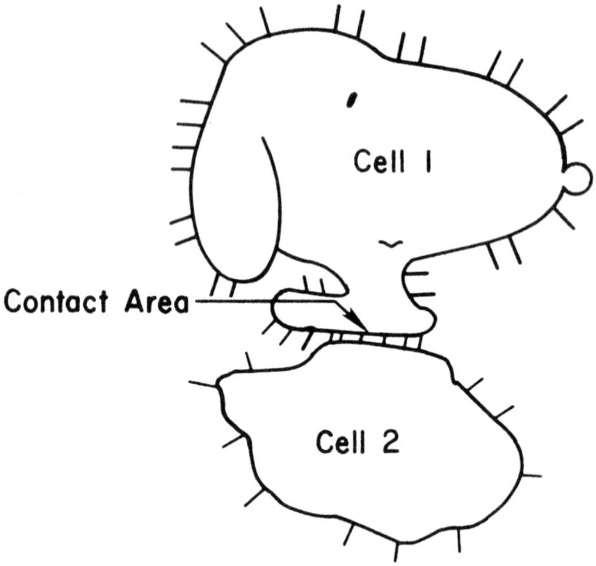

Fig. 2. Two cells in contact over an area are stuck together by intercellular bonds. Additional receptors will diffuse into the contact area and accumulate by binding to complementary receptors on the other cell, thus forming a contact cap. Evidently in this Figure the top cell is potentially phagocytic.

will accumulate. For numerical estimates, we may let K be an equilibrium constant for reactants in solution and measure N_2, say, as a <u>local</u> concentration per unit volume. For example we saw earlier that Ig molecules on a B cell have a local concentration $\simeq 4 \times 10^{-4}$M, so that if $K = 10^5$, 10^6, or 10^7 M^{-1}, $N_b/N_1 = 40$, 400, or 4×10^3 respectively. Thus for these examples, the concentration of bound receptors in the contact area would greatly exceed the concentration of receptors elsewhere on the first cell. Whether such an accumulation is reasonable depends on N_1 and other considerations beyond the scope of this article. There is obviously a limit to the number of receptors that a unit area of membrane can accomodate.

In Equation (4) I assumed that most receptors are free so that N_1 and N_2 are the intial receptor densities. It is easy to relax this requirement as is shown in the Appendix. Moreover it is there shown how these considerations can be applied to a cell which is stuck to a surface. In particular, I show how one can estimate whether the cell's receptors may become largely concentrated adjacent to the surface to which the cell is bound.

How fast may this receptor accumulation take place? Suppose that the contact area is a disc of radius r and that receptors on the cell under consideration have diffusion constant D. Suppose also that every receptor which enters the contact disc sticks there. (This assumption may overestimate the rate of receptor accumulation and will be discussed later.) With these assumptions, the mean time, T_d, for diffusion of a receptor into the disc is[23,24]

$$T_d = \frac{2R^2}{D} \ln \frac{R}{r} \tag{5}$$

where R is the cell radius. The rate at which receptors enter the disc, J_d, will be the total number, $4\pi R^2 N$ where N is the number of receptors per unit area, divided by T_d, so that

$$J_d = \frac{2\pi DN}{\ln(R/r)} \tag{6}$$

Hence the time, T_2, to double the initial number of receptors in the disc will be

$$T_2 = \frac{\pi r^2 N}{J_d} = \frac{r^2}{2D} \ln(R/r) \quad . \tag{7}$$

If, for exampble, $D = 10^{-10}$ cm^2/sec and $R \simeq 4\mu m$ then T_2 = 2, 12, or 70 sec respectively for r = 0.1, 0.3, or 1 μm. Thus receptors can accumulate very rapidly in a contact area.

Our assumption that every receptor which enters the disc sticks, requires that the mean time, T_r, for reaction of receptors in the contact area should be short compared to T_2. The reaction time can be obtained from Eq. (3). If we are considering the reaction time for receptors on cell 1 and let N_{2f} be the density of free receptors on cell 2, then Eq. (3) implies

$$T_r \simeq \frac{1}{2\pi\varepsilon(D_1+D_2)N_{2f}} \quad . \tag{8}$$

For $D_1 = D_2 = 10^{-10}$ cm^2/sec, $\varepsilon = 0.1$, Eq. (8) gives $T_r \simeq 80/N_{2f}$ sec where N_{2f} is in receptors per μm^2. Thus $T_r < T_2$ if the density of receptors on the opposing cell is large enough. For the above parameters, $T_r < T_2$ if

$$N_{2f} > \left[2\pi\varepsilon r^2 \ln(\tfrac{R}{r})\right]^{-1} = \frac{1.6}{r^2 \ln(R/r)} \tag{9}$$

If the receptors on one of the cells (or a surface) are immobile, then whether or not there is a significant flow of mobile receptors on the other cell into the contact area will also depend on the relative number of receptors on the two cells. Let N_1 be the number of mobile receptors per unit area on one cell and N_2 be the number of fixed receptors per unit area on the opposite surface. If $N_2 \gg N_1$ and $KN_2 \gg 1$, then there will be a large accumulation of mobile receptors in the contact area before they saturate the fixed receptors. If $N_2 \ll N_1$ (or $KN_2 \ll 1$) there can be only a minor redistribution of mobile receptors.

When cells are stuck together by soluble ligands, similar considerations will be expected to apply, subject to the complications noted at the end of the preceding section. Thus if a cell becomes stuck to another cell, to a surface, or to itself by means of a soluble ligand, mobile ligand receptors and ligand molecules may rapidly accumulate in the contact area. Such local accumulations have been observed in electron micrographs[25].

Accumulation of receptors in a contact cap will evidently reinforce the binding in that area. Some possible biological functions for such reinforcement are obvious. It might prevent escape of food. A high concentration of receptors might also furnish a good area for attachment[29] of contractile cytoskeletal elements in preparation for ingestion. Finally if the cell decides to let go of the sticky object it may find it advantageous to have most of its receptors accumulated in a small area where they can be exfoliated or internalized en mass, as in capping.

IMPLICATIONS FOR CELLULAR RECOGNITION IN IMMUNOLOGY

We have seen that firm binding can be established between two cells or between cell and substrate, by a relatively few reversible receptor-receptor or receptor-ligand bonds. Under appropriate conditions such bonds will form rapidly when the two cells come into contact. Moreover, additional receptors may accumulate in the contact area and reinforce the binding. It thus appears that cells with complementary receptors or cells which have receptors for the same ligand or macromolecular complex can bind to each other rapidly and tightly, under "appropriate conditions" of receptor accessibility, mobility, number and affinity. These "appropriate conditions" have been characterized in preceding sections.

What are the implications of these considerations for immunology? First of all there are a variety of in vitro studies, rosette assays, haemagluttination assays, studies of macrophages bound to surfaces or of lymphocytes to marcrophages, for which these models may be more or less applicable. In addition there may be a few in vivo phenomena which can be clarified. I have already mentioned the number of bonds required for adhesion of lymphocytes to endothelial cells. Similar considerations may apply to the interactions of granulocytes with endothelial cells. In these situations there are clear advantages to rapid formation of bonds between cells.

In lymphoid tissue, relative cell velocities are much smaller[3] so that cells which come into contact with each other will probably remain in contact for a minute or more. Moreover, lymphoid tissues will sometimes contain high concentrations of soluble antigen or immune complexes. Under such conditions, it seems likely that tight binding between cells is often brought about by these soluble ligands[21], even though, as noted earlier bond formation will be inhibited by antigen excess as well as antigen deficiency. Their binding will presumably render the cells relatively immobile, in agreement with the observation that for a day or two after injection of antigen into a lymph node, there is a marked decrease in the output of specific antigen reactive lymphocytes[3]. However lymphoid cells may have active mechanisms for terminating adhesive contacts. Such phenomena as B cell capping or the eating of B cell caps by macrophages[10] may represent a class of actions which lymphoid cells may take to terminate adhesive contacts through internalization or exfoliation of the sticky material. This point of view suggests that adhesion and contact cap formation between lymphoid cells may be a rather general signal for activation of the cells to do something.

Finally, let us consider the T cell recognition system. This appears to involve products of the major histocompatibility locus plus complementary receptors functioning in a cooperative fashion with an antigen recognition system which may be similar to that of B cells[26,27]. Since so little is known about the molecular components of the T cell recognition system, I shall confine my remarks to general deductions from earlier portions of this paper, and assume a dual recognition rather than an altered self model for recognition[26-28]. The cytotoxic T cell system will also be emphasized.

It is commonly supposed that a main function of T cells is to recognize small modifications in other cells which may be induced by viral infection or whatever and to kill the modified cells. Recognition of cell-bound antigen may have to take place in an environment in which there is a high concentration of soluble or extracellular antigen. Therefore a first question is how to design a system which is sensitive to cell bound antigen but insensitive to soluble antigen? An easy solution is possible if the local concentration of cell bound antigen substantially exceeds the concentration of soluble antigen, for in this case T cells having relatively low affinities for antigen such that their receptors are not saturated by soluble antigen will still be able to bind to and recognize the cell-bound antigens. This solution has at least two defects, namely T cells having the best (highest affinity) receptors for antigen would be ineffective in the presence of high concentrations of soluble antigen and the system could be subverted by an intracellular parasite which expressed a low concentration of antigens on the cell surface. These objections are not absolute since even a T cell with initially saturated receptors could bind to antigen on a target cell provided that the cells were in contact a long enough time. Moreover it is always possible, in principle, for an intracellular parasite to defeat the immune system by expressing no antigens on the cell surface.

A more sophisticated recognition system might require that the T cell first ascertain that it was adjacent to a cell before completely exposing its antigen binding receptors. Perhaps receptors which recognized histocompatibility antigens could be used for this purpose, yet it would seem inappropriate to establish tight and sustained binding between all histocompatible cells since this would render all lymphoid cells immobile. Therefore any such binding should be transient and last only a few minutes or less.

In any case, let us consider two interacting cells which we may call target and effector (T) cell. Suppose that the target cell has on its surface N(H) histocompatibility antigens, H, per unit area and N(X) other antigens, X, per unit area. For example, X might be a viral antigen or chemical determinant such as DNP. Let the effector cell have on its surface N(rH) and N(rX) complementary receptors per unit area. I assume that all these cell surface molecules are mobile. In addition, the effector cell will have H antigens and the target cell may have rH receptors but their interaction need not concern us. The key assumption of our model is that a cytotoxic signal from effector to target cell requires clustering of the rX receptors on the effector cell and of the H antigens on the target cell, together in a contact cap. In this view, clustering of H antigens on the target cell may be thought of as sensitizing the cell to being killed by the effector cell.

Allogeneic killing furnishes the simplest example, for in this case the target cell has an altered histocompatibility antigen, H^1, which we assumed to be recognized by rX. If K is the equilibrium constant for the interaction of H^1 with rX, then both receptors will be clustered in the contact area if $KN(H^1) >> 1$ and $KN(rX) >> 1$, and a killing interaction will result.

For the killing of self cells which have been modified by acquiring the antigen X, interaction of the X antigens with rX receptors will produce clustering of the receptors provided that $K_X N(X) >> 1$, where K_X is the equilibrium constant for this reaction. Similarly, interaction of H with rH, with equilibrium constant K_H will produce clustering of H provided that $K_H N(rH) >> 1$. Thus if these two inequalities were fulfilled, a killing signal would be delivered to the target cell.

It appears to me that this minimal model, especially if combined with assumptions concerning the selection and specificity of the rH and rX receptors as in ref. (27), can explain many of the features of T cell cytotoxicity. In addition, the model can be modified in various ways so as to solve some of the problems raised earlier concerning T cell recognition.

For example, it may be objected that histocompatible lymphoid cells do not stick together well by H-rH interactions. We could surmount this objection by an ad hoc assumption that such interactions are effective only after the cells have been brought close together by X-rX interactions or we could invent a molecular mechanism which might accomplish this. Suppose, in particular, that the rX and rH receptors are physically linked in the membrane of the effector cell[26] so that a clustering of the rX receptors by interaction with X would induce a co-clustering of rH receptors. Since both receptors are coded by the genome of the target cell, their interaction can be readily postulated. Suppose also that prior to clustering, the interaction of rH with H is weak ($K_H N(rH) << 1$) so that a negligible clustering of H antigens occur in any contact area. Nevertheless, if the rH receptors are clustered by a strong X-rX interaction, the resulting concentration of rH receptors will be strong enough to cluster the H antigens provided that $K_x N(X) K_H N(rH) >> 1$. Thus if rH and rX are linked on the effector cell, the X-rX interaction can greatly augment the number of H-rH bonds in the contact cap. One could go even further and postulate that rH receptors become exposed only after they have been clustered by the X-rX interaction.

Another objection noted earlier was that cell bound X antigens may have to be recognized in the presence of soluble X antigens which might saturate the receptors, rX. In principle this problem could be solved by an inversion of the preceding argument, namely postulate a linkage of rX and rH and assume that rX is not fully exposed until clustered by the H-rH interaction. However I don't see that you can run the argument both ways and the previous application is a little more appealing to me.

It has been noted that H antigens often co-cap with other antigens[22]. If X were linked to H on the target cell this would further enhance the interactions in the contact cap. However since X may be a viral antigen, it seems risky to rely on linkage of X with H for an effective cytotoxic system.

The killer T cell is likely to encounter various other lymphoid cells which it should not kill but, which have mobile antigen X bound to their own surface receptors. If such cells form contact caps with the killer T cell, these caps will contain concentrations of the potential target cells own receptors for X which could be assumed to negate the killing signal.

Finally, it is of interest to consider what would happen if the receptors rX on a killer T cell happened, by mutation or otherwise to recognize the self antigen H. Such a cell would naturally kill other cells expressing H as in allogeneic killing. However, if the cell formed a contact cap with itself, as mentioned earlier it would autodestruct thereby eliminating the problem.

These speculations concerning T cell recognition systems have been based on the assumption that receptor clustering in contact caps is a common event which forms an important means for intercellular signalling. The nature of the signal, whether killing, mitogenesis, suppression or something else, will presumably depend on the receptors involved. The particular model which I have described, appears to explain many aspects of T cell killing and illustrates how quantitative and testable predictions can be made concerning receptor aggregation.

ACKNOWLEDGEMENT

I am indebted to Micah Dembo for clarifying discussions on dual recognition models.

APPENDIX: RECEPTOR REDISTRIBUTION IN CONTACT AREAS

Consider two cells which are in contact, as shown in Fig. 2, over an area a. When the binding has reached equilibrium, the density of free receptors will be constant over each cell, both in the contact area and outside it. Let these values be N_{1f} and N_{2f} resepctively. Within the contact area, there will be additional bound receptors such that[14]

$$N_b = KN_{1f}\ N_{2f} \tag{A1}$$

If the surface receptor densities were N_1 and N_2 prior to contact and the cells have areas A_1 and A_2, then conservation of receptors implies

$$A_1N_1 = A_1N_{1f} + aN_b \tag{A2}$$

$$A_2N_2 = A_2N_{2f} + aN_b \tag{A3}$$

Equations (A1)-(A3) may be solved for N_b in terms of K, N_1, N_2, $\varepsilon_1 = a/A_1$, and $\varepsilon_2 = a/AN_2$. In particular we solve (A2) for N_{1f} and (A3) for N_{2f} and substitute in (A1) to obtain

$$\varepsilon_1\varepsilon_2N_b^2 - N_b(\varepsilon_2N_1 + \varepsilon_1N_2 + K^{-1}) + N_1N_2 = 0 \tag{A4}$$

which has the solution

$$N_b = \frac{\varepsilon_2N_1+\varepsilon_1N_2+K^{-1} - \sqrt{(\varepsilon_2N_1+\varepsilon_1N_2+K^{-1})^2 - 4\varepsilon_1\varepsilon_2N_1N_2}}{2\ \varepsilon_1\varepsilon_2}\ . \tag{A5}$$

Note that for ε_1 and ε_2 very small, equation (A4) or (A5) reduces to equation (4). When they are not small, equation (A5) gives the concentration of bound receptors in the contact area. Together with equations (A2) and (A3) this can be used to obtain the depletion of receptors on the rest of the cell surface.

It is also of interest to consider the redistribution of receptors on a cell which has become stuck to a surface. Under what conditions will the cell become depleted of receptors on its free surface? Suppose the density of receptors on the surface is $N_2(X)$ and X denotes position on the surface. As before, at equilibrium let N_{1f} represent the constant density of free receptors everywhere on the cell surface.

Equation (A1) still holds and in addition

$$N_2(X) = N_{2f}(X) + N_b(X)\quad , \tag{A6}$$

which together with (A1) gives

$$N_b(X) = \frac{KN_{1f}N_2(X)}{1 + KN_{1f}} \quad . \tag{A7}$$

Also

$$N_1(X) = N_{1f} + N_b(X) \quad , \tag{A8}$$

so that substituting (A7) in (A8),

$$N_1(X) = N_{1f}\left(1 + \frac{KN_2(X)}{1+KN_{1f}}\right) . \tag{A9}$$

If we now integrate this equation over the entire surface of the cell we obtain the total number of receptors on the cell, which is the initial density, N_{10}, times the cell area, A. Hence,

$$N_{10} = N_{1f}\left(1 + \frac{K\overline{N}_2}{1+KN_{1f}}\right) \tag{A10}$$

where $\overline{N}_2 \equiv A^{-1}\int N_2(X)dS$. Solving (A10) for N_{1f}, we find,

$$N_{1f} = \frac{K(N_{10}-\overline{N}_2) -1+ \sqrt{(K(N_{10}-\overline{N}_2)-1)^2+4KN_{10}}}{2K} \tag{A11}$$

This equation tells us the concentration of free receptors on the stuck cell in terms of the concentrations of receptors on the unstuck cell, N_{10}, and on the surface, $N_2(X)$, and the equilibrium constant K. If for example $\overline{N}_2 = N_{10}$, so that there are equal numbers of binding sites on the cell and on the surface area of attachment, then

$$KN_{1f} = \frac{\sqrt{1+4KN_{10}} -1}{2} \tag{A12}$$

For small values of KN_{10}, $N_{1f} \simeq N_{10}$ while for larger values, $KN_{1f} = \sqrt{KN_0}$.

REFERENCES

1. Immunological Reviews (1978). 40.
2. Katz, D.H. and Benacerraf, B., eds. (1976) The Role of Products of the Histocompatibility Gene Complex in Immune Responses, Academic Press, New York.
3. Bell, G. (1978) in Theoretical Immunology, Bell, G.I., Perelson, A.S., and Pimbley, G.H. eds., Marcel Dekker Inc., New York, pp. 341-378.
4. de Sousa, M. (1976) in Receptors and Recognition A2, Cuatrecasas, P. and Graves, M.F. eds., Chapman & Hall, London, pp. 105-163.
5. Zinkernagel, R.M., Callahan, G.N., Althage, A., Cooper, S., Klein, P.A., and Klein, J. (1978) J. Exp. Med. 147: 882-896.
6. Feldmann, M. (1978) in Theoretical Immunology, Bell, G.I., Perelson, A.S. and Pimbley, G.H. eds., Marcel Dekker, New York, pp. 281-301.

7. Taussig, M.J., Munro, A.J., Campbell, R., David, C.S. and Staines, N.A. (1975) J. Exp. Med. 142: 694-700.

8. Tada, T., Taniguchi, M. and David, C.S. (1976) J. Exp. Med. 144: 713-725.

9. Oláh, I., Röhlich, P. and Töro, I. (1975) Ultrastructure of Lymphoid Organs, An Electron-Microscopic Atlas, Lippincott, Philadelphia.

10. Griffin, F.M., Jr., Griffin, J.A. and Silverstein, S.C. (1976) J. Exp. Med. 144: 788-809.

11. Sanderson, C.J. and Glauert, A.M. (1977) Proc. Roy. Soc. Lond. B 198: 315-323.

12. Beckett, R.C., Bankhurst, A.D. and Williams, R.C. (1978) Arch. Allergy Appl. Immunol.

13. Singer, S.J. and Nicholson, G.L. (1972) Science 175: 720.

14. Bell, G.I. (1978) Science, 200: 618-627.

15. Pecht, I. and Lancet, D. (1976) in Chemical Relaxation in Molecular Biology, Rigler, R. and Pecht, I. eds., Springer-Verlag, Heidelberg.

16. Parsegian, V.A. (1973) Ann. Rev. Biophys. Bioeng. 2: 221.

17. Nir, S. and Anderson, M. (1977) J. Memb. Biol. 31: 1.

18. Korn, E.D. (1978) Proc. Nat. Acad. Sci. USA 75: 585-599.

19. Dembo, M. Goldstein, B., Sobotka, A.K. and Lichtenstein, L.M. (1978) J. Immunol. in press.

20. Elson, E.L., Schlesinger, J., Koppel, D.E., Axelrod, D. and Webb, W.W. (1976) in Membranes and Neoplasia: New Approaches and Strategies, Marchesi, V.T. ed., Less, New York.

21. Bell, G.I. Manuscript in preparation.

22. Bourguignon, L.Y.W., Hyman, R., Trowbridge, I. and Singer, S.J. (1978) Proc. Nat. Acad. Sci. USA 75: 2406-2410.

23. Adam, G. and Delbrück, M. (1968) in Structural Chemistry and Molecular Biology, Rich, A. and Davidson, N. eds., W.H. Freeman & Co., San Francisco, see also Berg, H.C. and Purcell, E.M. (1977) Biophys. J. 20: 193-219.

24. Bell, G.I. (1975) in Mammalian Cells: Probes and Problems, Richmond, C.R., Peterson, D.F., Mullaney, P.F. and Anderson, E.L. eds. (CONF-731007) U.S. Dept. Commerce pp. 275-276, see also Buas, M. (1977) Ph.D. thesis, University of Maryland, Physics Dept.

25. Singer, S.J. (1976) in Surface Membrane Receptors, Interface between Cells and Their Environment, Bradshaw, R.A., Frazier, W.A., Merrell, R.C., Gottlieb, D.I. and Hogue-Angeletti, R.A., eds, Plenum, New York pp. 1-24.

26. Langman, R.E. (1978) Rev. Physiol. Biochem. Pharmacol. in press.

27. von Boehmer, H., Haas, W. and Jerne, N. (1978) Proc. Nat. Acad. Sci. USA 75: 2439-2442.

28. Zinkernagel, R.M. and Doherty, P.C. (1974) Nature, 251, 547-548.

29. Bourguignon, L.Y.W. and Singer, S.J. (1977) Proc. Nat. Acad. Sci. USA 74: 5031-5035.

DISCUSSION

DR. WEBB: George, I am concerned about one thing. In calculating the forces, you have assumed an efficiency of adding theoretical individual bond strengths of about a third, whereas, as you well know, in ordinary plastic or brittle materials, we find when we measure the strengths of materials that one finds something more like one ten-thousandth, or perhaps if you are lucky, one-thousandth of the theoretical bond strengths. How do we manage to be so much more efficient in the cellular system?

DR. BELL: Certainly for most solids, the material strength is not that of the individual bonds but is rather determined by the stress which will nucluneate and propagate fractures along some imperfections in the material. For single crystals, for example carbon fibers, or high class steel one can approach the theoretical bond strength.

But I think we have sort of an ideal situation for maximizing strength because of the fluidity of the system. That is, if there is local stress placed upon one set of bonds, presumably the cell membrane will be deformed in order to accomodate that. There is also the feature that the bonds which we are considering are already nearly two orders of magnitude weaker than covalent or other electronic bonds, so that they do represent weak links.

To some extent, though, your objection is valid. One should worry about the fact that some of the bonds will be stressed more than others in most experimental situations. For example let's consider shearing forces tending to remove cells from surfaces in a flow; these would tend to stress some of the bonds, let's say on the tail end of the cell, more relative to those on the front end.

So I think that for detailed comparison with experimental results, one may have to worry about the distribution of stresses in the bonding pattern that you think is set up, and maybe you will find that you can get the cells apart or off a surface more easily.

For example, in the adhesion of the endothelial cells and lymphocytes, when one takes scanning electron micrographs, he sees the lymphocytes have microvilli sticking out; a microvillus sticks in a pit on the endothelial cell and it is very much as if there is a rope holding the lymphocyte locally.

DR. WEBB: Unfortunately, the same difficulty appears in the theory of the strength of rubber, the breaking of bonds which effectively reduces the strength, with all of that accommodation.

DR. VARESIO: What happens in your model if you consider that the receptor may exist in a soluble form in the micro-environment around the cells, such as the immunoglobulin or the histocompatibility antigens which may be shedded by the cells?

DR. BELL: That will interfere with the adhesion by blocking the ligand or by binding to sites on the opposing cell. If one knows what the receptor is he can introduce it into the theory. If he doesn't know, then he is obviously not in quite as good shape.

DR. SINGER: We have been thinking along very similar lines, as you know, George, that specific cell-cell interaction might be something like capping. As we were suggesting the first day of the meeting, if capping on a specific cell by a multivalent ligand may be accompanied by cytoskeletal rearrangements -- in particular, as I showed the first day, one of the markers that we have been quite confident about accompanying or capping by antibodies on isolated cells is the accumulation of alpha-actinin underneath these cells.

If the interaction between two different cells was a kind of mutual capping of their respective recognition molecules into the site of contact, and if in fact perhaps the significance of capping is physiologically that it is a means whereby two cells interact with one another, this denoting their mutual surface receptors, one might anticipate that if this is very analogous to capping induced by a soluble multivalent ligand, that alpha-actinin ought to be accumulated under the region where the two cells make contact.

Now, Benny Geiger in our lab has been doing some preliminary experiments, and I don't have the slides here, but he has made B cells that are immunized to sheep red cells and has formed rosettes in which the red cells are accumulated under appropriate conditions in one pole of the immune B lymphocyte, and has shown that alpha-actinin is accumulated immediately under the lymphocyte membrane where it comes into contact with the erythrocyte.

So I think the notion that cell-cell interaction between cells in which there are recognition molecules on one cell that con recognize the membrane antigens on the other involves some kind of mutual capping may have some importance.

Of course, the function of the cytoskeleton in such a process may be to lend extra stability to the bonding involved in the two cells and perhaps lengthen the time in which the cells remain in contact to allow for one or another kind of physiological interaction.

DR. RAFF: George, may I just ask you a question about your last model? One of the things that has plagued the dual recognition model, which I suppose is slightly the more popular model at the moment, the MHC involvement in T cell recognition, is that if there is on every T cell a receptor that recognizes their own MHC glycoproteins, then why don't you see those cells rosetting, for example, with any cell in your own body which has those MAC glycoproteins displayed on it?

Now, you have taken the two receptors, one for self, one for X, and put them together. I don't see how you get around that objection, why those receptors are together. You still have one receptor with some affinity for self. Why, when you increase the avidity enormously by making it multicontact binding to cells, you still don't get rosette formation?

You have put it together in terms of signalling, clustering and so on. But still there is a physical question: Why do those cells, if they have low affinity receptors for self, not form rosettes in view of avidity considerations?

DR. BELL: As you say, we don't want histocompatible cells to stick to each other. It seems to me that co-clustering alleviates this problem a little bit, but when one looks at it quantitatively, possibly not enough.

You can make an additional assumption that the rH receptors were not fully expressed or unmasked until there has been an X-rX interaction and this requirement would per se help things out. Or you could note that rosetting requires a sustained adhesive contact and postulate that H-rH interactions are transient (eg. cleared by an enzyme) unless stabilized by X-rX interactions.

Now, I must say that at times I have thought of trying to use this argument the other way around and say that since particularly the T cell recognition system is probably for looking at cellular antigens, maybe the T cell should first check if it is near a cell and then worry about exposing its rX receptors, but it seems to me the first way, namely, don't exposure your rH receptors until you have interacted with something, might be a little more reasonable.

DR. LAUFFENBURGER: How long, George, on the average do you expect your cells to be stuck together as they pass by?

DR. BELL: Well, these cells have relative velocities, as I said, on the order of a tenth micron per second if they are moving through lymphoid tissue, so if there is sticky stuff around and they once form an adhesive contact, I think they are going to have to do something to get rid of it, like internalize receptors or shed them. They are not going to come apart spontaneously once they have formed a significant number of bonds and I think bonds will form quite frequently.

DR. LAUFFENBURGER: I was wondering if perhaps it might be interesting to consider the lymph nodes as some sort of affinity chromatograph column in which you get differential residence times depending on if you induced some sort of subpopulation to become more sticky as it passed through.

DR. BELL: There is some experimental evidence on this, namely, if you put specific antigen into a lymph node and you collect the outcoming lymphocytes as a function of time and test them for capacity to recognize that antigen, you find that they are specifically eliminated from the efferent lymph for a period of a day or two. Meanwhile there is all sorts of cell proliferation and action going on in the lymph nodes and the antigen recognizing cells start coming out eventually.

DR. PAUL: George, one of the enigmas in the cellular interaction systems which are dominated by the MHC is that although one can show that the target cell or the collaborating cell has the MHC antigen on its surface, an anti-MHC antibody will block such functions such as cell killing. It is rather difficult to actually demonstrate that the antigen is available, and I mean that in several ways.

For example, in many cases killing of virus-infected cells can be blocked by antibody to MHC, but not by antibody to the viral antigen. Or the activation of T cells by antigen-pulsed macrophages can be blocked very efficiently to the Ia antigen. But antibody to antigen to DNP guinea pig albumin is without effect even when added at milligrams per ml of purified antibody, and one can trypsinize the cell and presumably remove the antigen, and yet not block the stimulatory activity of that macrophage.

It is almost as if, no matter how painful it seems to us to imagine, one cannot find the antigen which is stimulating the cells on the surface of the stimulatory cell. That I have always regarded as a central enigma, and I wondered if you have been aware of that, or alternatively if you have tried to incorporate it into your thinking on the subject.

DR. BELL: I haven't really been aware of it. I agree it is quite puzzling.

DR. RAFF: Although, Bill, you know there are a number of possible explanations for that, and one is burial, but another is that T cells seem to see sequence determinants on proteins, and antibodies are largely against conformational determinants, and that fits very nicely with the idea that T cells see digested antigens.

But recently, Peter Dogherty, in making monclonal antibodies against influenza virus, now has two clones where the antibody inhibits the T cell killing, and it is K restricted or D restricted, depending on the antibody. So it may well be that that will disappear when one has selected antibodies against sequence determinants.

DR. PAUL: It could well be. On the other hand, it is clear that the activation of T cells by dinitrophenolated antigens distinguishes very precise details of the nitrophenol group. For example, you can show that DNP can be distinguished from TNP, from paranitrophenol, from orthonitrophenol. And that distinction is so precise that it is very difficult to believe that the DNP group isn't actually being detected by the binding site. But it does not, perhaps, make up all the site.

But that precise variation is appreciated by the cell. On the other hand, you take an anti-DNP antibody and plaster it on that cell, and it ought to inhibit the binding a lot more than just moving a nitro group around. It has no effect at all.

Allen Rosenthal has recently used a hibrodoma reagent which is directed presumably against the same site on insulin that the T cell sees, and it has no blocking activity. So I don't think the dilemma will be resolved quite so easily.

DR. BELL: There are also some quantitative questions. We mentioned the possibility of very high local concentrations of antigen, and I presume it is often difficult to achieve such high concentrations in solution. So if there is enough time for interaction between the two cells so that antigen which is initially covered up can become free agian, why, there may be still interaction.

DR. SINGER: There are some other peculiar things that we may have to take into account that could possibly be involved in the kinds of things Bill was just saying. There is indication, certainly, on macrophages that -- I guess you just said it in one word, Marty, that there is a buried state for an antigen that has been attached to a cell surface. You can strip the cell surface of a macrophage of a particular antigen and then show, as I guess it was Luhr who showed, that there is with time a re-expression of that antigen as if it was in some perhaps vesicular state underneath the membrane and called forth after the stripping of the surface antigen was produced.

So maybe you have the capcity to re-express an antigen even though you have blocked all the antigen that was accessible at the surface.

DR. RAFF: There is one point, I think, that should come out in this kind of discussion in consideration, and that is it is hard enough to solve the problem of how soluble peptides bind the receptor signal to cell. Now when you are talking about a cell-cell interaction signalling where the receptor and the lignad are cell-bound membrane proteins, for example, or one a glycolipid, then we are really in the dark. I mean we have no idea, for example, how proteins are shed from membranes, and all membrane proteins are turning over and no one has a clue how the degradation process goes on, by and large.

So now when you take two membrane proteins or a carbohydrate and a membrane protein and have them interact at cell surface we really have no idea what the sequellae of that will be in terms of those cells obviously won't stay together forever. Those bonds have to be broken. Does the breaking involve pulling things out of membranes, or proteolysis or whatever?

It seems to me that these are problems that could be studied and have not been, so it is not too surprising that we are in the dark about this kind of model in detail.

Subject Index